Maarten Hooijberg · Practical Geodesy

Springer
Berlin
Heidelberg
New York
Barcelona
Budapest
Hong Kong
London
Milan
Paris
Santa Clara
Singapore
Tokyo

Maarten Hooijberg

Practical Geodesy

Using Computers

With 69 Figures and 21 Tables

Springer

Dr. Maarten Hooijberg
Doeverensestraat 27
NL-4265 JM Genderen
The Netherlands

ISBN 3-540-61826-0 Springer-Verlag Berlin Heidelberg New York

Cataloging-in-Publication data applied for

Typesetting: Camera ready by author
Production Editor: B. Schmidt-Löffler
Cover Design: design & production GmbH, Heidelberg

SPIN: 10551689 32/3136 - 5 4 3 2 1 0 - Printed on acid-free paper

Contents

Table of Contents

List of Figures

List of Tables

1. Preamble

1.1 Welcome to the World of Geodesy

Historical Outline

Towards the end of the seventeenth century hypothetical consideration about the shape of the Earth had given way to the first serious attempt to determine it by measurement.

An arc of longitude between Paris and Amiens in France, the distance measured by Picard in 1669-1671, enabled Christiaan Huygens and Isaac Newton to confirm their hypothesis of gravitation. All attempts to determine the shape of the Earth by measurement by French and Spanish Scientists in Peru (1735-1744) and by French and Swedish Scientists in Lapland (1736-1737) had given proof of the Earth's oblateness from arcs measured (Airy, 1830).

Johann Heinrich Lambert (1728-1777) had the facility for applying mathematics to practical questions that are not only interesting, but are still in use among geodesists. "... *It is no more than just, therefore, to date the beginning of a new epoch in the science of map projection from the appearance of Lambert's work ...* " (Craig, 1882).
The geoid was devised by Carl F. Gauss in 1828. He masterminded the conformal projection from the plane to the sphere. His concepts had been applied by Oscar Schreiber to the computation of a network in the Kingdom of Hannover. Schreiber changed the character of the arc measurement into a projection and grid system in 1866 (Laplace, 1799; Schreiber, 1866). One of the main problems of determining "the Figure of the Earth" was the lack of knowledge about the determination of longitude. Therefore, Sir George B. Airy was closely involved in remeasuring longitudes using the electric telegraph. Friedrich R. Helmert published his basic paper on deflection of the plumbline in 1886.

"Invar" was an important discovery in 1896. This metal could be rolled into tapes for measurements.
After the end of World War I, the "Väisälä Light Interference Comparator I" was designed and developed by Yrjö Väisälä. Using the Comparator for baseline measurements of up to 200 metres resulted in an accuracy of one part in 10^7. It was put into practical use by the Finnish Geodetic Institute in 1929 (Väisälä, 1923, 1930; Kukkamäki, 1954, 1978). Great improvements were being made in the design of theodolites by Heinrich Wild of Switzerland. This resulted in increased accuracy and speed of observations, combined with rugged instrument construction (Seymour, 1980; Strasser, 1966).

The UTM Grid System was devised by the US Army Map Service (AMS) and adopted by the US Joint Chiefs of Staff as the official US Army grid in 1947. Consequently, the concept of the co-ordinated grid had to receive its proper place. Co-ordinates were to be redefined in an operational sense as a function of a world-wide transverse Mercator system (Ayres, 1995).

Historically, World War II is the dividing line between conventional and electronic surveying systems. Except for small tasks associated with plane surveying, the popularity of measuring distances using tapes, stadia, subtense bars, and similar equipment is eroding as more convenient tools and techniques pass into common use (Leick, 1990).
One of the first opto-electronic distance measuring (EDM) instruments to be produced for geodetic purposes was the Geodimeter, a bulky instrument and labourious to use. It was developed by Ragnar Schöldström, and patented by Erik Bergstrand, geodesist of the Swedish Geographical Survey Agency in 1947.
A few years later it was followed by the magnetic-electronic distance measuring instrument, the Tellurometer, a system developed by T.L. Wadley of the South African Council for Scientific Research (Wadley, 1957).

Initially, progress using traditional triangulation techniques was slow, but the arrival of EDM instruments caused a revolution in surveying methods. A great variety of electronic surveying systems and methods were developed, depending on the mission to be accomplished: circular systems, hyperbolic systems, baseline measuring systems, radar altitude measuring systems, and sound navigation and ranging (sonar). These tools were applied to geodetic and hydrographic surveys, oceanographic research, navigation of ships and aeroplanes, including microwave Instrument Landing Systems (Laurila, 1976; Luse, 1985).

Triangulation procedures were abandoned in favour of trilateration networks, i.e. geodetic control by EDM instruments and first order traversing by theodolite, which were applied more and more after 1960. However, these methods do not measure positions, which is a crucial limitation (Bomford, 1977).
New geodetic information was added by the Army Map Service (AMS) in 1958, at the time when the first satellites went up. Discussions have led to the insight that the value of ellipticity, obtained only a year or so after the first artificial satellite had been launched, was fixed at f = 1/298.3. This is practically the same as had been put forward by F.R. Helmert in 1906 and T.N. Krassovsky in 1940 (Fischer, 1959).

The Doppler satellite system has an outstanding history. The initial idea was due to G.C. Weiffenbach and W.H. Guier at the Applied Physics Laboratory of John Hopkins University. In 1957/58, they made measurements of the Doppler shift exhibited on signals received from the artificial satellites, such as the Sputniks, Explorers and Vanguards, showing that satellite orbits could be deduced from them to a reasonable degree of accuracy.

F.T. McClure found that the roles could be inverted. If the orbits of satellites could be determined so accurately, then distance measurements at stations from any other known position could be used to fix unknown positions on the earth's surface. This system became known as NAVSAT (Navy Navigation Satellite System, or NNSS), but generally named the "Transit" Doppler system.

The period of greatest activity in this field started in the 1960s, during which time the whole task of providing a world-wide geodetic control network was achieved by corresponding efforts of certain scientific geodetic agencies (Buchar, 1962; Kaula, 1962; Schmidt, 1966; Williams, 1966).
Using radio frequencies, Doppler observations were made towards the Laser satellites Diadem II, Diadem III, Explorer 22, and the Laser Flashlight satellites such as Anna 1B, AJISAI, Etalon, GEOS A, GEOS B and Stella. Another system favoured the use of electronic distance measurement to track a comparatively small reflecting satellite. This was exemplified by Sequential Collation of Ranges (SECOR). The SECOR - a type of microwave EDM system - was used to establish an equatorial control network around the world.
Photographing large balloon satellites like the Echo I, Echo II, Passive Geodetic Earth Orbiting Satellite (PAGEOS) against the background of stars by ballistic cameras, such as Antares (France), SBG Zeiss Jena, NAFA (USSR), PC-1000, Hewitt (UK), Baker-Nunn, Wild BC-4 (USA), was another concept. Because of the designation of the latter camera, this project was called the BC-4 Triangulation. By 1973, the BC-4 data source was replaced by more accurate surveys based on the methods of Doppler satellite tracking (Bomford, 1977).

Satellite tracking has yielded much information about the gravity potential of the earth, and led eventually to mapping of the geoid throughout the world. Consequently, the trend is to describe new figures initially in geophysical terms. Determinations of the earth's figure from the time of GRS67 onwards are made in the truly *"Earth-Centred, Earth-Fixed"* (ECEF) geocentric co-ordinate system and based upon the theory of an equipotential ellipsoid (Arnold, 1970; Maling, 1992; Seeber, 1993).

Determination of the propagation velocity of light. The International Union of Geodesy and Geophysics (IUGG) adopted the value of c=299 792.458 km/sec in 1983 (Price, 1986; Rotter, 1984).
Basic research of the propagation velocity of light goes back to 1676: Römer measured c=215 000 km/sec. Through the years, many determinations of the velocity of light have been made with various methods: Weber and Kohlrausch measured c=310 800 km/sec (in 1857), Maxwell c=284 300 km/sec (in 1860), and Albert A. Michaelson c=299 910 km/sec (in 1879). In addition, Michaelson was awarded the Nobel prize for his pioneering work in *Interferometry* techniques in 1907 (Logsdon, 1992).

In recent years the Global Navigation Satellite System (GLONASS), Global Positioning System (NavStar or GPS), including Very Long Baseline Interferometry (VLBI), Satellite Laser Ranging (SLR), Lunar Laser Ranging (LLR) provided the world-wide ability to fix continuous positions with an accuracy equivalent to or better than conventional geodetic surveys whatever weather conditions.

The Doris tracking data of the Topex/Poseidon oceanographic satellite were used to calculate the co-ordinates of station positions, velocities and earth rotation parameters (IERS 1995).

In Practice

Only small area of a few hundred square kilometres can be accurately mapped and surveyed without a framework, because no difficulty is encountered due to earth curvature. Continuous mapping of a large area without one will result in disorder. Therefore, a high accuracy framework is imperative for geodetic and hydrographic surveys, for example in the southern part of Africa where several disparate datums have been established, between Eastern and Western Europe, and the North Sea with maritime frontiers between several countries: each with a different datum, origin and projection system (Bordley / Calvert, 1986).

Timesaving and practicality in manual skill are key words in selecting the right type of tools for projects undertaken. Today's practising surveyor cannot meet the competition if equipped only with traditional methods from the nineteenth century. Using GPS or Inertial Surveying Systems (ISS), transporting the equipment to a site is sufficient for position fixing. However, a "black box" instrument does not remove the requirement for a trained surveyor. It asks for a certain level of knowledge and therefore adequate tuition is required for certification because of those basic ideas being hidden behind the automated operation of such instruments.

Development of completely new instruments such as Advanced Very High Resolution Radiometer (AVHRR) imagery by airborne radio sounding and CycloMedia (Beers, 1995) for terrestrial photogrammetry using digital panoramic images for rapid acquisition of data will be imperative for the coming years.

Local surveys are always done with respect to the local datum. Because satellite systems operate globally, a global datum is adopted, and position fixes must be transformed from the global datum to the appropriate local datum and vice versa. Consequently, relating position fixes and previously established co-ordinated reference marks to the same mathematical framework using conversions, least squares adjustments and transformations remain essential.

The Formulae

Friedrich R. Helmert's celebrated book "Die Mathematischen und Physikalischen Theorieen der Höheren Geodäsie" was published in 1880 (Part 1) and 1884 (Part 2). It remained for many years the most authoritative work in German on the subject. In 1880, Alexander Ross Clarke published the book "Geodesy", which was an equally important work in the UK and for the British Commonwealth Territories.
Their formulae are still found in textbooks and were designed for use with tables and logarithms.

The computation of planar co-ordinates of all points was done using logarithms before the appearance of the electronic computer. Theoretically, the availability of eight-place logarithm and interpolation tables for the Gauss-Krüger transverse Mercator projection system made the use of 3° wide zones in Germany essential.
Later, a rotary-machine method was introduced, again based on the logarithmic method of computation; not strictly exact, but the differences between stations in any particular area were sufficiently correct for all practical purposes.

The question about the accuracy of computed data is nevertheless very important. Using a computer to the fullest extent means that many existing formulae must be rewritten in a form which fits the computer best. Vincenty: "Many apparently unrealisable programmes can be written for a computer if more thinking is given to recasting the equations." See also (Hooijberg, 1979, 1996; Vincenty, 1976b).

Methods and Algorithms

The basic equations used in algorithms in this book were obtained from (Boucher, 1979; NIMA, 1991; Meade, 1987; Moritz, 1992; Vincenty, 1984a), used by (Floyd, 1985; Stem, 1989a).
The Meridional Arc calculations are based on formulae by Klaus Krack, the Isometric Latitude calculations are based on formulae by Ralph Moore Berry, and are optimised by the author.

Some more algorithms can be found in (Claire, 1968; Clark, 1976; Burkholder, 1985; C&GS SP No. 8, 1952; C&GS SP No. 65-1 Part 49, 1961; DA, 1958; Krack, 1981, 1982; Pearson II, 1990; Rune, 1954; Thomas, 1952).

All methods and algorithms are elucidated by selected examples found in the literature.

The Programs

The procedures comprise different cases, each involving one or more surveying conversions and/or transformations, such as:

- arc-to-chord or (t - T) correction (δ) |[1]
- bi-linear interpolation
- conversion of co-ordinates from system to system
- datum transformation
- geodetical to grid co-ordinate conversion and vice versa
- grid scale factor (k) |[1]
- length of the meridional arc
- meridian convergence (γ) |[1]
- properties of a geodetic line.

In the wider context of computing, personal computer-based geodetic applications are models of great virtue when it comes to ease of use and flexibility. The user with any experience of mainframe systems will know that, while they do their specific tasks well once the user has learned their particular quirks and ways of working, that is usually all such a system can do. To get them to do anything else is usually risking crashing the system.

Many techniques are available for doing co-ordinate transformations, some offer mathematical ease at the sacrifice of accuracy. At the other extreme, some are accurate under all conditions.
The universal programs with 2-D, and 3-D formulae listed in this book were developed on an IBM compatible personal computer. The reader is assumed to be acquainted with the general principles of geodesy and associated jargon, which is essential for proper use of the programs. Those who are conversant with the extended use of Mathematical Geodesy will find that certain refinements in explanation have been neglected for the sake of simplicity. Such readers are cordially invited to study the papers referred to and sources of reference quoted in the text.

This book makes no attempt to teach the BASIC language.

A Future in Geometrical Geodesy

Every scientist stands on the shoulders of his predecessors. Friedrich R. Helmert (1843-1917) was well endowed with all those qualities that make a successful geodesist. He was the professor of Geodesy at the Technical University at Aachen, director of the Prussian Geodetic Institute in Potsdam, and of the Central Office of the "Internationale Erdmessung". Through his work, geodesy has experienced discoveries, conceptual inventiveness, and achievement of technical ingenuity, which still have their effect (Torge, 1991).
In his fundamental monograph, Friedrich R. Helmert established geodesy as a proper science, devoted to the measurement of the shape of the earth and its gravity field in a three dimensional Euclidean space and time.

It should be noted that the remarkable increase in quality and quantity of the results obtained from the geodetical techniques during the last twenty-five years is because they were based on about 150 years of development of theory. Despite the expected universal use of GPS surveying in the future, there will still be situations in which angular measurement by theodolite or distance measurement with an EDM instrument will be more economical (Leick, 1990; Torge, 1991).

---

[1] Using the properties of a projection.

As ongoing geodetic research provides increasingly better estimates of geoidal heights, absolute geoidal heights' accuracies are being improved as more data are collected.

It is as *A.R. Robbins* of Geodesy and Surveying at Oxford University mentioned in 1978:

"The Revolution in Geodesy has only just begun and we are privileged to be taking part in it"

What about the future? It is hoped that the *development* of the mathematical geodesy will provide subjects - such as a refined geoid, prediction of permanent earth tides, earth quakes, and tectonic plate movements - early the next century.

Consider the phrase "Heads I win, tails you lose". This is the simplest possible illustration of probability of occurrence. The organisation of tomorrow must prevent problems before they can occur and, more important, find ways to change and improve geodetic processes with the support of electronic information technology.

Geodesy must cover costs. Geodesy involves working closely with colleagues in the adaptation of the future. This future requires strategic thinkers who can retain an overall perspective while dealing with the day to day practicalities. Although the techniques of mathematical geodesy are separate subjects, within the scope of this book brief references are made to this type of technique.
Material for further thought is outlined by *celebrated pioneers* - such as Vidal Ashkenazi, Willem Baarda, Arne Bjerhammer, Erik W. Grafarend, Juhani Kakkuri, - and quoted verbatim in the text.

In future the reader may be called to work on them to provide geodetic expertise.

This Book

The book "Practical Geodesy" has been written for those who, either for practical purposes or simply because they enjoy experimenting, wish to calculate geodetic positions.

Why is another treatise on geodesy desirable? It is launched because a title is needed that addresses the geodetic community's request for accurate, practical information - few of the textbooks that are available satisfy these needs.
It can be argued that the subject of geodesy is better documented than other fields of Earth Sciences.
Very little has been published - despite insistent demands for documentation - about conversion from a particular projection to another projection system, the transformation from a specific geodetic datum to another, together with demonstration by "global" examples.

The material contained in this book has been specially chosen and covers the practical aspect of geodesy and many details, which are so often left to the imagination. An up-to-date publication should embody the latest ideas in geodetic computation techniques while a computer of modest capacity can cope with the algorithms developed through the years.

With these points of view in mind the author has compiled this publication to take stock of available data (datums, ellipsoids, units, etc.), to focus on applications and to illuminate spatial developments. It would help the user to take advantage of the geodetic revolution and allay anxieties about the changes that this industry will inevitably bring.

The turning point came as GPS - (Grafarend, 1995b): the *Global Problem Solver* - sparked a Klondikian rush for various solutions that attracted everyone from the inventive genius to the best minds of the age.
In 1981 Baarda said: *"Looking back over the twenty-five years needed to establish a theory, one is surprised at the long and winding path that had to be followed before an essentially very simple line of thought could be formulated".* (Baarda, 1981) is right, though. This book has evolved through the author's twenty-five years of advising and writing on geodetic methods and computations, while the author was on the staff of a major Dutch-British company in the field of world-wide exploration and mining of mineral resources. In 1974 the first

"HP45 topopack" - about compact pocket calculator algorithms - was produced in which user requests were incorporated, prepared for use by practising land surveyors and hydrographers.
Following this topopack the author subsequently introduced topopacks for the HP67/97/41CV pocket calculators, the HP85/9845B desktop calculators, and now this publication.
It is devoted to creating a better understanding consistent with the current state-of-the-art topics of geodetic surveying.

Every effort has been made to present a practical approach to computing in which the how-to-do-it approach is stressed to provide thoughtful advice, to create inspiration and confidence among students, GIS-consultants, hydrographers, land surveyors, including the out-of-touch professionals with an engagement in the geodetic and hydrographic profession. Furthermore, for the benefit of the advanced geodesist, the publication points at the latest technological developments such as geoids, vertical datums, permanent earth tides, and universal multi-dimensional databases for more ambitious arrangements. The author was very fortunate to receive help with ideas, data, algorithms, criticism, corrections and response that contributed significantly to a systematic presentation of the subjects discussed in this book.

Did the author lay The Golden Rule? Not quite; for as shown hereafter, writing this publication does not depend on just one individual but on non-political co-operation between the author and scientists, colleagues of distinguished agencies, research institutes, and universities in the field of geodesy: "the World of Geodesy!"

It is the author's sincere wish that it will perhaps encourage and help those already practising geodesy to achieve an even higher standard of efficiency, and arouse the interest of others, resulting often in some new idea, or at least material for further thought.

Land van Heusden en Altena, Summer 1997
Maarten Hooijberg

1.2 Organisation of this Book

This book is divided into 12 chapters:

Chapter 1	"Preamble" describes this publication
Chapter 2	"BASIC" guidelines and recasting algorithms
Chapter 3	Definitions and description of datums, ellipsoids, geoids, origins and reference systems
Chapter 4	Geodesic. Calculation of the direct and inverse problems
Chapter 5	Description of conformal projections in general
Chapter 6	Description of the Gauss-Krüger / transverse Mercator projection
Chapter 7	Description of the Lambert conformal conical projection
Chapter 8	Description of the oblique Mercator projection
Chapter 9	Description of datum transformation and bi-linear interpolation
Chapter 10	Description of miscellaneous applications
Chapter 11	Appendix, examples and the programs
Chapter 12	Bibliography, and Indices

Chapter 11 contains following programs:

1. ELLIDATA.BAS – calculation of spheroid parameters and constants
2. REFGRS00.BAS – calculation of geometric parameters
3. BDG00000.BAS – geodesic - calculation of the direct problem
4. GBD00000.BAS – geodesic - calculation of the inverse problem
5. GK000000.BAS – Gauss-Krüger projection conversion
6. LCC00000.BAS – Lambert conformal conical projection conversion
7. OM000000.BAS – oblique Mercator projection conversion
8. TRM00000.BAS – about datum transformation
9. RDED003*.BAS – about bi-linear interpolation

1.3 Acknowledgements

Many years ago, Brigadier Martine Hotine, UK Director of Military Survey, Chief of the Geographical Section General Staff and Adviser on Surveys to the UK Secretary of State for Technical Co-operation, quoted Field-Marshal Lord Montgomery about *co-operation* as follows:

"It ain't the guns or armament
Or the money they can pay,
It's the close co-operation
That makes 'em win the day.
It ain't the individual
Or the army as a whole,
But the everlastin' teamwork
Of every bloomin' soul".

The author expresses his appreciation for providing valuable comments and substantial corrections to improve this publication:

Erik W. Grafarend, University of Stuttgart, Department of Geodetic Science, Stuttgart, *Germany*

Special acknowledgement is due to:

James E. Ayres, Scientific Advisor of the National Imagery and Mapping Agency (formerly DMA), US Department of Defense, Washington DC, *United States,* and his staff for their assistance about the information on Datums and miscellaneous Grid Systems.

Thaddeus Vincenty, Geodetic Research and Development Laboratory, NOAA (retired), Silver Spring, Maryland, *United States,* who influenced all sections on errors.

The author gratefully appreciates all the help and support provided, without which this publication could not have been written:

Earl F. Burkholder, Consulting Geodetic Engineer, Circleville, Ohio, *United States*
Emmett L. Burton, National Imagery and Mapping Agency, US Department of Defense, Washington, DC, *United States*
Carl E. Calvert, Ordnance Survey, Southampton, *United Kingdom*
J.F. Codd, Ordnance Survey of Northern Ireland, Belfast, *Ireland*
M.J. Cory, Ordnance Survey of Ireland, Dublin, *Ireland*
Bjørn G. Harsson, Geodetic Institute, Norwegian Mapping Authority, Hønefoss, *Norway*
Juhani Kakkuri, Finnish Geodetic Institute, Masala, *Finland*
Klaus Krack, Federal Armed Forces University of Munich, Neubiberg, *Germany*
Michel Le Pape, Institut Géographique National (IGN), St. Mandé, *France*
Washington Y. Ochieng, University of Nottingham (IESSG), Nottingham, *United Kingdom*
Jean M. Rüeger, University of New South Wales, Sydney, *Australia*
Henning Schoch, Institut für Angewandte Geodäsie (IfAG), Leipzig, *Germany*
Aurelio Stoppini, Universita degli Studi Facoltà di Ingegneria, Perugia, *Italy*
Govert L. Strang van Hees, Netherlands Geodetic Commission, Delft, *The Netherlands.*

The author would like to thank for supporting this book:

Dirk J. Bakker, Rijkswaterstaat - Dir. Noordzee, Rijswijk (Z.H.), *The Netherlands*
Carlos van Cauwenberghe, Min. van de Vlaamse Gemeenschap, Oostende, *Belgium*
A.W. van Dam, College of Advanced Technology, Amsterdam, *The Netherlands*
Hans Dessens and Addie Ritter, DUT Library, Delft, *The Netherlands*
Ning Jinsheng, Wuhan Technical University of Surveying and Mapping, Wuhan, *P. R. of China*
Dieter Schneider, Federal Office of Topography (L+T), Berne, *Switzerland*
Luciano Surace, Istituto Geografico Militare (IGM), Florence, *Italy*
Pierre Voet, Nationaal Geografisch Instituut (NGI), Brussels, *Belgium*
Walter Welsch, Federal Armed Forces University of Munich, Neubiberg, *Germany*
Alan F. Wright, Global Surveys Limited, Birmingham, *United Kingdom.*

Contributions of several colleagues *in Asia, Europe, the United States* - including the services of colleagues of NOAA, National Geodetic Survey, Silver Spring, Maryland - are also gratefully acknowledged.

Various drawings are reproduced by courtesy of NIMA, Washington, DC, *United States of America.*

Views expressed are those of the author.

2. BASIC Guidelines and Algorithms

BASIC is a simple, easy-to-use computer programming language with English-like statements and mathematical notations.
To use the BASIC language, you must load it into the memory of your computer.
To achieve compatibility, you must type the following command and press RETURN:

BASIC [Filename] /D

Filename is the name of a BASIC program file.
/D allows certain functions to return double-precision results.
The functions affected are ATN, COS, EXP, LOG, SIN, SQR, and TAN.

2.1 Double-Precision Form for Numeric Constants

Numeric constants can be integers, single-precision or double-precision numbers. Integer constants are stored as whole numbers only. Single-precision numeric constants are stored with seven digits, although only six may be accurate.

Note

Double-precision numeric constants are stored with 17 digits of precision, and printed with as many as 16 digits. A double-precision constant is any numeric constant with either <u>8 or more digits</u>, *the exponential form* <u>using D</u> *or* <u>a trailing number sign</u> *(#), e.g. -1.09432D-06, 3490.0# and 7654321.1234.*

Variables are the names that were chosen to represent values used in a BASIC program. The value of a variable may be assigned specifically, or may be the result of calculations in your program. If a variable is assigned no value, BASIC assumes the variable's value to be zero.

The following are the arithmetic operators recognised by BASIC, and they appear in order of precedence:

(^) Exponentiation, (-) Negation, (*) Multiplication,
(/) Floating-point Division, (+) Addition and (-) Subtraction.

2.2 Error Messages

Error messages that you might encounter while using BASIC. If, during the evaluation of an expression a division by zero is encountered, the *"Division by zero"* error message appears, machine infinity with the sign of the numerator is supplied as the result of the division and execution continues.

If the evaluation of an Exponentiation results in zero being raised to a negative power, the *"Division by Zero"* error message appears, positive machine infinity is supplied as the result of the Exponentiation, and execution continues.

If overflow occurs, the *"Overflow"* error message appears, machine infinity with the algebraically correct sign is supplied as the result and execution continues. The errors that occur in overflow and division by zero will not be trapped by the error trapping function.

2.3 Conversion of BASIC Programs

A concern in preparing this publication was which algebraic language to use. The advantage of conversational features of the BASIC language was chosen in order to be compatible with as many systems as possible.

Note

The structure of the program listings is not fully utilised in order to permit wider use of the routines. The input / output formats are kept as simple as possible.

Therefore, transference of the routines is possible to languages such as PowerBASIC, FORTRAN, C^{++}, or an spreadsheet, and independent of any computer Operating System (OS).
The program listings are preceded by examples with an intermediate output, because situations may arise in which a program fails to operate as expected.

A further important factor is the design of the equations. Many early computing programs were not written in a suitably economical form. The nested form of an equation e.g. for calculation of the Meridional Arc distance is equally valuable. See examples in [2.4]. Operations within parentheses are performed first. Inside the parentheses, the usual order of precedence is maintained.

It follows that the programs without parentheses, or those containing all the SIN(x) terms, spent a certain amount of computing time without furthering progress in the calculations. Not only are the equations easier to write in an appropriate programming language, they depend upon progressively raising terms to higher powers, and therefore reduce the risk of underflow or overflow conditions (Hooijberg, 1979, 1996; Vincenty, 1971).

2.4 Recasting Algorithms

Methods to calculate the radius of the rectifying sphere for the International Ellipsoid, scale factor $k_0 = 1$:

Method 1 (Helmert, 1880):

$$r = a (1 - n) (1 - n^2) (1 + n^2 / 4 (9 + 25 / 16 n^2 (9 + n^2 / 4 (49 + 81 / 16 n^2)))) \quad (2.01)$$

Calculation time: 0.97 ms (Figure 1)
Result: 6 367 654.50005 76 m

Method 2 (Krack, 1982):

$$r = a (1 - n) (1 - n^2) (1+9 / 4 n^2 (1+25 / 16 n^2)) \quad (2.02)$$

Calculation time: 0.67 ms
Result: 6 367 654.50005 76 m

Method 3 (Krack, 1983). The most significant aspect of this method is its speed combined with accuracy:

$r = a (1 + n^2 / 4) / (1 + n)$ used by programs of this publication (3.01) (2.03)

Calculation time: 0.30 ms
Result: 6 367 654.50005 67 m

The examples 1, 2, 3, 4 and 5 with formulae (2.04- 2.13) are *incomplete* "BASIC" programs and refer to the calculation of the Meridional Arc (G_m) according to the formula:

$\delta G_m = \rho_m \delta B$ (Helmert, 1880).

All angles are expressed in *radians.*

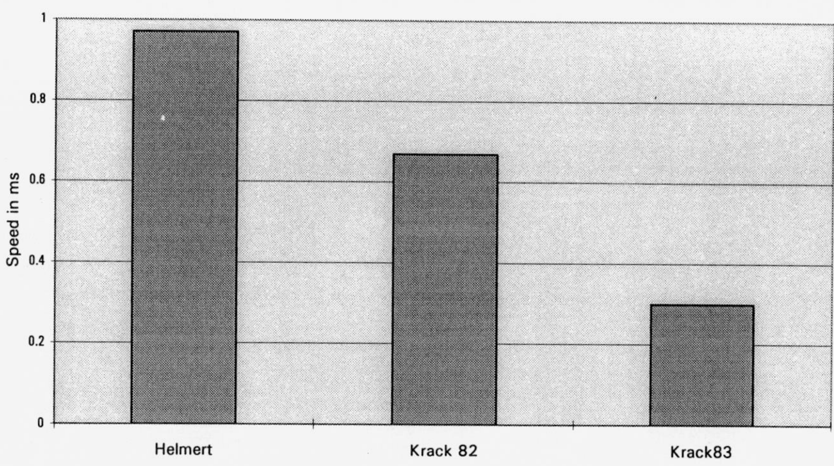

Figure 1: Comparison of computing time to calculate the radius of the rectifying sphere

Example 1 (J/E/K, 1959). See the calculation of the Meridional Arc, Direct; Inverse by iteration:

```
E2=C/2*(-3/4*E3+15/16*E3^2-525/512*E3^3+2205/2048*E3^4-72765/65536*E3^5)      (2.04)
GM=R0*LAT+E2*SIN(2*B)+E4*SIN(4*B)+E6*SIN(6*B)+E8*SIN(8*B)                    (2.05)
```

Calculation time: 11.0 ms (direct)
Calculation time: 16.7 ms (inverse by iteration), see Figure 2

Example 2 (Krack, 1982), solving Direct and Inverse without iteration; equations in brackets:

```
E2=C/2*E3/4*(-3+E3/4*(15+E3/32*(-525+E3/4*(2205-72765/32*E3))))              (2.06)
GM=R0*LAT+E2*SIN(2*B)+E4*SIN(4*B)+E6*SIN(6*B)+E8*SIN(8*B)                    (2.07)
```

Calculation time: 10.4 ms (direct)
Calculation time: 10.4 ms (inverse)

Example 3 (Hooijberg, 1996), solving Direct and Inverse with one sin(x)= f(C):

```
E2=C*(((((-86625/8*E3+11025)/64*E3-175)/4*E3+45)/16*E3-3)/4*E3)              (2.08)
GM=R0*LAT+(((E8*C+E6)*C+E4)*C+E2)*C1*SQR(1#-C)                               (2.09)
```

Calculation time: 6.0 ms (direct)
Calculation time: 6.0 ms (inverse)

Example 4 (Krack, 1983), solving Direct and Inverse *without fractions,* with one sin(x)= f(C):

```
E1=-N*(36+N*(45+39*N))                                                       (2.10)
GM=R0*(LAT+SQR(1-C)*C1/12*(E1+C*(E2+C*E3)))                                  (2.11)
```

Calculation time: 3.2 ms (direct)
Calculation time: 3.2 ms (inverse)

Example 5. NGS 5 (Stem, 1989a) solving Direct and Inverse with fractions and one sin(x)= f(C):

```
U2=-3*N/2+9*N^3/16, etc.
U0=2*(U2-2*U4+3*U6-4*U8)                                                     (2.12)
GM=R0*(LAT+(SQR(1-C*C)*C)*(U0+C*C*(U2+C*C*(U4+U6*C*C))))                     (2.13)
```

Calculation time: 12.8 ms (direct)
Calculation time: 12.8 ms (inverse)

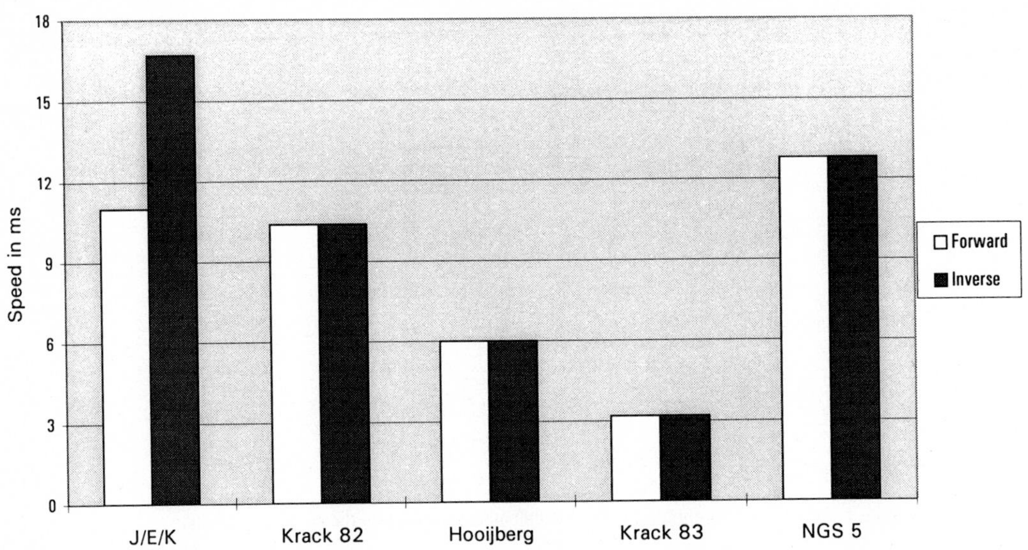

Figure 2: Comparison of computing time to calculate the meridional arc

Notes

For treatment of algorithms, see (Glasmacher / Krack, 1984).

Furthermore the interested reader may consult Grafarend:

"Various geodetic problems such as the free non-linear geodetic boundary value problem, and the problem of non-linear regression, demand the inversion of an univariate, bivariate, trivariate, in general multivariate homogeneous polynomial of degree n.

A new algorithm is discussed using the Symbolic Computer Manipulation algorithm"

(Grafarend, 1996a).

2.4.1 Accuracy and Precision

The term accuracy refers to the closeness between calculated values and their correct or true values. The further a calculated value is from its true value, the less accurate it is. Conversely, calculated values may be accurate but not precise if the calculated values are well distributed about the true value, but are significantly different from each other. Calculated values will be both precise and accurate if the values are very closely grouped around the true value. Precision is expressed in terms of the mean of the squares of the errors (Bomford, 1977).

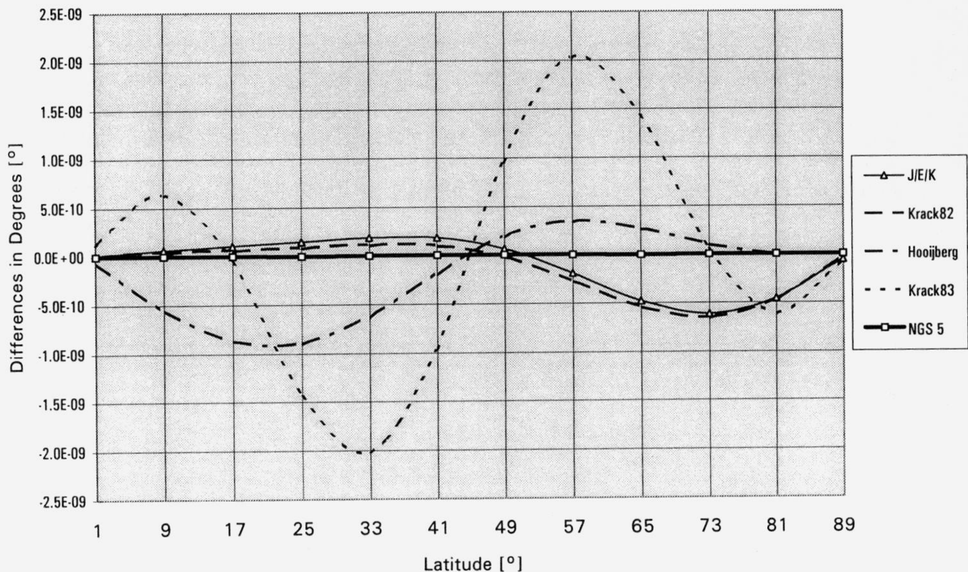

Figure 3: Round-trip error of the meridional arc in degrees

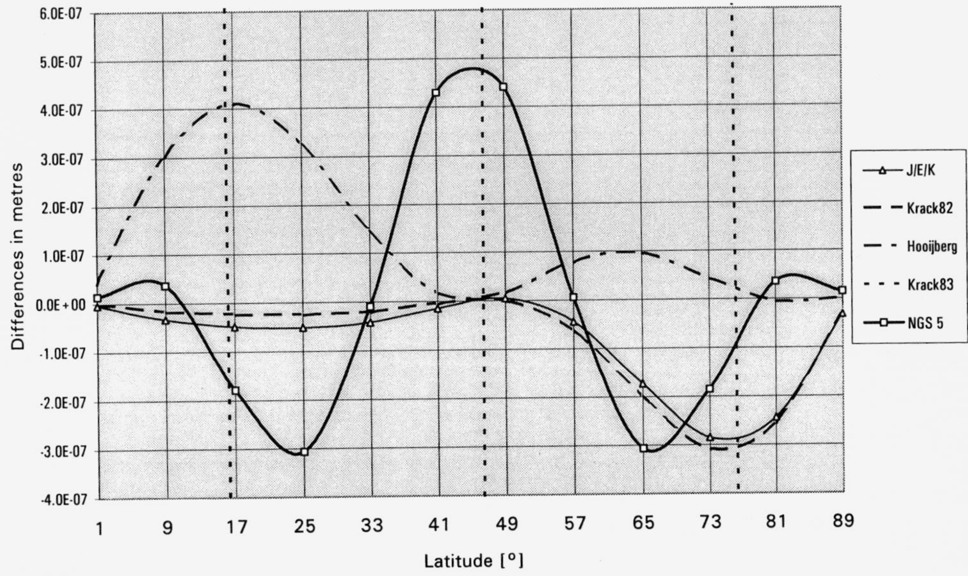

Figure 4: Round-trip error of the meridional arc in m

2.4.2 Errors

The term error can be considered as referring to the difference between a given calculation and a true value of the calculated quantity.

The "round-trip errors" are the differences in degrees or in metres between the starting and ending co-ordinates, illustrated in Figure 3 and Figure 4, and are calculated as follows:

- Latitudes (B^R) are converted to the Meridional Arc distances (G_m), which are converted back to Latitudes
- Meridional Arcs are converted to Latitudes, which are converted back to Meridional Arcs.

The round-trip error of (J/E/K, 1959) is:	$3 * 10^{-7}$ m for G=f(B), and	$6 * 10^{-10}$ ° for B = f(G)
The round-trip error of (Krack, 1982) is:	$3 * 10^{-7}$ m for G=f(B), and	$6 * 10^{-10}$ ° for B = f(G)
The round-trip error of (Hooijberg, 1996) is:	$4 * 10^{-7}$ m for G=f(B), and	$9 * 10^{-10}$ ° for B = f(G)
The round-trip error of (Krack, 1983) is:	$2 * 10^{-4}$ m for G=f(B), and	$2 * 10^{-9}$ ° for B = f(G)
The round-trip error of NGS 5 is:	$3 * 10^{-7}$ m for G=f(B), and	$3 * 10^{-12}$ ° for B = f(G).

---

3. Datums and Reference Systems

3.1 The Figure of the Earth

Clairaut, 1743: "Théorie de la figure de la terre tirée des principles de l'hydrostatique".

The mathematical Figure of the Earth, a term which in modern usage is usually applied to the classical definition of the geoid, is defined as the equipotential (level) surface of the earth's gravitational field. This surface, on average, coincides with mean sea level (MSL) in the open undisturbed ocean or its hypothetical extension under the land masses so as to encircle the earth (Bomford, 1977; Clarke, 1880; Fischer, 1845; Listing, 1873; Torge, 1991).

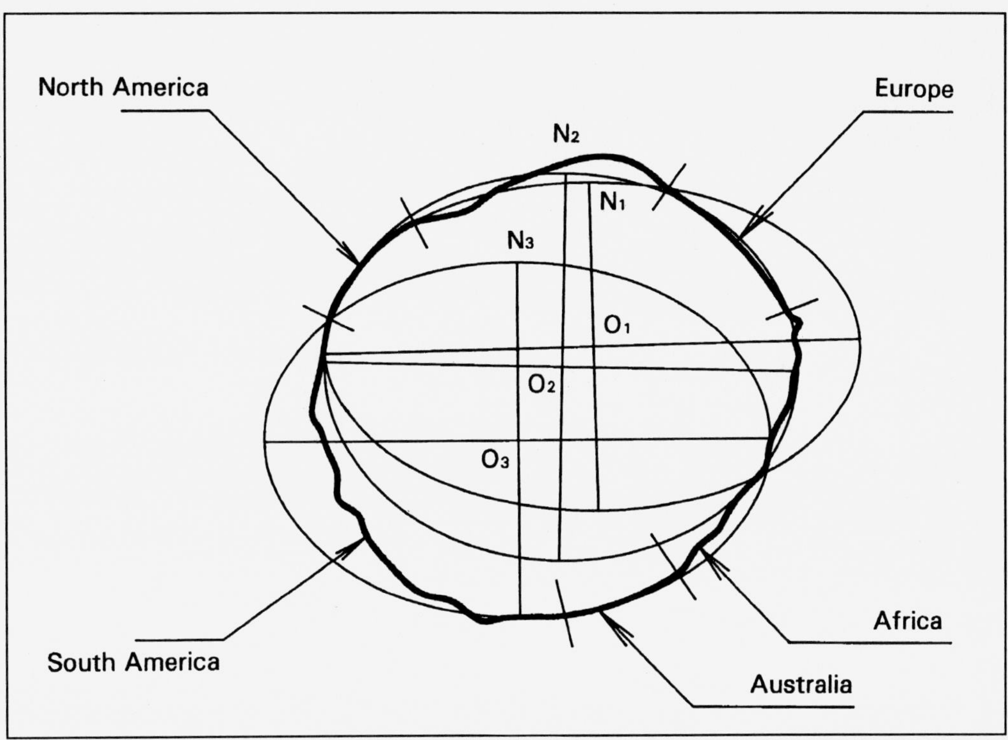

Figure 5: The geoid and three reference ellipsoids

To obtain complete and accurate results in the geodetic calculations, it is necessary to know the form of the geoid. Clarifying that the MSL is not an equipotential surface may be necessary and in its simplest definition it would comprise a mean of sea level surface approximated and observed over 18.67 years |[2].

Nevertheless, the difference between the Vertical Datum of Sweden and Finland (RH70 - N60) is 0.08 m, between the Vertical Datum of Sweden and FRG (RH70 - HN76) is 0.16 m, and between the Vertical Datum of Sweden and Poland (RH70 - H60) is 0.37 m (Ekman, 1995a, 1995b; Sjöberg, 1993).

In Figure 5 an ellipsoid that fits very well, for instance, in Australia, does not necessarily fit in America or Europe. Some of these ellipsoids and areas in use are given in Table 2: Datums and origins.

---

[2] Rotation of the lunar orbital plane. Although the moon completes one orbital pass in one month, it takes 18,67 years for a complete circuit of the orbital plane around the ecliptic pole (Doodson, 1960).

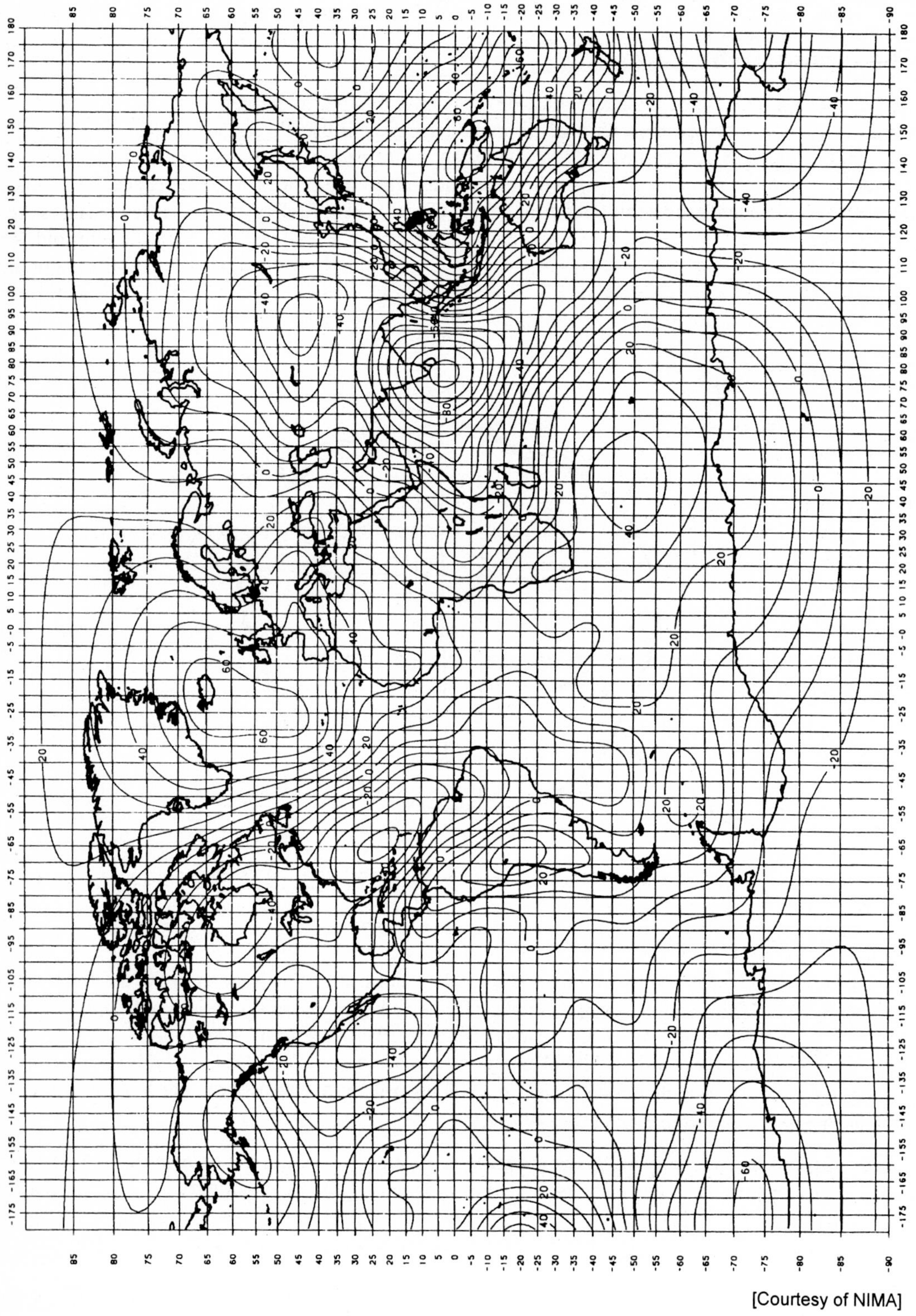

[Courtesy of NIMA]

Figure 6: WGS 84 geoid referenced to the WGS84 ellipsoid in m

As a surface perpendicular to the direction of gravity, the geoid manifests the gravitational forces of the earth, which vary the irregular intensity and direction from place to place due to the irregular mass distribution in the earth. These forces also affect the orbits of satellites and the trajectories of missiles. The geoid cannot be used as a computational surface (Hirvonen, 1962).

Using a geometrical figure to describe the geoidal surface of the Earth, the ellipsoid is chosen as an approximation for the shape of the geoid. The details of the irregular geoid are described by the separations from the chosen computational biaxial reference ellipsoid at specific points.

"Geoid undulations" is a word to describe the separation (geoidal height=N) between the two surfaces. In a mathematical sense, the geoid is also defined as so many metres above (+N) or below (- N) the ellipsoid. The overall size and shape of the geoid is expressed in these approximations by two parameters, e.g. the major axis and the flattening. The geoid can be depicted as a contour chart which shows the deviations of the geoid from the ellipsoid selected as the mathematical figure of the Earth (Doodson, 1960).

Classical Geoid and Quasi Geoid

Defining a geoid requires adopting either a *classical geoid* or a *quasi geoid.* Calculating orthometric heights using the actual gravity field will define the classical geoid. Calculating normal heights using the normal gravity field will define a quasi geoid, which is not exactly an equipotential or a level surface.
The difference between the classical geoid and the quasi geoid is approximately proportional to the square of the height above mean sea level (MSL). At MSL the difference is almost zero, it increases to about 0.1 m for a height of one thousand metres, and extends to about one metre for a height of three thousand metres (Ekman, 1995b; Heiskanen, 1967; Torge, 1989).

Geodetic control data are collected and adjusted on an adopted regional, national, continental or world-wide datum. Thus, there is little difference between angles and distances measured on the topographic surface of the Earth and their geodetic counterparts represented on the ellipsoid.

In the early 1950s, AMS (Army Map Service, now NIMA) added new information about the *southern hemisphere*:

- A field party in the Sudan completed a missing link in the triangulation along the Meridian 30° E through Africa (Fischer, 1959, 1972)
- The Inter American Geodetic Survey (IAGS), now a component of NIMA, completed a long triangulation arc through Central and South America.

These two long arcs from way in the north to way down south, more than 100° long, were analysed at AMS and a new Figure of the Earth was derived: Hough 1960.

Since 1957, after the launching of the artificial Earth satellites - the Sputniks by the USSR, the Vanguards by the USA - the flattening of the Earth was firmly established as 1 : 298.3.

The geoidal profile 30° E (Figure 7) is on various datums: ED50 (International Ellipsoid), WGS84, and Arc Datum, Adindan Datum (Clarke 1880 Modified ellipsoid).

A chosen ellipsoid together with its uniquely chosen location (a datum point) and orientation with respect to the geoid constitutes a geodetic reference datum. Consequently, a major goal of geodesy was the determination of the geoid of the period between 1900 and 1950. In classical geodetic practice, this is generally done so that the reference ellipsoid is a close fit to the geoid in the particular area of interest (Torge, 1991).

In Satellite Geodesy, the centre of the Reference Ellipsoid |[3] - is made to coincide with the Earth's gravitational centre. In *"Relativistic Geodesy, the geoid is the surface nearest to mean sea level on which precise clocks run with the same speed"* (Bjerhammer, 1986).

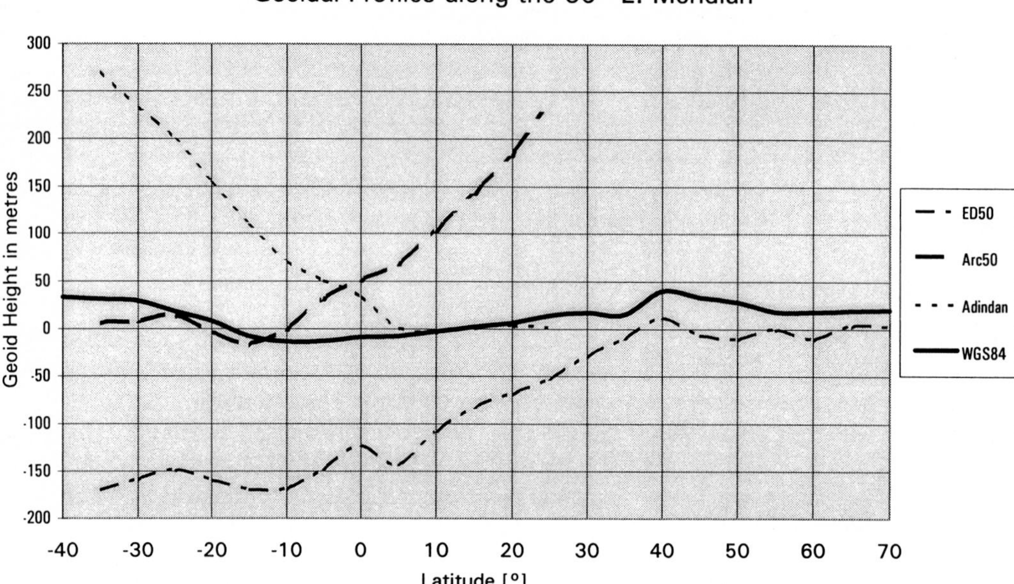

Figure 7: Comparison of various datums along the meridian from Finland to South Africa

Present world-wide mapping operations which use the geoid as a reference surface for elevations are dependent on an improved WGS84 geoid (Figure 6). The accuracy of the geoid is expected to have a horizontal accuracy better than one metre and a vertical accuracy better than two metres world-wide. Any improvement will have a beneficial impact on mapping and positioning operations, which demand an accurate geoid. As ongoing geodetic research provides increasingly better estimates of geoidal heights, absolute geoidal height accuracies are being improved as more data are collected.

Currently, there is no world-wide vertical world geodetic datum cover, but we expect to see such a capability early the next century. One of the actual goals of geodesy is, therefore, to investigate the geoidal surface (Ayres, 1995).

[3] NIMA stopped considering spheroids and ellipsoids as equivalent due to the fact that an spheroid is very complex surface while an ellipsoid is a simple two-degree surface.

3.2 Vertical Datum

In the classical (2+1)-dimensional geodesy and the 3-dimensional geodesy exists the problem of vertical control. For completeness the vertical datum and associated jargon is discussed in notes and its definition.

Notes About Projective Heights in Geometry and Gravity Space:

> Grafarend: *"With the advent of artificial satellites, global positioning is carried out in geodetic reference systems, in three dimensional Euclidean space at some reference epoch. In particular, systems are responsible for the materialisation of three-dimensional geodesy in an Euclidean space.*
>
> *In gravity space the triplet (X, Y, Z) is transformed into physical heights with respect to the reference equipotential surface, the geoid, at some reference epoch by a geodesic projection. The geodesic projection is performed by a curved line - the geodesic - as the plumbline / orthogonal trajectory with respect to a family of equipotential surfaces in curved gravity space"*
>
> (Grafarend, 1995c).

Permanent Tide in Positioning

A report by Poutanen et al provides the zero-crust hypothesis that corresponds to the theory already accepted in the resolution of IAG in 1983 for gravimetric works which is in tune with modern practice and technical development in the field of ECEF co-ordinate computation. In summary:

> *"Little has been published about the treatment of the tidal deformation of the Earth in GPS computation. Most ECEF co-ordinates are reduced to the* "nontidal" crust, *conventionally defined using physically meaningless parameters. However, the great demand for ever increasing accuracy and the need to combine the GPS based co-ordinates with other methods requires an agreeable way to handle the tides.*
>
> *The tide-generating potential of the Moon and the Sun result in deformation of the shape of Earth. This deformation can be divided into a periodic and a time-independent or permanent part. The periodic part should always be eliminated. Concerning the permanent part, the* nontidal geoid *is the equipotential surface of the gravity field of the Earth when the permanent tidal deformation is eliminated. A* zero geoid *refers to the gravity field of the Earth when the permanent tidal deformation is preserved.*
>
> *The gravity field of the Earth with its permanent tidal deformation preserved, plus the time-independent part of the tidal field generated by Sun and Moon refers to a* mean geoid. *Equally, the* nontidal crust *refers to the model of the Earth when the permanent tidal deformation is eliminated and the* zero crust (=mean crust) *refers to the model of the Earth when this deformation is preserved"*
>
> (Kakkuri, 1996).

For further reading, the interested reader may consult (Poutanen, 1996).

The Geopotential Value W_0

An attractive field for future research will be in line of thought about W_0 - *the geopotential value of the equipotential surface* - in which Grafarend has contributed so substantially. A study about the spheroidal "free air" potential variation may serve as a guide to solve the problem of W_0 *without* using the gravity field of a spheroidal equipotential surface.

In summary, Grafarend reports:

> The "Baltic Sea Level Project" situated between 53°N, 6°E and 66°N, 28°E is used as an example. Twenty-five GPS stations are determined in WGS84 3-D co-ordinates with its orthometric height at the Finnish Height datum N60 epoch 1993.4 in the vicinity of *primary mareographic stations*. Matching the 3-D co-ordinates of the primary mareographic stations with the GPS station co-ordinates was carried out using a new spheroidal free-floating-variation technique. Transformation spherical harmonic coefficients of the gravity potential results into spheroidal harmonic coefficients, computed for all GPS stations using a global model OSU91A of the gravitational potential field. Using the orthometric heights of the GPS stations with respect to the geoid reference datum, the gravity potential with respect to a global spheroidal model has been transformed to yield the *vertical geodetic* W_0 value.
>
> The mean of W_0 data: 6 263 685.58 kgal m, the geopotential (tide gauge) value W_0, was found which holds for the reference epoch 1993.4, including the RMS error: 0.36 kgal m. As a check the W_0 data of epoch 1993.4 are favourably compared with the spheroidal model gravity field of the Somigliana-Pizetti type.
> Variations of the miscellaneous W_0 data vs. 6 263 686.085 kgal.m |[4], *the* W_0 *value* of The Geodesist's Handbook (GH) 1992 (Burša, 1992), are shown in Figure 8.
>
> (Grafarend, 1997c).

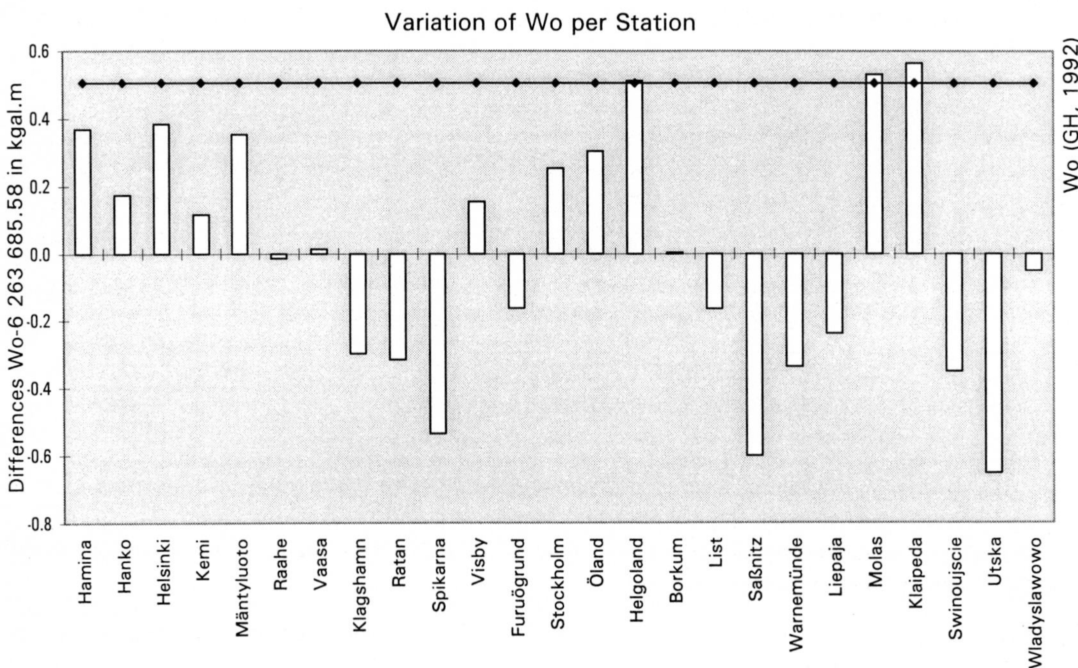

Figure 8: Variation of the various W_0 data vs. W_0 data of The Geodesist's Handbook 1992

[4] Correct W_0 value of The Geodesist's Handbook 1992

3.2.1 Definition of the Vertical Datum

In the determination of precise geodetic heights, either orthometric or normal heights are employed. Using C, geopotential number, heights are defined as follows:

> "Orthometric Height is the height $H_o = C/g$ of a point *P,* in which g is the mean gravity value assigned to the plumbline between the geoid and the point *P.*
>
> Normal Height is the height $H_n = C/\gamma$ of a point *P,* in which γ is the mean normal gravity value assigned to the normal plumbline between the ellipsoid and the point *Q,* where the normal potential is equal to the actual potential at the point of calculation *P.* (So-called *telluroid mapping* $P \rightarrow Q$*, [Figure 48])*" (Kakkuri, 1996).

Consequently, the ideal vertical datum is completely defined by the equipotential surface. The *ideal vertical datum* is the geoid. In that case C is equal to zero and consequently $H_o = H_n = zero.$ In practice, the vertical positions are given with respect to the local mean sea level of a determined tide gauge and epoch |[5]

(Kakkuri, 1995).

3.3 Linear Units of Measurement

Eratosthenes (276-195 BC) of Alexandria found that at the 21st of June - the time of the summer solstice - the rays of the sun descended vertically into a well in Syene (Assuan), whereas in Alexandria, roughly on the same meridian, they formed an angle γ with the direction of the plumb line (Airy, 1830; Torge, 1991; Wilford, 1981).
From the length of the shadow of an obelisk, called a gnomon, he determined this angle as γ= 7°.2, i.e. 1/50 of a complete circle. Eratosthenes was told that a camel caravan needed 50 days to make the journey - Syene to Alexandria - and that camels usually travelled 100 stadia a day. The distance would be about 5 000 stadia. With the assumed length of a stadium as 1/10 n mi, we obtain an earth circumference of 25 000 n mi or 46 300 km. This value departs from a mean spherical earth circumference of 40 000 km by 16% approximately. Further, the reader should note about the stadium that its value varies between 150 and 185 m in literature.

An essential aspect of geodetic measurement is the need to define the standard to which the measured lengths and the dimensions of the reference ellipsoid are related, such as the standard Bar O_1, the legal metre, the foot, the Russian double-toise, the Belgian toise, the Prussian standard toise, the British toise, the Indian- or Australian ten-foot standard bars, the Gunter's chain, Wiener Klafter, and finally, the international metre.

In March 19, 1791, the "Assemble constituante" of France sanctioned a project of "Commission Générale des Poids et Mesures" that a ten-millionth part of the earth's meridian quadrant should be adopted as the national standard of length, to be called the metre (Admiralty, 1965).

At its meeting in Paris of October 20, 1983 the Conférence Générale des Poids et Mesures (CGPM) redefined the metre as follows (Price, 1986; Rotter, 1984):

> *"The metre is the length of the path travelled by light in vacuum during a time interval of 1/299 792 458 of a second".*

In other words this definition gives a fixed value to the speed of light of 299 792 458 m s^{-1} exactly.

[5] For treatment of Telluroid or Terroid, see (Graaff-Hunter, 1960).

Unit	$=m_{INT}$	One m_{INT}=
International inch	0.0254 (exactly)	39.37
International foot	0.3048 (exactly)	3.28083 98950 13123
International yard	0.9144 (exactly)	1.09361 3298
British Imperial yard	0.91439 84155	1.09361 5193
British Imperial foot	0.30479 94718	3.28084 558
Foot of bar O_1	0.30480 07491	3.28083 18318
(South) African geodetic foot	0.30479 72654	3.28086 9335
Indian foot	0.30479 84100 66	3.28085 70097
Indian foot 1956 |[6]	0.3048 (exactly)	3.28083 98950 13123
US survey yard	0.91440 18288 03658	1.09361 11111 11111
US survey foot	0.30480 06096 01219	3.28083 33333 33333
US survey foot	12.00 / 39.37 (exactly)	39.37 / 12.00 (exactly)
Toise of Paris	1.94903 63098 24586 73212	0.51307 40740 74074 07407
Klafter 1871	1.89648 4	0.52729 156
Legal metre |[7] [3.4.2.2]	1.00001 33554 1	0.99998 66447 70

Table 1: Conversion of linear units

Note:

The original "toise of Paris" conversion is: 1 metre = 5 130 740 × 864 ÷ 10 000 000 = 443.295 936 Paris' lines. The conversion factor used in (Table 1): 443.296÷ 864
(J/E/K, 1959).

The notation "legal metre" varies with the country and is therefore only of local national significance.

3.4 Geodetic Reference Datum

Historically, as countries around the world began to map and survey the nearby land and sea, they developed local reference frames (datums) to provide a point of origin and orientation for the specific area or region. These local geodetic datums generally employed nongeocentric ellipsoids as the best-fitting figure of the Earth in the region being surveyed or mapped. A co-ordinate reference system generally consists of the arbitrary theoretical definition of a system of geodetic co-ordinates.

Classical Geodesy

A geodetic datum is usually given a distinctive name. It is concerned with the realisation of that reference system, through a series of conventionally agreed measures (Geodetic Glossary, 1986).

---

[6] (Bhattacharji, 1962)

[7] (Zeger, 1991)

Before 1960, defining a complete geodetic datum by five quantities was customary:

- latitude of an initial point
- longitude of an initial point
- azimuth of a line from this point
- 1st unique parameter semi-major axis (a) of a reference ellipsoid, and
- 2nd unique parameter, e.g. reciprocal flattening (f^{-1}) of a reference ellipsoid.

In addition, specification of the components of the deflection of the vertical at the initial point, or the condition that the minor axis of the ellipsoid be parallel to the Earth's axis of rotation granted one more quantity. The datum was *not complete* because the origin of the co-ordinate system remained free to shift in one dimension. The definition does not correspond to modern usage (Geodetic Glossary, 1986).

Satellite Geodesy

After 1960, the arrival of Earth orbiting satellites allowed the development of global geodetic reference frames for practical use. The first of these global reference frames was the US-Department of Defense (DoD) World Geodetic System of 1960 (WGS60). This was followed by improved systems in 1966 and 1972 and ended in development of the World Geodetic System of 1984 (WGS84), see [3.5]. This Conventional Terrestrial Reference Frame (CTRF) provides a coherent set of global models and definitions which form the basis for all current DoD mapping, charting, navigation and geodesy. WGS84 was realised through Doppler observations from the TRANSIT satellite system using a world-wide group of tracking stations (Kouba, 1994).

Parallel efforts by the scientific associations have resulted in the development of other highly accurate terrestrial reference frames, such as an annual series of International Earth Rotation Service (IERS) Terrestrial Reference Frames (ITRF) [11.5]. These frames are based on observations from Satellite Laser Ranging (SLR), Lunar Laser Ranging (LLR), Very Long Baseline Interferometry (VLBI) and the Global Positioning System (GPS or NavStar).

At least eight constants are needed to form a *complete geodetic datum:*

- three constants to specify the location of the origin of the co-ordinate system
- three constants to specify the orientation of the co-ordinate system, and
- two constants to specify the dimensions of the reference ellipsoid (Geodetic Glossary, 1986).

Geocentric datums are located by computations that take into account the Earth's gravitational field as determined from measurements on the surface or from analyses of the orbits of satellites.

The national triangulation network of every country involves a properly defined ellipsoidal geodetic datum, e.g. ED50, ED87. New geodetic reference frames, such as *US-DoD* WGS84 (or its predecessor WGS72), and *CIS* SGS90 (or its predecessor SGS85), were established to facilitate a world-wide use of efficient and precise geodetic satellite techniques in surveying. GRS80 is a terrestrial continental geodetic reference frame for, e.g. the European or North American continental mapping, navigation and positioning applications.

Traditional local vertical datums are based on many realisations of mean sea level (MSL). Currently, there is no world-wide vertical WGS84 cover, with a refined geoid defined to unify and tie together these local vertical datum systems. However, research is now underway to allow use of the ellipsoid or geoid, as measured by the GPS, as a common international terrestrial reference. Key to such an objective is a *refined geoid*
(Ayres, 1995).

A brief description of the geodetic datums and co-ordinate reference systems in the world follows this introduction (Hager, 1991).

Datums and Origins

Datum	Ellipsoid	Origin	Latitude	Longitude
Adindan	Clarke 1880 Mod.	Station 15 Adindan	22° 10' 07".110 N	31° 29' 21".608 E
American Samoa 1962	Clarke 1866	Betty 13 ECC	14° 20' 08".34 S	170° 42' 52".25 W
Arc-Cape 1950	Clarke 1880 Mod.	Buffelsfontein	33° 59' 32".000 S	25° 30' 44".622 E
Argentine	International	Campo Inchauspe	35° 58' 16".56 S	62° 10' 12".03 W
Ascension Island 1958	International	Mean of three Sta.	7° 57' 00".0 S	14° 23' 00".0 W
Amersfoort	Bessel 1841	O.L.-Vrouwe Tower	52° 09' 22".178 N	5° 23' 15".5 E
Australian Geodetic	ANS 1966	Johnston Geodetic	25° 56' 54".5515 S	133° 12' 30".0771 E
Bermuda 1957	Clarke 1866	Ft. George	32° 22' 44".36 N	64° 40' 58".11 W
Bern 1898	Bessel 1841	Bern Observatory	46° 57' 08".66 N	7° 26' 22".5 E
Betio Island 1966	International	1966 Secor Astro	1° 21' 42".03 N	172° 55' 47".9 E
Brussels	International	Royal Observatory	50° 47' 57".704 N	4° 21' 24".983 E
Camp Area	International	Camp Area Astro	77° 50' 52".521 S	166° 40' 13".753 E
Canton Astro 1966	International	1966 Canton Secor	2° 46' 28".99 S	171° 43' 16".530 W
Chua Astro Brazil	International	Chua Astro	19° 45' 41".16 S	48° 06' 07".56 W
Corrego Alegre Brazil	International	Corrego Alegre	19° 50' 15".14 S	48° 57' 42".75 W
European 1950	International	Potsdam-Helmert Twr	52° 22' 51".4456 N	13° 03' 58".9283 E
European 1987	International	Munich D-7835	48° 08' 22".2273 N	11° 34' 26".4862 E
Geodetic Datum 1949	International	Papatahi Trig.	41° 19' 08".900 S	175° 02' 51".000 E
Ghana	War Office	Leigon G.C.S 121	5° 38' 52".270 N	0° 11' 46".08 W
Graciosa Island	International	SW Base 1948	39° 03' 54".934 N	28° 02' 23".882 W
Geodetic Datum 1949	International	Papatahi	41° 19' 08".9 S	175° 02' 51" E
Gizo, Provisional DOS	International	Gux 1	9° 27' 05".272 S	159° 58' 31".752 E
Guam 1963	Clarke 1866	Togcha or Lee No. 7	13° 22' 38".490 N	144° 45' 51".560 E
Hjorsey 1955	International	Hjorsey	64° 31' 29".260 N	22° 22' 05".840 W
Hu-Tzu-Shan	International	Hu-Tzu-Shan	23° 58' 32".340 N	120° 58' 25".975 E
Iben Astro, Navy (Tk)	Clarke 1866	Iben Astro 1947	7° 29' 13".05 N	151° 49' 44".42 E
IGN France	Clarke 1880IGN	Paris 1922	48° 50' 14"0 N	2° 20' 14".025 E
Indian	Everest	Kalianpur 1895	24° 07' 11".26 N	77° 39' 17".57 E
Indonesia 1974	GRS67	Padang	0° 56' 38".414 S	100° 22' 08".804 E
Ireland 1965	Airy Modified	Donard Slieve	54° 10' 48".2675 N	5° 55' 11".8675 W
Isle Socorro Astro	Clarke 1866	Station 038	18° 43' 44".93 N	110° 57' 20".72 W
Johnston Island 1961	International	Johnston Island 1961	16° 44' 49".729 N	169° 30' 55".219 W
Kusaie Astro 1962/65	International	Allen Sodano Light	5° 21' 48".8 N	162° 58' 03".28 E
Liberia 1964	Clarke 1880	Robertsfield Astro	6° 13' 53".02 N	10° 21' 35".440 W
Luzon 1911	Clarke 1866	Balanacan	13° 33' 41".000 N	121° 52' 03".000 E
Midway Astro 1961	International	Midway Astro 1961	28° 11' 34".5 N	177° 23' 35".72 W
Merchich	Clarke 1880IGN	Merchich	33° 26' 59".672 N	7° 33' 27".295 W
Namibian	Bessel	Schwarzeck	22° 45' 35".82 S	18° 40' 34".549 E

(Table 2: Datums and origins, cont'd)

Datum	Ellipsoid	Origin	Latitude		Longitude	
Naparima	International	Naporima	10° 16' 44".86	N	61° 27' 34".62	W
Nigeria	Clarke 1880	Minna	9° 38' 08".87	N	6° 30' 58".76	E
North American 1927	Clarke 1866	Meades Ranch	39° 13' 26".686	N	98° 32' 30".506	W
NAD27 - Cape Canaveral	Clarke 1866	Central	28° 29' 32".364	N	80° 34' 38".77	W
NAD27, White Sands	Clarke 1866	Kent 1909	32° 30' 27".079	N	106° 28' 58".694	W
Old Bavarian	Bessel 1841	Munich	48° 08' 20".000	N	11° 34' 26".483	E
Old Hawaiian	Clarke 1866	Oahu West Base	21° 18' 13".890	N	157° 50' 55".800	W
OSGB36	Airy	Herstmonceux	50° 51' 55".271	N	0° 20' 45".882	E
Pico de las Nieves	International	Pico de las Nieves	27° 57' 41".273	N	15° 34' 10".524	W
Prv. S. American 1956	International	La Canoa	8° 34' 17".17	N	63° 51' 34".88	W
Prv. S. Chile 1963	International	Hito XVIII	64° 67' 07".76	S	68° 36' 31".24	W
Qornoq	International	Station 7008	64° 31' 06".27	N	51° 12' 24".86	W
Rauenberg	Bessel 1841	Potsdam-Helmert Twr	52° 22' 53".9540	N	13° 04' 01".1527	E
Riga	Bessel 1841	St. Peter Church (old)	56° 56' 53".9190	N	24° 06' 31".8980	E
Riga	Bessel 1841	St. Peter Church (low)	56° 56' 53".9350	N	24° 06' 31".9370	E
Roma	International	M. Mario	41° 55' 25".51	N	12° 27' 08".40	E
Sierra Leone 1960	Clarke 1880	DOS Astro SLX2	8° 27' 17".6	N	12° 49' 40".2	W
South American	GRS67	Chua (Brazil)	19° 45' 41".6527	S	48° 06' 04".0639	W
Swallow Islands	International	1966 Secor Astro	10° 18' 21".42	S	166° 17' 56".79	E
System 1932	Bessel 1841	Pulkovo Geodetic Obs.	59° 46' 18".71	N	30° 19' 38".55	E
System 1942	Krassovsky	Pulkovo Geodetic Obs.	59° 46' 18".55	N	30° 19' 42".09	E
HD1972	GRS1980	Szólóhegy	47° 17' 52".6156	N	19° 36' 09".9865	E
Tananarive (Antanunarivo)	International	Tananarive Obs. 1925 (Antananarivo)	18° 55' 02".10	S	47° 33' 06".45	E
Tokyo	Bessel 1841	Tokyo Obs. (old)	35° 39' 17".515	N	139° 44' 40".502	E
Viti Levu 1916	Clarke 1880	Monavatu (Lat. only)	17° 53' 28".285	S		
		Suwa (Lon. only)			178° 25' 35".35	E
Voirol 1960	Clarke 1880IGN	Voirol Observatory	36° 45' 07".9	N	3° 02' 49".45	E
Wake Island	International	Astro 1952	19° 16' 48".7	N	166° 38' 46".8	E
Yacare	International	Yacare	30° 35' 53".68	S	57° 25' 01".30	W
Yof Astro 1967	Clarke 1880 Mod.	Yof Astro 1967	14° 44' 41".62	N	17° 29' 07".02	W
Five ILS Stations						
CIO epoch 1900-05	Carloforte Obs.	Italy	39° 08' 08".941	N	-	
CIO epoch 1900-05	Gaithersburg Obs.	USA, MD	39° 08' 13".2022	N	-	
CIO epoch 1900-05	Kitab Obs.	CIS	39° 08' 01".850	N	-	
CIO epoch 1900-05	Misuzawa Obs.	Japan	39° 08' 03".602	N	-	
CIO epoch 1900-05	Ukiah Obs.	USA, CA	39° 08' 12".096	N	-	

(Geodetic Glossary, 1986)

(Courtesy NIMA)

Table 2: Datums and origins

Datums and Associated Reference Ellipsoids

Courtesy (NIMA, 1991).

Local Geodetic Datum	Reference Ellipsoid	Country	Area: Africa
North Sahara 1959	Clarke 1880	Algeria	
Voirol1960	Clarke 1880	Algeria	
Arc1950	Clarke 1880	Botswana	
Adindan	Clarke 1880	Burkina Faso	
Point 58	Clarke 1880	Burkina Faso	
Arc1950	Clarke 1880	Burundi	
Adindan	Clarke 1880	Cameroon	
Minna	Clarke 1880	Cameroon	
Pointe Noire 1948	Clarke 1880	Congo	
Ayabelle Lighthouse	Clarke 1880	Djibouti	
Old Egyptian 1907	Helmert 1906	Egypt	
European 1950	International 1924	Egypt	
Adindan	Clarke 1880	Ethiopia	
Massawa	Bessel 1841	Ethiopia (Eritrea)	
M'Poraloko	Clarke 1880	Gabon	
Leigon GCS 121	Clarke 1880	Ghana	
Dabola	Clarke 1880	Guinea	
Bissau	International 1924	Guinea-Bissau	
Arc1960	Clarke 1880	Kenya	
Arc1950	Clarke 1880	Lesotho	
Liberia 1964	Clarke 1880	Liberia	
Arc1950	Clarke 1880	Malawi	
Adindan	Clarke 1880	Mali	
Merchich	Clarke 1880	Morocco	
Schwarzeck	Bessel	Namibia	
Point 58	Clarke 1880	Niger	
Minna	Clarke 1880	Nigeria	
Adindan	Clarke 1880	Senegal	
Afgooye	Krassovsky 1940	Somalia	
Sierra Leone 1960	Clarke 1880	Sierra Leone	
Cape - Arc1950	Clarke 1880	South Africa	
Adindan	Clarke 1880	Sudan	
Arc1950	Clarke 1880	Swaziland	
Arc1960	Clarke 1880	Tanzania	
Carthage	Clarke 1880	Tunisia	
Voirol 1960	Clarke 1880	Tunisia	
European 1950	International 1924	Tunisia	
Arc1950	Clarke 1880	Zaire	
Arc1950	Clarke 1880	Zambia	
Arc1950	Clarke 1880	Zimbabwe	

(Table 3: Local datum and associated reference ellipsoids, cont'd)

Local Geodetic Datum	Reference Ellipsoid	Country Area: Atlantic Ocean
Antarctica	Clarke 1880	Antarctica
Camp Area Astro	International 1924	Antarctica, McMurdo Camp Area
North American 1927	Clarke 1866	Antigua
Antigua Island Astro 1943	Clarke 1880	Antigua, Leeward Islands
Ascension Island 1958	International 1924	Ascension Island
Observatorio Meteorologico 1939	International 1924	Azores (Corvo and Flores Islands)
Graciosa Base SW 1948	International 1924	Azores (Faial, Graciosa, Pico)
Sao Braz	International 1924	Azores (Sao Miguel, Santa Maria Islands)
Cape Canaveral	Clarke 1866	Bahamas, Florida
North American 1927	Clarke 1866	Bahamas (incl. San Salvador Islands)
Bermuda 1957	Clarke 1866	Bermuda Islands
Pico de las Nieves	International 1924	Canary Islands
L. C. 5 Astro 1961	Clarke 1866	Cayman Brac Island
Deception Island	Clarke 1880	Deception Island (Antarctica)
Sapper Hill 1943	International 1924	East Falkland Island
North American 1927	Clarke 1866	Greenland (Hayes Peninsula)
Qornoq	International 1924	Greenland (South)
Hjorsey 1955	International 1924	Iceland
Porto Santo 1936	International 1924	Madeira Islands
Montserrat Island Astro 1958	Clarke 1880	Montserrat (Leeward Islands)
Fort Thomas 1955	Clarke 1880,	Nevis, St. Kitts (Leeward Islands)
Porto Santo 1936	International 1924	Porto Santo
Puerto Rico	Clarke 1866	Puerto Rico
Selvagem Grande 1938	International 1924	Salvage Islands
Graciosa Base SW 1948	International 1924	Sao Jorge, Terceira
ISTS 061 Astro 1968	International 1924	South Georgia Islands
Astro DOS 71/4	International 1924	St Helena Island
Naparima BWI	International 1924	Trinidad and Tobago
Tristan Astro 1968	International 1924	Tristan da Cunha
Puerto Rico	Clarke 1866	Virgin Islands

Local Geodetic Datum	Reference Ellipsoid	Country Continent: Asia
Herat North	International 1924	Afghanistan
Indian	Everest 1830	Bangladesh
Indian	Everest 1830	Bhutan
Timbalai 1948	Everest	Brunei Darussalam
Indian 1960	Everest 1830	Con Son Island (Vietnam)
Timbalai 1948	Everest	East Malaysia (Sabah, Sarawak)
Hong Kong 1963	International 1924	Hong Kong
Indian	Everest 1956	India, Nepal
Indonesian 1974	GRS67	Indonesia
Bukit Rimpah	Bessel 1841	Indonesia (Bangka and Belitung Islands)
Gunung Segara	Bessel 1841	Indonesia (Kalimantan)
Djakarta (Batavia)	Bessel 1841	Indonesia (Sumatra)
European 1950	International 1924	Iran

(Table 3: Local datum and associated reference ellipsoids, cont'd)

Local Geodetic Datum	Reference Ellipsoid	Country	Continent: Asia
Tokyo	Bessel 1841	Japan	
Pulkovo 1942	Krassovsky 1940	Lao, People's Democratic Republic of	
Macau	International 1924	Macau	
Indian	Everest 1830	Myanmar	
Pulkovo 1942	Krassovsky 1940	Mongolia	
Tokyo	Bessel 1841	Okinawa	
Indian	Everest	Pakistan	
Luzon 1911	Clarke 1866	Philippines	
Pulkovo 1942	Krassovsky 1940	China, People's Republic of	
Xian 1980	IAG75	China, People's Republic of	
Pulkovo 1942	Krassovsky 1940	Korea, People's Republic of	
Kertau 1948	Everest 1948	Singapore	
South Asia	Fischer 1960 Modified	Singapore	
Tokyo	Bessel 1841	South Korea	
Kandawala	Everest 1830	Sri Lanka	
Hu-Tzu-Shan	International 1924	Taiwan	
Indian 1954	Everest 1830	Thailand	
Indian 1975	Everest 1975	Thailand	
Hu-Tzu-Shan	International 1924	Taiwan	
Indian 1960	Everest 1830	Vietnam	
Pulkovo 1942	Krassovsky 1940	Vietnam	
Kertau 1948	Everest 1948	West Malaysia	

Local Geodetic Datum	Reference Ellipsoid	Country	Continent: Australia
Australian Geodetic 1966	Australian National	Australia	
Australian Geodetic 1984	Australian National	Australia	
Australian Geodetic 1966	Australian National	Tasmania	
Australian Geodetic 1984	Australian National	Tasmania	

Local Geodetic Datum	Reference Ellipsoid	Country	Area: Central America
North American 1927	Clarke 1866	Barbados	
North American 1983	GRS80	Barbados	
North American 1927	Clarke 1866	Barbuda	
North American 1983	GRS80	Barbuda	
North American 1927	Clarke 1866	Belize	
North American 1983	GRS80	Belize	
North American 1927	Clarke 1866	Caicos Islands	
North American 1983	GRS80	Caicos Islands	
North American 1927	Clarke 1866	Canal Zone	
North American 1983	GRS80	Canal Zone	
North American 1927	Clarke 1866	Costa Rica	
North American 1983	GRS80	Costa Rica	
North American 1927	Clarke 1866	Cuba	
North American 1983	GRS80	Cuba	

(Table 3: Local datum and associated reference ellipsoids, cont'd)

Local Geodetic Datum	Reference Ellipsoid	Country Area: Central America
North American 1927	Clarke 1866	Dominican Republic
North American 1983	GRS80	Dominican Republic
North American 1927	Clarke 1866	El Salvador
North American 1983	GRS80	El Salvador
North American 1927	Clarke 1866	Grand Cayman
North American 1983	GRS80	Grand Cayman
North American 1927	Clarke 1866	Guatemala
North American 1983	GRS80	Guatemala
North American 1927	Clarke 1866	Honduras
North American 1983	GRS80	Honduras
North American 1927	Clarke 1866	Jamaica
North American 1983	GRS80	Jamaica
North American 1927	Clarke 1866	Nicaragua
North American 1983	GRS80	Nicaragua
North American 1927	Clarke 1866	Turks Islands
North American 1983	GRS80	Turks Islands

Local Geodetic Datum	Reference Ellipsoid	Country Continent: Europe
Pulkovo 1942	Krassovsky 1940	Albania
Potsdam	Bessel 1841	Austria
European 1950	International 1924	Austria
European 1987	International 1924	Austria
European 1950	International 1924	Belgium
European 1987	International 1924	Belgium
Pulkovo 1942	Krassovsky 1940	Bulgaria
European 1950	International 1924	Channel Islands
European 1987	International 1924	Channel Islands
Pulkovo 1942	Krassovsky 1940	Commonwealth of Independent States (CIS)
European 1950	International 1924	Cyprus
S-JTSK	Bessel 1841	Czech, Slovakia, prior to Jan 1, 1993
Pulkovo 1942	Krassovsky 1940	Czech, Slovakia
European 1950	International 1924	Denmark
European 1987	International 1924	Denmark
Pulkovo 1942	Krassovsky 1940	Estonia (Eesti)
ETRS89	GRS80	Estonia (Eesti)
European 1950 (KKJ)	International 1924	Finland
European 1987	International 1924	Finland
EUREF89	GRS80	Finland
Paris 1922	Clarke 1880IGN	France
European 1950	International 1924	France
European 1987	International 1924	France
Pulkovo 1942	Krassovsky 1940	Federal Republic of Germany (East)
Potsdam	Bessel 1841	Federal Republic of Germany (West)
European 1950	International 1924	Federal Republic of Germany (West)

(Table 3: Local datum and associated reference ellipsoids, cont'd)

Local Geodetic Datum	Reference Ellipsoid	Country	Continent: Europe
European 1987	International 1924	Federal Republic of Germany (West)	
European 1950	International 1924	Gibraltar	
European 1987	International 1924	Gibraltar	
European 1950	International 1924	Greece	
ETRF89	GRS80	Greece	
Pulkovo 1942	Krassovsky 1940	Hungary	
Donard 1965	Airy Modified	Ireland	
European 1950	International 1924	Ireland	
European 1987	International 1924	Ireland	
European 1950	International 1924	Italy (incl. Sardinia, Sicily)	
European 1987	International 1924	Italy (incl. Sardinia, Sicily)	
Roma 1940	International 1924	Italy (incl. Sardinia)	
Roma 1983	International 1924	Italy (incl. Sardinia)	
Pulkovo 1942	Krassovsky 1940	Latvia (Latvija)	
LKS1992	GRS80	Latvia	
Pulkovo 1942	Krassovsky 1940	Lithuania (Lietuva)	
ETRS89	GRS80	Lithuania	
European 1950	International 1924	Luxembourg	
European 1987	International 1924	Luxembourg	
European 1950	International 1924	Malta	
Amersfoort	Bessel 1841	Netherlands	
European 1950	International 1924	Netherlands	
European 1987	International 1924	Netherlands	
European 1950	International 1924	Norway	
European 1987	International 1924	Norway	
Oslo	NGO1948	Norway	
EUREF89	GRS80	Norway (>1996)	
Pulkovo 1942	Krassovsky 1940	Poland (Polska)	
European 1950	International 1924	Portugal	
European 1987	International 1924	Portugal	
Pulkovo 1942	Krassovsky 1940	Romania	
European 1950	International 1924	Scotland	
European 1987	International 1924	Scotland	
European 1950	International 1924	Shetland Islands	
European 1987	International 1924	Shetland Islands	
Potsdam	Bessel 1841	Slovenia (Slovenija)	
European 1950	International 1924	Spain	
European 1987	International 1924	Spain	
European 1950	International 1924	Sweden	
European 1987	International 1924	Sweden	
RT1990	Bessel 1841	Sweden	
SWEREF93	GRS80	Sweden	
Berne CH-1903^{+}	Bessel 1841	Switzerland	
European 1950	International 1924	Switzerland	
European 1987	International 1924	Switzerland	
CHTRS	GRS80	Switzerland	
OS Great Britain 1936	Airy 1830	UK (incl. Isle of Man, Shetland Islands)	

(Table 3: Local datum and associated reference ellipsoids, cont'd)

Local Geodetic Datum	Reference Ellipsoid	Country	Continent: Europe
European 1950	International 1924	UK	
European 1987	International 1924	UK	
OSGRS80 Great Britain	GRS80	UK	
Potsdam	Bessel 1841	Yugoslavia	
Local Geodetic Datum	**Reference Ellipsoid**	**Country**	**Area: Indian Ocean**
Anna 1 Astro 1965	Australian National	Cocos Islands	
ISTS 073 Astro 1969	International 1924	Diego Garcia	
Kerguelen Island 1949	International 1924	Kerguelen Island	
Tananarive Observatory 1925	International 1924	Malagasy Rep. (Madagascar)	
Mahe 1971	Clarke 1880	Mahe Island	
Reunion	International 1924	Mascarene Islands	
GAN 1970	International 1924	Republic of Maldives	
Local Geodetic Datum	**Reference Ellipsoid**	**Country**	**Area: Middle East**
Ain el Abd 1970	International 1924	Bahrain	
European 1950	International 1924	Iraq	
European 1950	International 1924	Israel	
European 1950	International 1924	Jordan	
European 1950	International 1924	Kuwait	
European 1950	International 1924	Lebanon	
Oman	Clarke 1880	Oman	
Nahrwan	Clarke 1880	Oman (Masirah Island)	
Qatar National	International 1924	Qatar	
Nahrwan	Clarke 1880	Saudi Arabia	
Ain el Abd 1970	International 1924	Saudi Arabia	
European 1950	International 1924	Saudi Arabia	
European 1950	International 1924	Syria	
Nahrwan	Clarke 1880	United Arab Emirates	
Nahrwan	Clarke 1880	Yemen	
Pulkovo 1942	Krassovsky 1940	Yemen	
WGS72	Global Definition	Yemen	
Local Geodetic Datum	**Reference Ellipsoid**	**Country**	**Continent: North America**
North American 1927	Clarke 1866	Alaska (incl. Aleutian Islands)	
North American 1983	GRS80	Alaska (incl. Aleutian Islands)	
North American 1927	Clarke 1866	Canada	
North American 1983	GRS80	Canada	
North American 1927	Clarke 1866	ConUS	
North American 1983	GRS80	ConUS	
Cape Canaveral	Clarke 1866	Florida	
North American 1927	Clarke 1866	Mexico	
North American 1983	GRS80	Mexico	

(Table 3: Local datum and associated reference ellipsoids, cont'd)

Local Geodetic Datum	Reference Ellipsoid	Country Continent: North America
North American 1927	Clarke 1866	Newfoundland, Nova Scotia

Local Geodetic Datum	Reference Ellipsoid	Country Area: Pacific Ocean
American Samoa 1962	Clarke 1866	American Samoa Islands
Kusaie Astro 1951	International 1924	Caroline Islands
Easter Island 1967	International 1924	Easter Island
Bellevue (IGN)	International 1924	Efate and Erromango Islands
Santo (DOS) 1965	International 1924	Espirito Santo Island
Kusaie Astro 1951	International 1924	Fed. States of Micronesia
Viti Levu 1916	Clarke 1880	Fiji Islands (Viti Levu)
GUX 1 Astro	International 1924	Guadalcanal Island
Guam 1963	Clarke 1866	Guam
North American 1983	GRS80	Hawaii, Maui, Oahu, Kauai and Niinau
Old Hawaiian	Clarke 1866	Hawaii, Maui, Oahu, Kauai and Niinau
Astro Beacon "E" 1945	International 1924	Iwo Jima
Johnston Island 1961	International 1924	Johnston Island
Astronomical Station 1952	International 1924	Marcus Island
Wake-Eniwetok 1960	Hough 1960	Marshall Islands
Midway Astro 1961	International 1924	Midway Islands
DOS 1968	International 1924	New Georgia Islands (Gizo Island)
Geodetic Datum 1949	International 1924	New Zealand
Chatham Island Astro 1971	International 1924	New Zealand (Chatham Island)
Canton Astro 1966	International 1924	Phoenix Islands
Pitcairn Astro 1967	International 1924	Pitcairn Island
Astro Tern Island (FRIG) 1961	International 1924	Tern Island
Wake Island Astro 1952	International 1924	Wake Atoll

Local Geodetic Datum	Reference Ellipsoid	Country Continent: South America
Campo Inchauspe 1969	International 1924	Argentina
South American 1969	South American 1969	Argentina
Prov. South American 1956	International 1924	Bolivia
South American 1969	South American 1969	Bolivia
Corrego Alegre	International 1924	Brazil
South American 1969	South American 1969	Brazil
Prov. South American 1956	International 1924	Chile (near 19°S)
South American 1969	South American 1969	Chile
Prov. Sth Chilean 1963, Hito XVIII	International 1924	Chile (near 53°S)
Bogota Observatory	International 1924	Colombia
Prov. South American 1956	International 1924	Colombia
South American 1969	South American 1969	Colombia
Prov. South American 1956	International 1924	Ecuador
South American 1969	South American 1969	Ecuador (incl. Baltra, Galapagos)
Prov. South American 1956	International 1924	Guyana
South American 1969	South American 1969	Guyana

(Table 3: Local datum and associated reference ellipsoids, cont'd)

Local Geodetic Datum	Reference Ellipsoid	Country Continent: South America
Chua Astro	International 1924	Paraguay
South American 1969	South American 1969	Paraguay
Prov. South American 1956	International 1924	Peru
South American 1969	South American 1969	Peru
Zanderij	International 1924	Suriname
Naparima	South American 1969	Trinidad and Tobago
Yacare	International 1924	Uruguay
Prov. South American 1956	International 1924	Venezuela
South American 1969	South American 1969	Venezuela

Geodetic Datum	Reference Ellipsoid	Area: World-wide
DXZ78	Global Definition	People's Republic of China
DXZ88	Global Definition	People's Republic of China
SGS85	Global Definition	Commonwealth of Independent States
SGS90	Global Definition	Commonwealth of Independent States
WGS60	Global Definition	United States of America
WGS66	Global Definition	United States of America
WGS72	Global Definition	United States of America
WGS84	Global Definition	United States of America, world-wide aeronautical and hydrographical charts

(Courtesy NIMA)

Table 3: Local datum and associated reference ellipsoids

Note:

Most regional or local geodetic datums, e.g. ED50, ED79, ED87 are or will be superseded in future by another geodetic datum, such as ETRF89, with ellipsoid GRS80

(Ehrnsperger, 1989, OS, 1995; Poder, 1989; Vermeer, 1995).

3.4.1 Defined Ellipsoids

The mathematical reference surface, where the reductions take place, is called an oblate ellipsoid |[8] *or* reference ellipsoid*: an ellipse that rotates around its* polar, semi-minor axis *(Bigourdan, 1912; Tardi, 1934).*

An ellipsoid is determined by the Figure of the Earth or datum, such as the "International" Reference Ellipsoid by John F. Hayford in 1909. It is defined by the equatorial radius of the Earth or semi-major axis, $a = 6\ 378\ 388$ *m and the reciprocal flattening,* $f^{-1} = 297$ *(Hristov, 1968).*

[8] NIMA stopped considering spheroids and ellipsoids as equivalent due to the fact that a spheroid is very complex surface while an ellipsoid is a simple two-degree surface. Consequently, throughout this book the word "ellipsoid" is used to prevent confusion

In geodesy, several reference ellipsoids are used (Strasser, 1957):

Reference Ellipsoid [9]	Semi-Major Axis	Unit	Reciprocal Flattening
Airy 1830	***6 377 563.3964***	***m***	***299.32496 459***
Airy 1830	20 923 713	feet-bar O_1	299.32496 459
Airy Modified 1965	***6 377 340.189***	***m***	***299.32496 459***
Andrae 1876	6 377 104.43	m	300
APL 4.5 1965	6 378 137	m	298.25
Australian National *(IAU65)*	***6 378 160***	***m***	***298.25***
Bessel 1841 ORIGINAL	3 272 077	toise	299.15282
Bessel 1841	***6 377 397.155***	***m***	***299.15281 285***
Bessel *(Namibia)*	***6 377 483.865***	***m***	***299.15281 285***
Bessel NGO1948	6 377 492.0176	m	299.1528
Clarke 1858	6 378 360.706	m	294.26
Clarke 1866	***6 378 206.4***	***m***	***294.97869 82***
Clarke 1866	20 925 832.16	US survey feet	294.97869 82
Clarke 1880DoD	***6 378 249.145***	***m***	***293.465***
Clarke 1880IGN	6 378 249.2	m	293.46602 13
Clarke 1880G	6 378 249.14533	m	293.465
Clarke 1880G	20 926 202	imperial feet	293.465
Clarke 1880 Arc Modified	6 378 249.145326	m	293.46630 76
Clarke 1880 Palestine	6 378 300.79	m	293.46630 7656
Danish National Grid	6 377 103.965	m	300
Delambre 1810	6 378 428	m	311.5
Engelis 1985	6 378 136.05	m	298.2566
Everest 1830	317 014.1182	indian chains	300.8017
Everest 1830 *(India)*	***6 377 276.3458***	***m***	***300.8017***
Everest 1830	6 974 310.6	indian yards	300.8017
Everest 1830	20 922 931.8	indian feet	300.8017
Everest 1956 *(India)*	***6 377 301.243***	***m***	***300.8017***
Everest 1948 *(West Malaysia, Singapore)*	***6 377 304.063***	***m***	***300.8017***
Everest 1969 *(West Malaysia)*	***6 377 295.664***	***m***	***300.8017***
Everest *(Pakistan)*	***6 377 309.613***	***m***	***300.8017***
Everest *(Brunei, East Malaysia)*	***6 377 298.556***	***m***	***300.8017***
Fischer 1960 / Mercury	6 378 166	m	298.3
Fischer 1960 Modified	***6 378 155***	***m***	***298.3***
Fischer 1968	6 378 150	m	298.3

(Table 4: Reference ellipsoids, cont'd)

[9] NIMA stopped considering spheroids and ellipsoids as equivalent due to the fact that a spheroid is very complex surface while an ellipsoid is a simple two-degree surface.

Reference Ellipsoid [10]	Semi-Major Axis	Unit	Reciprocal Flattening
Fischer 1968 (MMD)	6 378 160	m	298.24716 74
GEM 10 1979	6 378 139	m	298.257
GRS67	6 378 160	m	298.24716 75
GRS80 *or* ***New International***	***6 378 137***	***m***	***298.25722 21008 827*** [11]
Hayford 1909	6 378 388	m	297
Helmert 1906	***6 378 200***	***m***	***298.3***
Hough 1960	***6 378 270***	***m***	***297***
IAG 1975	6 378 140	m	298.25722 2
Indonesian 1974 *(GRS67)*	***6 378 160***	***m***	**298. 24716 75**
International 1924	***6 378 388***	***m***	***297***
Kaula 1961	6 378 163	m	298.24
Krassovsky 1940	***6 378 245***	***m***	***298.3***
Maupertuis 1738	6 397 300	m	191
NWL-9D 1965	6 378 145	m	298.25
NWL-10D	6 378 135	m	298.26
Plessis 1810	6 376 523	m	308.64
Shdanov 1893	6 377 717	m	299.7
SGS85 (СГС85)	6 378 136	m	298.257
SGS90 (СГС90)	6 378 136	m	298.257
South American 1969 *(GRS67)*	***6 378 160***	***m***	***298.25***
Struve	6 378 297	m	294.73
Walbeck 1817	6 376 896	m	302.78
War Office (McCaw) 1924	6 378 300.58	m	296
WGS60	6 378 165	m	298.3
WGS66	6 378 145	m	298.25
WGS72	***6 378 135***	***m***	***298.26***
WGS84	***6 378 137***	***m***	***298.25722 3563***
WGS84 (G730)	6 378 137	m	298.25720 1157

Table 4: Reference ellipsoids

National Units

Colonel Alexander R. Clarke recognised also that the relation between the various national units has to be determined first when various national arcs are combined in deducing the elements of an ellipsoid. Therefore, he compared several European standards with the British Imperial Yard in 1863-1870. He considered these relations in his computations of elements from 1866 onwards. The *unit of his elements* are unique, contrary to those computed before (Strasser, 1957).

[10] Table 4 lists, in bold type and italics, reference ellipsoids associated with WGS84 (NIMA, 1991).

[11] (Burkholder, 1984)

3.4.2 Defining Parameters

3.4.2.1 An Era in the Science of Geodesy

Until the time of C.F. Gauss and F.W. Bessel, computer scientists had to simply judge as they best they could how and when to use supernumerary angles. The principle of least squares |[12] showed that a system of corrections ought to be applied to the reduction of the horizontal angles and calculation of the triangulations |[13].

The reduction of the supernumerous observations, in the past a very laborious process, led to complex calculations. The first grand development of this principle is contained in the book entitled "Gradmessung in Ostpreußen und ihre Verbindung" by Friedrich W. Bessel, published in 1838. It marks an era in the science of geodesy. The defining fundamental parameters of the "Bessel 1841" ellipsoid are the length of the equatorial radius of the Earth, $a = 3272077^T.14$ |[14], $b = 3261139^T.33$ ($= a / 1.00335\ 39847$), thus reciprocal flattening $f^{-1} = 299.152\ 82$.

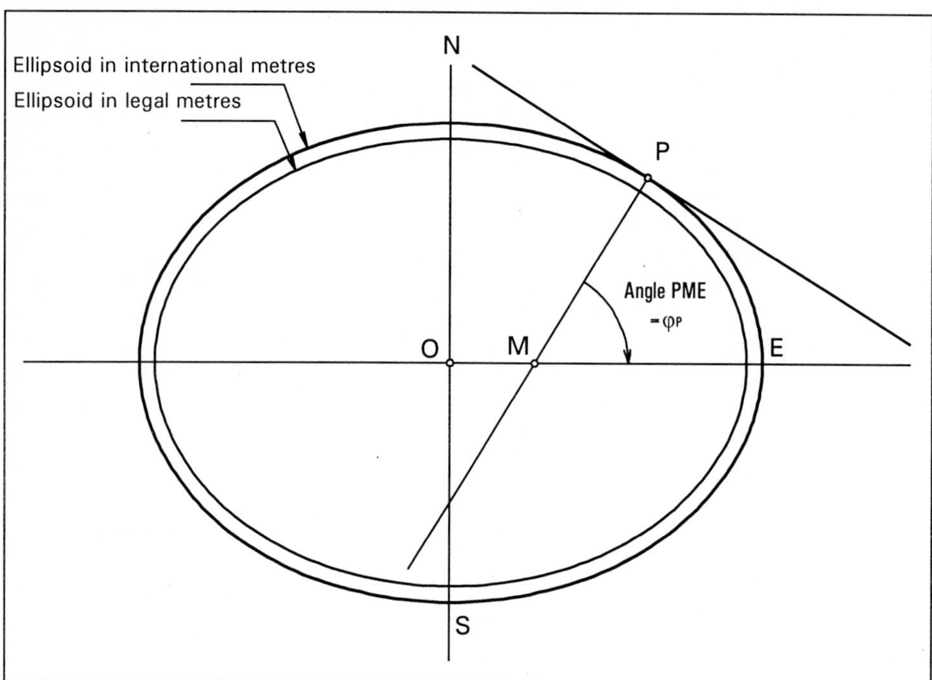

Figure 9: Bessel ellipsoids in legal metres and international metres

3.4.2.2 Legal Metres or International Metres

The rotational ellipsoid is created by rotating the meridian ellipse about its minor axis. Using two geometrical parameters, the shape of the ellipsoid is described: the semi-major axis a and the semi-minor axis b. Generally, b is replaced by one of a number of quantities which is more suitable for series expansions: the (geometrical) flattening f, or the first and second eccentricities e and e', respectively.

The conversion from toise |[14] to *legal metres* using the appropriate conversion factor gives the *original* defining parameters of the Bessel ellipsoid as, see Table 1 and Table 5 (König, 1951; Krack, 1995; Torge, 1991).

---

[12] Methods of H.J. Walbeck, later on C.F. Gauss and E.J.C. Schmidt

[13] See [3.7], Footnotes

[14] Toise of Paris

Bessel 1841		
log a	=	6.80464 34637 (unique (J/E/K, 1959))
log b	=	6.80318 92839 (unique)
a	=	6377397.15507 604|96 89970 31762 |[15]
b	=	6356078.96289 778|47 19996 70996
a / b	=	1.00335 39847 9|1992 86789
1 / f	=	299.15281 28533 4|0587 66096
f	=	0.00334 27731 81578 77|108
e^2	=	0.00667 43722 306|14 05990
e'^2	=	0.00671 92187 979|70 65522
c	=	6398786.84814 66|733 41962 56064

Table 5: Derived ellipsoidal parameters of Bessel 1841 according to Helmert

Notation	a | b	e^2	e'^2	f	n
$a =$					$\frac{c(1-n)}{1+n}$
$b =$		$a(1-e^2)^{1/2}$	$\frac{a}{(1+e'^2)^{1/2}}$	$a(1-f)$	
$\frac{a}{b} =$				$\frac{1}{1-f}$	
$\frac{b}{a} =$		$(1-e^2)^{1/2}$	$\frac{1}{(1+e'^2)^{1/2}}$	$1-f$	$\frac{1-n}{1+n}$
$c =$	$\frac{a^2}{b}$	$\frac{a}{(1-e^2)^{1/2}}$	$a(1+e'^2)^{1/2}$	$\frac{a}{1-f}$	
$e^2 =$	$\frac{a^2-b^2}{a^2}$		$\frac{e'^2}{1+e'^2}$	$f(2-f)$	$\frac{4n}{(1+n)^2}$
$e'^2 =$	$\frac{a^2-b^2}{b^2}$	$\frac{e^2}{(1-f)^2} = \frac{e^2}{1-e^2}$	$\frac{f(2-f)}{(1-f)^2}$		$\frac{4n}{(1-n)^2}$
$f =$	$\frac{a-b}{a}$	$1-(1-e^2)^{1/2}$	$1-(1+e'^2)^{-1/2}$		$\frac{2n}{1+n}$
$n =$	$\frac{a-b}{a+b}$	$\frac{1-(1-e^2)^{1/2}}{1+(1-e^2)^{1/2}}$	$\frac{(1+e'^2)^{1/2}-1}{(1+e'^2)^{1/2}+1}$	$\frac{f}{2-f}$	

Table 6: Defining parameters and associated constants

[15] Calculated using program "BIGCALC" to 50 decimal places

3.4.2.3 Parameters and Associated Constants

Semi-major axis, a = 6 377 397.155 m_{LEG} , b = 6 356 078.963 m_{LEG}, and f^{-1} = 299.152 812 85.

These parameters are used in many countries in Continental Europe and Asia as m_{INT}, international metres. Germany passed an *International Metre Act* in 1893 (Zeger, 1991; J/E/K, 1959).

However, in southern Africa (Namibia) the defining parameters of the Bessel ellipsoid are given in *really* international metres as (Rens, 1990):

Semi-major axis, a = 6 377 483.865 m_{INT}, b = 6 356 165.38297 m_{INT}, and again f^{-1} = 299.15281 285.

After entering an equatorial radius of the Earth, semi-major axis= a, and its reciprocal flattening = f^{-1}, program ELLIDATA.BAS [11.7.1] lists the defining parameters and associated constants for that particular ellipsoid for use in geodetic applications and calculates the meridional arc distance (Figure 9) and [3.4.4].

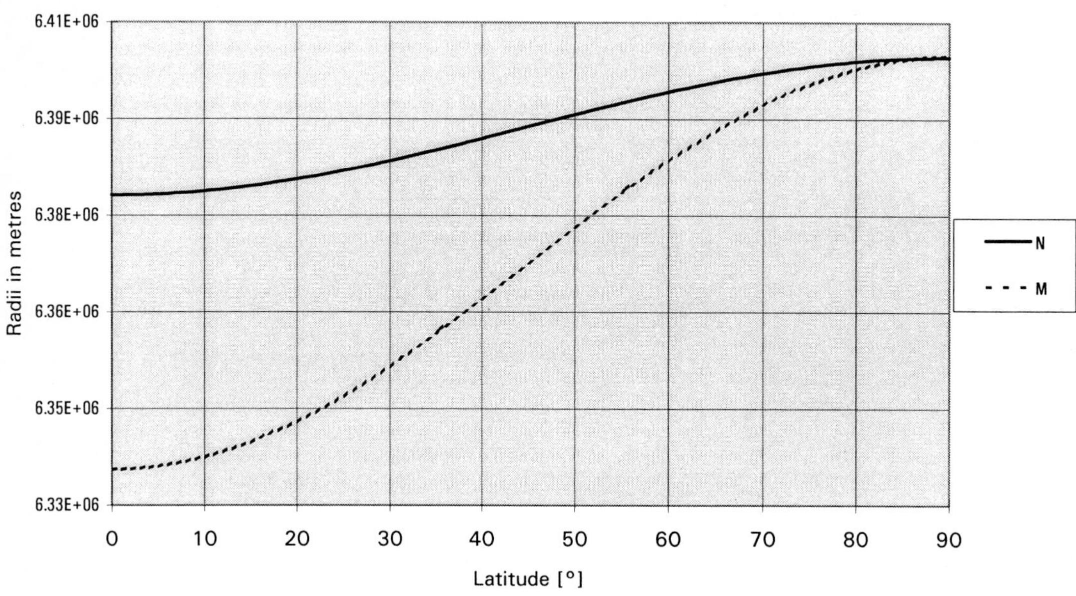

Figure 10: Radii of curvature in Prime Vertical and in the Reference Meridian

The radii N and M (Figure 10) have the following features (J/E/K, 1959):

- N =ν; (3.11) >= M in any latitude
- M =ρ; (3.12) = N at the poles (= c), and reach their maximum values
- N = a on the equator, and M and N have their minimum values
- M = a in latitude 54°.8, approximately
- M = b in latitude 35°.3, approximately.

3.4.3 Spheroidal Mapping Equations

Symbols and Definitions

All angles are expressed in *radians* (Krack, 1983). See also program ELLIDATA.BAS [11.7.1]. For the geometrical parameters, see Table 6.

Geocentric Equipotential Primary Constants

a	= semi-major axis or equatorial radius of the earth, see [3.5.5.2]	
GM	= value × 10^8 m^3 s^{-2}	G×M product, geocentric gravitational constant
J_2	= value ×10^{-8}	dynamic form factor, un-normalised form
$C_{2.0}$	= value × 10^{-6}	dynamic form factor, normalised form
ω	= value × 10^{-11} rad s^{-1}	angular velocity of the earth

Classical Ellipsoid Constants

b	semi-minor axis of the ellipsoid
f^1	reciprocal flattening of the ellipsoid
f	flattening of the ellipsoid
k_0	scale factor assigned to the meridian
φ	parallel of geodetic latitude, positive north
S_0	meridional distance from the equator to φ_0, multiplied by the scale factor
e^2	first eccentricity squared
e'^2	second eccentricity squared
n	second flattening
W	first auxiliary quantity
V	second auxiliary quantity
M	=(ρ) = radius of curvature in the the meridian (Figure 10)
N	=(ν) = radius of curvature in the prime vertical (Figure 10)
c	polar radius of curvature
r	radius of the rectifying sphere, having the same meridional length as that of the spheroid
R_M	arithmetic mean radius of a and b
R_G	geometric mean radius of a and b
R_α	radius of curvature in azimuth α
R_Q	sphere of equal volume as the spheroid

Derived Constants

Compute constants for ellipsoid as given below, or use (6.04 - 6.09):

$$r = a\,(1 + n^2/4)\,/\,(1 + n) \tag{3.01}$$

$$f = 1\,/\,f^1 \tag{3.02}$$

$$c = a\,/\,(1 - f) \tag{3.03}$$

$$a\,/\,b = 1\,/\,(1 - f) \tag{3.04}$$

$$n = f\,/\,(2 - f) \tag{3.05}$$

$$e^2 = f\,(2 - f) \tag{3.06}$$

$$e'^2 = e^2\,/\,(1 - f)^2 \tag{3.07}$$

$$b = a\,(1 - f) \tag{3.08}$$

Input: geodetic co-ordinates of a point P (B_i^R)

$$W = (1 - e^2 \sin^2 B_i^R)^{1/2} \tag{3.09}$$

$$V = (1 + e'^2 \cos^2 B_i^R)^{1/2} \tag{3.10}$$

$$N = a\,/\,W \quad = c\,/\,V \tag{3.11}$$

$$M = c\,/\,V^3 \tag{3.12}$$

R_M	$= (M\,N)^{1/2} \quad = c / V^2$	(3.13)
c	$= a^2 / b$	(3.14)
R_α	$= c / (V + (V^3 - V) \cos^2 \alpha) \quad = (M\,N) / (M \sin^2 \alpha + N \cos^2 \alpha)$	(3.15)
R_{MAB}	$= (a + b) / 2$	(3.16)
R_{GAB}	$= \sqrt{(a\,b)}$	(3.17)
R_Q	$= a\,(1 - f / 3 - f^2 / 9)$	(3.18)

Meridional Arc formula

Forward Constants:

$$E1 = - n (36 + n (45 + 39 n))$$
$$E3 = - 280 n^3 \qquad (3.19)$$
$$E2 = 90 n^2 - E3$$

Inverse Constants:

$$F1 = n (36 + n (- 63 + 93 n))$$
$$F3 = 604 n^3 \qquad (3.20)$$
$$F2 = 126 n^2 - F3$$

Direct Computation

Input: geodetic co-ordinates of a point P (B_i^R)
Output: G_m of a point P (B_i^R)

$$G_m = k_0\, r\, (B_i^R + \sin B_i^R \cos B_i^R / 12\, (E1 + \cos^2 B_i^R (E2 + \cos^2 B_i^R\, E3))) \qquad (3.21)$$

Inverse Computation

Input: G_m of a point P (B_i^R)
Output: geodetic co-ordinates of a point P (B_i^R)

$$\omega = G / (r^* k_0) \qquad (3.22)$$
$$B_i^R = (\omega + \sin \omega \cos \omega / 12 (F1 + \cos^2 \omega (F2 + \cos^2 \omega\, F3))) \qquad (3.23)$$

□

Derived functions for the WGS84 ellipsoid are given, see Table 7.
For further reading, see (Agajelu, 1987; Baeschlin, 1948; J/E/K, 1959).

3.4.4 Parameters and Associated Constants for the WGS84

Name	Constants and Magnitudes	Explanation	Formulae
$\pi / 4$	= .7853981633974483		
degrad	= .0174532925199433	conversion factor degrees into radians	
Reference Ellipsoid: WGS84, a geocentric equipotential ellipsoid (NIMA, 1991)			
a	= 6 378 137 m	semi-major axis or equatorial radius of the earth	
GM	$= 3\,986\,005{*}10^8\ m^3\, s^{-2}$	geocentric gravitational constant	
J_2	= calculated value (3.29):	dynamic form factor, un-normalised form	
$C_{2.0}$	$= - 484.166\,85{*}10^{-6}$	dynamic form factor, normalised form	
ω	$= 7\,292\,115{*}10^{-11}\ rad\ s^{-1}$	angular velocity of the earth	

cont'd

cont'd

Name		Derived Geometric Constants and Magnitudes	Explanation
b	=	6356752.31424518	semi-minor axis
f^{-1}	=	298.257223563	reciprocal flattening
k_0	=	1	scale factor on central meridian
f	=	3.352810664747481D-03	flattening of meridional ellipse
c	=	6399593.625758493	polar radius of curvature
a / b	=	1.003364089820976	
n	=	1.679220386383705D-03	ratio of length
e^2	=	6.694379990141317D-03	first eccentricity squared
e	=	8.181919084262149D-02	first eccentricity of meridional ellipse
$1 - e^2$	=	.9933056200098587	
$\sqrt{(1 - e^2)}$	=	.9966471893352525	
e'^2	=	6.739496742276435D-03	second eccentricity squared
$1 + e'^2$	=	1.006739496742276	
$\sqrt{(1 + e'^2)}$	=	1.003364089820976	
r	=	6367449.145822624	radius of the rectifying sphere
E1	=	-6.057900872590829D-02	coefficient of meridional arc - direct
E2	=	2.551061090413531D-04	coefficient of meridional arc - direct
E3	=	-1.32580949715567D-06	coefficient of meridional arc - direct
F1	=	6.027472805828684D-02	coefficient of meridional arc - inverse
F2	=	3.524324588751549D-04	coefficient of meridional arc - inverse
F3	=	2.859960486721517D-06	coefficient of meridional arc - inverse
R_{MAB}	=	6367444.65712259	arithmetic mean radius
R_{GAB}	=	6367435.679716192	geometric mean radius
R_Q	=	6371000.804881453	radius of sphere of same volume
Lat (deg)	:	45	*input:* arbitrary latitude (B_i)
N	=	6388838.290121148 = ν	radius of curvature in the prime vertical
M	=	6367381.815619549 = ρ	radius of curvature in the meridian
R_M	=	6378101.030201018	mean radius of c and V
R_α	=	6378092.007544452	radius of curvature in azimuth (α)
G_m	=	4984944.377977213	*output* = G = length of meridional arc
Lat	=	45.00000000002307	round-trip latitude (B_i)
Lat (deg)	:	60	*input:* arbitrary latitude (B_i)
N	=	6394209.173847894 = ν	radius of curvature in the prime vertical
M	=	6383453.857229078 = ρ	radius of curvature in the meridian
R_M	=	6388829.252275326	mean radius of c and V
R_α	=	6391516.948369661	radius of curvature in azimuth (α)
G_m	=	6654072.819442124	*output* = G = length of meridional arc
Lat	=	59.99999999804641	round-trip latitude (B_i)

Table 7: WGS84 constants and parameters

Program ELLIDATA.BAS uses the parameters a, f, and k_0 for any reference ellipsoid as input and calculates the parameters and associated constants according to Table 6.

3.5 Geodetic Reference Systems

Positioning by navigation satellites is carried out in geodetic reference systems |[16], in three dimensions (3-D). The geocentric system of co-ordinates has been called "Earth-Centred, Earth-Fixed" (ECEF) and utilises a rectangular cartesian co-ordinate system having three mutually perpendicular *X, Y, Z*-axes with the origin located at Earth's centre of mass. This applies to the TRANSIT system, which was realised in the 1960s, and NavStar, USAF "PROGRAM 621B", has been in development since 1973.
Here is how it started (Parkinson, 1996).

3.5.1 Mercury Datum of 1960

Prior to the advent of specifically geodetic satellites, geodesists from the AMS, now NIMA, developed an astro-geodetic world system, using all available data, along with an early determination of reciprocal flattening f^1= 298.3 from observations of Sputnik I, II and the Vanguards. This system was selected by NASA to position the Mercury project tracking stations, and became known as the Mercury Datum of 1960 (Seppelin, 1974).

3.5.2 World Geodetic System of 1960

The Department of Defense (DoD) in the late 1950s generated a geocentric reference system to which different geodetic networks could be referred to and compatibility was established between the co-ordinates of sites of interest. Efforts of the US Army, Navy and Air Force were combined leading to the development of the World Geodetic System of 1960 (WGS60). In accomplishing WGS60, a combination of surface gravity data, astro-geodetic data and results from HIRAN and SHORAN surveys were used to obtain a best-fitting ellipsoid for the significant datum areas. The sole contribution of satellite data to the development of WGS60 was the value for the ellipsoid's reciprocal flattening f^1=298.3 which was obtained from the nodal motion of a satellite. At the same time, the semi-major axis was determined as a=6 378 165 m in 1958.

3.5.3 World Geodetic System of 1966

The World Geodetic System Committee (WGSC) developed a replacement for WGS60 in January 1966. Using additional data, such as surface gravity observations, results from triangulation and trilateration networks, Doppler observations, optical satellite data and improved techniques, the WGSC devised WGS66.

The defining parameters of the World Geodetic System of 1966 (WGS66) ellipsoid were the reciprocal flattening f^1=298.25, determined from satellite data and the semi-major axis a=6 378 145 m, determined from a combination of TRANSIT and astro-geodetic data involving a geoid match technique. A geopotential coefficient set was selected as the WGS66 EGM (gravity model). Using the set in a spherical harmonic expansion resulted in the world-wide WGS66 Geoid, a geoid referenced to the WGS66 ellipsoid.

The numeric values of the fundamental parameters were soon rendered out of date by the rapidly improving results from *satellite geodesy*.

3.5.4 World Geodetic System of 1972

Since the development of WGS66, large quantities of additional optical and electronic satellite data had become available from both TRANSIT and optical satellites, gravity, triangulation and trilateration surveys, high precision traverses and astronomical surveys. The WGSC developed a replacement for WGS66 in 1970.

---

[16] See [3.7] Footnotes

The electronic satellite data consisted of:

- *Baker-Nunn camera* data of the Smithsonian Astrophysical Observatory (SAO)
- *Doppler data* available from the numerous sites established by *geodetic receivers* (Geoceivers)
- *Doppler data* provided by US Navy and co-operating non-DoD satellite tracking stations
- Electronic satellite data was provided by the *SECOR* (Sequential Collation of Ranges) equatorial control network
- Optical satellite data provided by the *Wild BC-4* camera system.

In addition, greater capabilities had been developed in both computer hardware and software. Further, ongoing research in improved computational procedures and error analyses resulted in superior methods and a greater facility for handling and combining data.

In 1975 the WGSC completed the development of the DoD World Geodetic System of 1972 (WGS72). Selected satellite, surface gravity and astro-geodetic data available through 1972 from both DoD and other sources was used in a unified WGS solution (a large scale least squares adjustment).

TRANSIT. Geopotential coefficients and station co-ordinates have been determined using sampled Doppler observations of satellites at various inclinations. An accurate set of station co-ordinates has been derived using sampled or integrated Doppler observations from the Navy Navigation Satellite System (NNSS).

WGS72 Parameters. In determining the WGS72 ellipsoid and associated parameters, the Committee decided to adhere to the approach used by the International Union of Geodesy and Geophysics (IUGG) in establishing the *Geodetic Reference System of 1967* (GRS67). Accordingly, an equipotential ellipsoid of revolution, i.e. a surface on which all values of the potential are equal, was taken as the form for the WGS72 ellipsoid.

Different combinations of TRANSIT tracking and astro-gravimetric datum orientation stations provided values for the semi-major axis ranging from 6 378 133.6 to 6 378 136.6 metres. The value adopted by the WGS Committee for the semi-major axis of the WGS72 ellipsoid is a = 6 378 135 ±5 metres and the WGS72 ellipsoid reciprocal flattening has been rounded to f^{1}=298.26 (Ihde, 1981; Seppelin, 1974).

3.5.5 World Geodetic System of 1984

The *World Geodetic Reference System of 1984 (WGS84)* is a Conventional Terrestrial System (CTS), realised by modifying the *Navy Navigation Satellite System* (NNSS, NAVSAT or TRANSIT), Doppler reference frame (NSWC 9Z-2) in origin and scale, and rotating it to bring its reference meridian into coincidence with the Bureau International de l'Heure (BIH) Terrestrial System (BTS), epoch 1984.0, defined Zero Meridian.

One principal purpose of a World Geodetic System is to provide the means by which local geodetic datums can be referenced to a *single geocentric system.* However, the number of local geodetic datums, or local horizontal datums requiring such referencing is extensive.

3.5.5.1 Mathematical Relationship

The *WGS84 Terrestrial Reference Frame* is the frame of a standard earth rotating at a constant rate around a Conventional Terrestrial Pole (CTP). However, the universe is in motion, the earth is nonstandard, and events occur in an instantaneous world. Therefore, the WGS84 co-ordinate System, which is identical to the BIH-defined CTS in its definition, must be related mathematically to an Instantaneous Terrestrial System (ITS) and to a Conventional Inertial System (CIS).

The relationship between the CIS, ITS and the WGS84 co-ordinate System, can be expressed as (NIMA, 1991):

- the origin is the Earth's centre of mass
- the Z-axis is the direction of the Conventional Terrestrial Pole (CTP) for polar motion, as defined by the Bureau International de l'Heure (BIH) for epoch 1984.0 on the basis of the co-ordinates adopted for the BIH stations
- the X-axis is the Intersection of the WGS84 Reference Meridian Plane and the plane of the CTP's Equator, the Reference Meridian being the Zero Meridian defined by the BIH for epoch 1984.0 based on the co-ordinates adopted for the BIH stations
- the Y-axis completes a right-handed, Earth-Centred, Earth-Fixed (ECEF) orthogonal co-ordinate system, measured in the plane of the CTP Equator, 90° East of the X-axis.

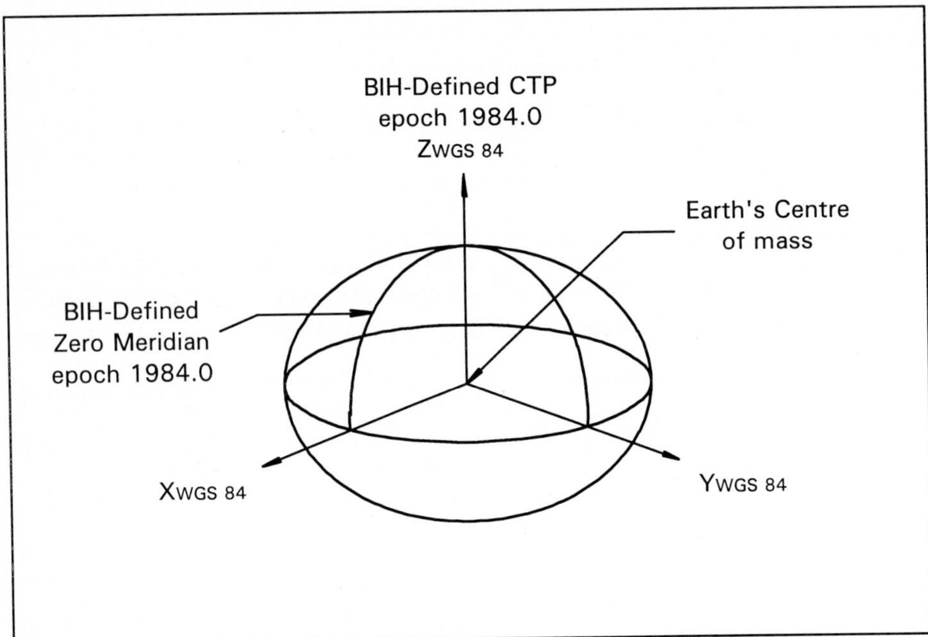

Figure 11: WGS84 co-ordinate system definition

3.5.5.2 Fundamental Geometrical Constants WGS84 and Others - World-wide

Equipotential ellipsoids of revolution are determined by a set of four constants, adopted by the International Union of Geodesy and Geophysics, and authorised by the IAG e.g. for:

GRS67, GRS80, IAG75 and WGS84 (Table 8).

Parameter	Notation	Units	Ellipsoid GRS67	Ellipsoid IAG75
Semi-major axis	a	m	6 378 160	6 378 140
Geocentric gravitational constant	GM	$10^8\ m^3\ s^{-2}$	$3986030*10^8$	$3986005*10^8$
Angular velocity of the Earth	ω	$10^{-11} rad\ s^{-1}$	$7292115.1467*10^{-11}$	$7292115*10^{-11}$
Dynamic form factor, un-normalised form	J_2	10^{-8}	$108270*10^{-8}$	$108263*10^{-8}$
Dynamic form factor, normalised form	$C_{2.0}$	10^{-6}	-	-
Parameter	**Notation**	**Units**	**Ellipsoid GRS80**	**Ellipsoid WGS84** |[17]
Semi-major axis	a	m	6 378 137	6 378 137
Geocentric gravitational constant	GM	$10^8\ m^3\ s^{-2}$	$3986005*10^8$	$3986005*10^8$
Angular velocity of the Earth	ω	$10^{-11}\ rad\ s^{-1}$	$7292115*10^{-11}$	$7292115*10^{-11}$
Dynamic form factor, un-normalised form	J_2	10^{-8}	$108263*10^{-8}$	|[18]
Dynamic form factor, normalised form	$C_{2.0}$	10^{-6}		$-484.16685*10^{-6}$

Table 8: Constants for GRS67, IAG75, GRS80 and WGS84 ellipsoids

GRS Constants

Geodetic Reference Systems (GRS) are based on the theory of the geocentric equipotential ellipsoid, defined by constants. The International Union of Geodesy and Geophysics (IUGG) has chosen the following set of conventional ellipsoid constants (Moritz, 1992):

a	= m ______	semi-major axis or equatorial radius of the earth
GM	= value $* 10^8 m^3 s^{-2}$ ______	geocentric gravitational constant
J_2	= value $* 10^{-8}$ ______	dynamic form factor, un-normalised form
$C_{2.0}$	= value $* 10^{-6}$ ______	dynamic form factor, normalised form
ω	= value $* 10^{-11} rad\ s^{-1}$ ______	angular velocity of the earth.

The formulae

The following relation exists between the zonal sphere function J_2 and the eccentricity e:

$$e^2 = 3\ J_2 + (4/15)(\omega^2 * a^3 / GM)(e^3 / 2q_0), \text{ in which} \quad (3.24)$$

$$2q_0 = (1 + 3/e'^2))\arctan(e') - (3/e') \text{ and} \quad (3.25)$$

$$e' = e(1 - e^2)^{-1/2} \quad \text{second eccentricity} \quad (3.26)$$

Equipotential ellipsoid parameters:

$$e^2 = a^2 - b^2 / a^2 \quad \text{square of the first eccentricity} \quad (3.27)$$

$$f = 1 - (1 - e^2)^{1/2} \quad \text{flattening} \quad (3.28)$$

Table 9: Closed formulae to calculate the square of the first excentricity e²

[17] Analogous to the BIH Defined Conventional Terrestrial System (CTS), or BTS, epoch 1984.0

[18] Calculated J_2=108 262,99890 51944*10^{-8}

Using basic equation (3.27), and taking (3.25) and (3.26) into account, (3.24) calculates iteratively e^2, and (3.28) furnishes the flattening (f). See program REFGRS00.BAS [11.7.2] (Moritz, 1968-1992).

3.5.5.3 Revision of WGS84

Revision of WGS84 Terrestrial Reference Frame. A careful review and evaluation of the four WGS84 defining parameters indicated that the WGS84 EGM value is the one that warrants revision (Kouba, 1994; Malys / Slater, 1994).

NIMA has replaced the WGS84 EGM value by the International Earth Rotation Service (IERS, 1992), Standard value of 3 986 004.418 * 10^8 m^3 s^{-2} (mass of the atmosphere included). The reference frame enhancements embodied in WGS84 (G730) have brought the operational DoD reference frame into coincidence with the ITRF92 at a level of approaching 10 cm.

Please note that the change (0.582 * 10^8 m^3 s^{-2}) to the WGS84 EGM is within the original 1 sigma (σ) uncertainty for this parameter (0.6 * 10^8 m^3 s^{-2}) as identified in (NIMA, 1991). All other WGS84 parameters are to remain at their current values.

This refined WGS84 reference frame, along with the improved EGM value have been given the designation WGS84 (G730) which refers to GPS week 730, the first full week of the calendar year 1994.

The revision is needed *only for specific high-accuracy DoD applications* such as *precise orbit determination.*

The values for a and f, calculated by the formulae of Table 9 in program REFGRS00.BAS [11.7.2], are entered in e.g. program ELLIDATA.BAS, [11.7.1] to derive the associated constants.

Derived functions for the reference ellipsoids GRS80, and WGS84 are given in Table 10.

Reference Ellipsoid: Name	GRS80 Constants and Magnitudes	WGS84 Constants and Magnitudes
a	**6 378 137**	**6 378 137**
b	6 356 752.31414 0347	6 356 752.31424 518
c	6 399 593.62586 4032	6 399 593.62575 8493
f^{-1}	**298.25722 21008 827**	**298.25722 3563**
f	3.352810681183638D-03	3.352810664747481D-03
n	1.679220394629406D-03	1.679220386383705D-03
e^2	6.694380022903416D-03	6.694379990141317D-03
e	8.181919104283185D-02	8.181919084262149D-02
e'^2	6.739496775481622D-03	6.739496742276435D-03
r	6 367 449.14577 0252	6 367 449.14582 2624
N (=ν for $40^{2/3}$ °)	6 387 222.31141 441	6 387 222.31136 9852
M (=ρ for $40^{2/3}$ °)	6 362 551.38275 4965	6 362 551.38283 1662

Table 10: Calculated values for GRS80 and WGS84 for comparison

3.5.6 Geodetic Reference System of 1980

IUGG Resolution No. 7, specifies that the Geodetic Reference System of 1980 (GRS80) be geocentric. The orientation of the system is specified in the following way:

> The rotation axis of the reference ellipsoid is to have the direction of the Conventional International Origin for Polar Motion (CIO), the Zero Meridian as defined by the Bureau International de l'Heure (BIH) is used.

To this definition there corresponds a rectangular co-ordinate system XYZ whose origin is the geocentre, whose Z-axis is the rotation axis of the reference ellipsoid, defined by the direction of CIO, and whose X-axis passes through the Zero Meridian according to the BIH (Heiskanen, 1967; Ihde, 1991; Moritz, 1968-1992).

3.5.7 North American Datum of 1983

The establishment of the North American Datum of 1983 (NAD83), international in scope, was the result of co-operation between the *National Geodetic Survey* of the United States, the *Geodetic Survey of Canada* (GSC), the *Danish Geodetic Institute* (representing Greenland), and by the *Inter American Geodetic Survey* (IAGS) (representing Mexico and Central America). The geodetic data were compiled by the National Imagery and Mapping Agency's *Hydrographic Topographic Centre.*

NAD83 is historically the third official US datum, and supersedes the North American Datum of 1927 (NAD27). The vertical datum is called the North American Vertical Datum of 1988 (NAVD88). NAVD88 supersedes the National Geodetic Vertical Datum of 1929 (NGVD29) (Osterhold, 1993; Zilkoski, 1992).

Datum Redefinition with respect the Geodetic Reference System of 1980

This effort, consisting of the accumulation, validation, automation, and simultaneous least squares adjustment of horizontal survey information, involved 1 785 772 geodetic observations connecting 266 436 control stations within the United States, Canada, Mexico, and Central America. Greenland, Hawaii, and the Caribbean islands were connected independently by Doppler observations and the application of Very Long Baseline Interferometry (VLBI).

The computations were done with respect to the ellipsoid of the Geodetic Reference System of 1980 (GRS80), recommended by the International Association of Geodesy (Schwartz, 1989).

Datum Differences

The WGS84 ellipsoid differs very slightly from the GRS80 ellipsoid which was used for ETRF89 and NAD83. The differences can be seen in Table 10: Calculated values for GRS80 and WGS84 for comparison. This results in two sets of co-ordinates at all common points. These differences arise because NIMA used the normalised form of the coefficient of the second zonal harmonic of the gravity field as a fundamental geodetic constant, while GRS80 had used the un-normalised form.

Furthermore, the normalised value used by NIMA was obtained by using the mathematical relationship:

$$C_{2.0} = -J_2 / (5)^{1/2} \tag{3.29}$$

and rounding the result to eight significant figures (NIMA, 1991).

Readjustment

The readjustment project included the computation of geoidal heights and deflection of the verticals at 193 241 control points, computed by the method of astro-gravimetric levelling, using almost 1 400 000 gravity points and 5 000 observed astro-geodetic deflections (Schwartz, 1989).

Besides the expected latitude and longitude co-ordinate unknowns, the solution included scale factor unknowns for many groups of Electronic Distance Measuring (EDM) instruments.

Additional parameters related the co-ordinate systems of the terrestrial observations, Doppler observations, and the VLBI data to the final co-ordinate system. The Greenland networks are brought into NAD83 through the GPS and Doppler observations.

The horizontal survey network in Mexico, Central America, and the Caribbean islands, exclusive of Puerto Rico and the US Virgin Islands consisted of 1 884 stations established by first-order triangulation and traverse methods. Observations among these stations included 9 970 directions, 82 Laplace azimuths, 55 base lines (Invar and Geodimeter) and there were 4 000 km of traverse. After that they were treated as a part of the US network (Schwartz, 1989; Sodano, 1958).

Helmert blocking

The computer, and advances in higher mathematics, permitted the simultaneous adjustment of observations and the resolution of unknowns for NAD83.

The last stage of validation was the continental adjustment itself. The first linearisation and solution established that the normal equations could be solved and that the network therefore was properly connected.
New surveys and their processing accounted for less than 25% of the total cost. Using the Helmert block adjustment computation the 928 735 simultaneous linear normal equations accounted for less than 10% of the total cost (Schwartz, 1989).

3.5.8 Soviet Geocentric System of 1985

Geodetic Reference Systems of the CIS

Geodetic Reference Systems (GRS) are based on the theory of the geocentric equipotential ellipsoid, defined by constants. The CIS|[19] Global Navigation Satellite System (GLONASS) provides continuous, world-wide access to precise position, time and velocity since May 1995. The operation of GLONASS is quite similar to the US Global Positioning System (GPS).

Both satellite navigation systems, GPS and GLONASS, enable users to determine their position and navigation calculations in geocentric global co-ordinate systems. GPS utilises the WGS84, whereas GLONASS satellite information is given in *Soviet Geocentric System of 1985* (SGS85), and its successor SGS90 in 1995. The Soviet Geocentric System origin and orientation of the X-, Y- and Z-axes are specified in the GLONASS ICD |[20].

These co-ordinate systems are based on models of the elliptical shape of the earth's surface. The origin of each Cartesian co-ordinate system is thought to be at the Earth's gravitational centre. The Z-axis of the SGS85, and its successor SGS90, is based on the mean direction of the earth's rotation at the time between 1900 and 1905. The X-axis is aligned to the ellipsoidal Zero Meridian, while the Y-axis completes the system as a right-hand orthogonal system. Note that the values of latitude of 5 ILS stations (Table 2) define the mean pole of 1903.0, which is identical with the more commonly defined mean pole of epoch 1900-05 of G. Cecchini.

Co-ordinate System and Definitions

The ephemerides, transmitted by each Satellite of the GLONASS system as part of the operational information, describe the location of the phase centre of the transmitting antenna of this satellite in the Earth-bound SGS90 geodetic system as follows (Figure 12).

The Soviet Geocentric System of 1990 (SGS90 or CГC90) represents the rectangular co-ordinate system as:

- the co-ordinate origin is located at the Earth's centre of mass
- the Z-axis is the direction of the Conventional International Origin (CIO), directed towards the mean North Pole in the *mean epoch 1900-1905,* as defined in resolutions of the International Astronomical Union and the International Association of Geodesy
- the X-axis of the system lies in the plane of the terrestrial Equator of the epoch 1900-1905. The Reference Meridian Plane XOZ is parallel to the Mean Greenwich Meridian. It defines the position of the Origin based on the longitude values adopted for the Bureau International de l'Heure (BIH) stations, and
- the Y-axis supplements the geocentric rectangular co-ordinate system to a right-handed, Earth-Centred, Earth-Fixed (ECEF) orthogonal co-ordinate system, measured in the plane of the terrestrial equator of the epoch 1900-1905, 90° East of the X-axis.

The geodetic latitude φ_P of point P is defined as the angle between the normal to the surface of an ellipsoid and the equatorial plane. The geodetic longitude λ_P is the angle between the plane of the Prime Meridian and the plane of the meridian passing through point P (positive direction of reckoning longitudes from the Prime Meridian eastwards) and the geodetic height H is the distance along the normal from the surface of an ellipsoid to point P. See also (Grafarend, 1995c).

---

[19] Commonwealth of Independent States, formerly the USSR

[20] Interface Control Document

System 1942

The "Pulkovo System 1942 of Survey Co-ordinates" (System 1942, CK42) is the reference system whose origin coincides with the centre of the Krassovsky ellipsoid. The direction of the axes of the CK42 deflects slightly (less than one second of arc) from the direction of the SGS85 axes.

Fundamental parameters of the earth ellipsoid correlated with SGS85, are given below in Table 11.

3.5.8.1 Fundamental Geometrical Constants SGS85 of the CIS

Constants Designation	Symbol	Magnitude	
Basic Geodetic Constants			
Angular velocity of the Earth's rotation	ω	$7\,292\,115 * 10^{-11}$	rad s^{-1}
Geocentric gravitational constant with correction for atmosphere	GM	$3\,986\,004.4 * 10^{-8}$	$m^3\,s^{-2}$
Geocentric gravitational constant of the atmosphere	GM_a	$0.35 * 10^{9}$	$m^3\,s^{-2}$
Parameters of the Earth Ellipsoid			
Semi-major axis	a	6 378 136.m	
Reciprocal flattening	f^{-1}	298.257	
Velocity of light (in a vacuum)	c	299 792 458.	m s^{-1}
Acceleration at the equator due to theoretical gravity	γ_e	978 032.8	mgal
Correction to acceleration due to theoretical gravity (γ) for the atmospheric attraction	$\delta\gamma_a$	- 0.9	mgal

Table 11: Constants for SGS85 ellipsoid

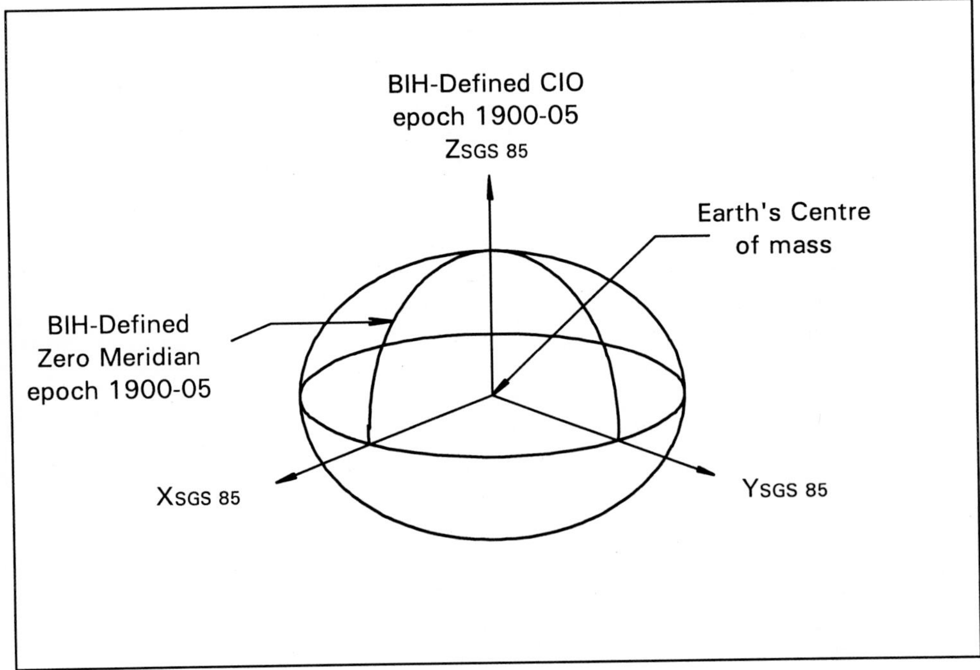

Figure 12: SGS85 and SGS90 co-ordinate system definition

Reference Ellipsoid: Name	Krassovsky 1940 Constants and Magnitudes	SGS85 Constants and Magnitudes
a	**6 378 245**	**6 378 136**
b	6 356 863.01877 3047	6 356 751.30156 8781
c	6 399 698.90178 2711	6 399 592.63853 1641
f^{-1}	**298.3**	**298.257**
f	3.352329869259135D-03	3.352813177896914D-03
n	1.67897918065816D-03	1.679221647182098D-03
e^2	6.693421622965943D-03	6.69438499958795D-03
e	8.181333401693115D-02	8.181922145552321D-02
e'^2	6.738525414683492D-03	6.739501819472925D-03
r	6 367 558.49687 4189	6 367 448.13949 0456
N (=ν for $40^{2/3}$ °)	6 387 329.16175 5153	6 387 221.31675 8546
M (=ρ for $40^{2/3}$ °)	6 362 661.36251 8751	6 362 550.37354 8071

Table 12: Calculated values for Krassovsky 1940 and SGS85 for comparison

3.6 Space-based Positioning Systems

3.6.1 SGS85 to WGS84 Transformations

GPS is referred to the DoD's (US Department of Defense) World Geodetic System of 1984 (WGS84) ellipsoid. The DoD has built in the capability to control the accuracy of the Standard Positioning Service (SPS) signals available to civil and international users by a combination of dithering the satellite clock and manipulating the ephemeris data.

The Global Navigation Satellite System (GLONASS) is managed and owned by CIS |[21], including Belarus, Khazakhstan, Russia, and the Ukraine. The authorisation to use the space system rests with Russian Space and Time Agency and the Russian Ministry of Defense. GLONASS is similar to the GPS in that it is a space-based navigation system providing global, 24 hour-a-day, all weather access to precise position, velocity and time information to a properly equipped user. Each GLONASS satellite continuously broadcasts its own precise position as well as less precise position information for the entire constellation. The transmitted GLONASS ephemeris message, the satellite position, its velocity and also its acceleration is directly given in the form of Cartesian Earth-Centred, Earth-Fixed (ECEF) co-ordinates and extrapolation terms in Euclidean space and time (Daly, 1992-1994; Kazantsev, 1992).

GLONASS was initially based on the CIS "Pulkovo System 1942 of Survey Co-ordinates" (Krassovsky) and the World Geodetic System of 1984 (WGS84) ellipsoid (Moskvin, 1989). Hereafter it was based on the Soviet Geocentric System of 1985 (SGS85) and as of 1995 on the Soviet Geocentric System of 1990 (SGS90).

Although the almanac parameters used for the position determination of GLONASS satellites are similar to those of the GPS (both are kinds of Keplerian parameters), the more precise ephemeris data of GLONASS and GPS differ completely in contents and timely updates. The values of the defining parameters of the SGS85 EGM gravity model and ellipsoid are slightly different from the WGS84 EGM.
A more difficult problem arises from the differences between the ECEF reference frames of WGS84 and SGS85 / SGS90. Although for the WGS84, the definition of individual parameters is well known and has been published, information about the SGS85 and SGS90 (or PZ90) is hard to obtain (Benhallam, 1996).

For a position determination using a combination of both satellite navigation systems, GPS and GLONASS, a uniform co-ordinate system is necessary. Some questions remain concerning the position reference ellipsoids used by GPS and GLONASS.

[21] The Commonwealth of Independent States, formerly the USSR

The Global Positioning System (GPS) became fully operational in 1994, and the Global Navigation Satellite System (GLONASS) in 1995 (Misra, 1992, 1994).

3.6.2 GPS and GLONASS Characteristics

Function	United States GPS	Soviet GLONASS
S/C number in the fully deployed system	21+3 standby	21+3 standby
Number of orbit planes	6	3
Orbit inclination	55°	64°.85
Orbit altitude	20 180 km	19 100 km
Revolution period orbit	approx. 12 hours	approx. 11 hours, 15 min
Orbital Parameters updated	satellite broadcast every hour	satellite broadcast ECEF position, velocity, acceleration updated every 30 minutes.
Ephemeris data representation	Kepler-like elements of S/C orbit parameters and extrapolating coefficients	9 parameters of S/C motion in the geocentric cartesian co-ordinate system
Geodetic co-ordinate system	WGS84 ECEF Frame	SGS90 ECEF Frame (1995)
Orbital Parameters updating	every hour	position, velocity, acceleration updated every 30 min
Phase-lock ranging signals	to GPS synchroniser	to GLONASS synchroniser
Method of S/C signal division	code division multiplexing	frequency division multiplexing
Almanac content	152 bit	120
Duration Almanac transmitting time	12.5 min	2.5 min
Frequencies in the L_1 range	1575.42 ±1.0 MHz	1602.5625 - 1615.5 ±0.5 MHz
Frequencies in the L_2 range	1227.6 MHz	1246.4375 - 1256.5 MHz
Number of code elements	1023	511
C/A-code frequency	1.023 MHz	0.511 MHz
P-code frequency	10.23 MHz	5.11 MHz
Cross talk level on 2 adjacent channels	- 21.6 dB	- 48 dB
Synchro code repetition period	6 s	2 s
Bit number in synchro code	8	30
Bi-binary coding of information	no	yes
Type of ranging code	C/A and P(Y) Codes	C/A and P Codes
Availability	Selective (S/A)	Non Selective
Time Synchronisation	GPS Time with UTC (USNO)	GLONASS Time with UTC (SU)
Ground track repeat period	1 sidereal day	8 sidereal days

Table 13: GPS and GLONASS characteristics

(Beser, 1992; Ivanov, 1992; Kazantsev, 1991; Seeber, 1993; Sluiter, 1995).

Local SGS85 to WGS84 Transformations

Investigations of local SGS85 to WGS84 transformations within narrow regions were published and led to the calculated local co-ordinate offsets. The 4th dimension (time offset) is included (Table 14):

Area	ΔX m	ΔY m	ΔZ m
Braunschweig	51	156	1

Area	Δφ m	Δλ m	ΔAlt m (corrected)
Minneapolis	- 52	31	85
Anchorage	1	10	2
Taipei	- 71	- 26	98

Table 14: Local SGS85 to WGS84 co-ordinate offsets

(Hartman, 1991a,1991b, 1992; Lechner, 1993, 1994; Gouzhva, 1991, 1995).

3.7 Footnotes

3.7.1 Reference Systems

Referring to (Ashkenazi, 1986b), a review paper with definitions and descriptions of the various types of co-ordinate systems and their mutual relationships, geodetic co-ordinates are discussed in [3.7.1.2].
This is followed in [3.7.1.3] by astronomic co-ordinates.

3.7.1.1 Reference System Errors

Ashkenazi: "Traditionally, the cartographer, the seafaring navigator and the geodetic surveyor have always expressed their co-ordinates in geographical terms, i.e. latitude and longitude, whereas the civil engineer and the landsurveyor preferred theirs in terms of projection grid co-ordinates, i.e. northings and eastings. Transformations between these various co-ordinate systems involve not only complex algebraical formulae, but also some very specific numerical parameters, which are appropriate for different countries and continents and which can only be determined empirically. Moreover, the treatment and interpretation of the different systems of co-ordinates may frequently involve some very *basic conceptual misunderstandings.* These include confusing astronomic latitudes and longitudes with their geodetic counterpart, treating projection northings and eastings as if they were ordinary plane co-ordinates and, in the case of positions derived from observations to *Transit / NavStar* satellites, applying the wrong set of transformation parameters or using inappropriate geoidal contour charts. These are typical examples of the sort of *common misconceptions* leading to gross errors and affecting even the most precisely determined absolute positions. Relative positioning, with respect to another point or a framework of points with known co-ordinates, eliminates some of the worst effects of these systematic sources of error, and is commonly used in geodetic surveying. However, instantaneous navigation (especially by using satellites) is most likely to be based on continuously determined, successive absolute positions and will therefore inevitable be affected by *reference system errors.* This is particularly important in the case of land surveying where much higher accuracies will be expected".

3.7.1.2 Geodetic Latitude and Longitude

A geodetic co-ordinate system is a 2-dimensional system used to define the unique location of a datum point on a mathematical model of the earth, the ellipsoid. Latitude represents distance north or south of the equator while longitude represents distance east or west from an arbitrarily accepted reference, the meridian, such as Ferro, Genoa, Greenwich, Oslo and Paris. Latitude and longitude are both given in sexagesimal or centesimal units.
If a three dimensional position is required, height above or below the ellipsoid at an arbitrary location is given in units of length. Elevation above or below mean sea level can and has been used instead of the ellipsoidal height where accuracy is not critical or geoidal heights are not available.

The geodetic position of a point on Earth is defined by the ellipsoidal co-ordinates of the projection of the two-dimensional point on to the surface of a geodetic reference ellipsoid, along the normal to that ellipsoid (Ashkenazi, 1991).

Geodetic Latitude φ_G of a point is the angle φ^P. The geodetic latitude (φ_G) is defined as the inclination of the normal to the ellipsoidal equatorial plane, and geodetic meridian as the plane through the normal and the minor axis of the reference ellipsoid, the angle PME (Figure 9), (Maling, 1992).
It follows that geodetic longitude (λ_G) of a point is the angle t, the geodetic longitude angle between its geodetic meridional plane and the IERS Reference Zero Meridian (IRM), measured in the plane of the geodetic equator, positive from 0° to 180°E, and negative from 0° to 180°W.

Note

No two points on the same reference Ellipsoid / Datum / Epoch can have identical geodetic co-ordinates.

3.7.1.3 Astronomic Latitude and Longitude

The astronomic latitude (φ_A or Φ) of a point is given by the angle between the vertical at that point and the equatorial plane. The astronomical meridian is defined as the plane passing through the vertical at that point and the spin axis of the Earth. The astronomic longitude (λ_A or Λ) of a point is the angle between two planes, one of which is the local meridian and the other the *Zero Meridian* at Greenwich.

Two points on the same reference ellipsoid may have identical astronomic co-ordinates. Therefore, astronomic co-ordinates do not therefore constitute a co-ordinate system in the geometrical sense.

3.7.2 The Geodetic Datum Problem

3.7.2.1 Datum Definition

A datum may be defined as a set of specifications for a three-dimensional co-ordinate system for a collection of positions on earth's surface:

- a two-dimensional reference surface to which the latitude and longitude co-ordinates are referred as well as the quantities which determine the origin used and orientation of the ellipsoid with respect to the Earth
- a three-dimensional cartesian co-ordinate system, the origin, orientation, and scale of which must fit the co-ordinates of physical points in the system. It is associated with every *geodetic datum*.

For both the reference surface and the co-ordinate system we must consider:

- issues of philosophy or principle [3.5.6; 3.5.8]
- issues of materialisation, how these attributes are achieved.

A common new datum is obtained by defining the transformation from each individual co-ordinate system to the final system (Chovitz, 1989; Schwartz, 1989; Wolf, 1987).

3.7.3 Datum and Ellipsoid

Guy Bomford:

> "The word *ellipsoid* can be used either with reference to the axes and flattening of the figure or to the three arbitrary constants at the origin as well. The word *datum* can also be used to refer only to the latter, or to the dimensions of the ellipsoid as well. For example, the International Ellipsoid (a and f), located by the *Potsdam 1950 Origin* (ξ_o, η_o, N_o) constitutes the *European Datum*" (Bomford, 1977).

3.7.4 Datum as Co-ordinate System

A system can also be specified by describing the relationship between it and another co-ordinate system. Attempting to match the Bureau International de l'Heure (BIH) Terrestrial System (BTS), NAD83, OSGRS80, EUREF89 and WGS84 co-ordinate systems are defined in terms of their relationship to the NSWC 9Z-2 co-ordinate system. These co-ordinate systems are almost identical because *Geodetic Agencies* were co-ordinating their efforts.

3.7.5 Datum as Co-ordinates

The datum is defined as an abstract, equipotential reference surface, usually followed by a definition which states that a horizontal geodetic datum is composed of adopted horizontal co-ordinates of a set of physical points. The adopted co-ordinates actually determine the origin and orientation of a datum. The concept of extending a datum by adding new points implies that a set of physical points are available from which the process is begun. These are the fundamental points that participate in the initial datum adjustment.

Charles R. Schwartz reports:

> *"New points that will be added are not. In most geodetic datums, the distinction between fundamental and non-fundamental points has been lost. Typically a new point surveyed to first-order accuracy and adjusted into the network has been treated as equal in usefulness to a fundamental first-order point, and superior to a second-order fundamental point. Unfortunately this common, but incorrect practice has often misled users as to the accuracy of a point's co-ordinates" (Schwartz, 1989).*

See also [11.4], Spatial Databases.

3.7.6 Overlapping Datums

Some physical points are fundamental to two datums, e.g. ETRF89 and WGS84. The co-ordinates of these points in the two systems may differ in precision to sub millimetre level.. Using Global Positioning System observations in the single point positioning mode, together with a satellite ephemeris given in the WGS84 co-ordinate system, these co-ordinates will be in WGS84 "raw" doppler co-ordinates. The user should do something about co-ordinate misclosures remaining in ETRF89: the co-ordinates receive adjustment corrections from integration with terrestrial observations. This results in two different sets of co-ordinates at common points.

3.7.7 Mixing Co-ordinates

The Datum Problem

The *lack of awareness* by the general user of the necessity to link co-ordinate values with a specific datum is the main problem associated with geodetic datums.

ETRF 89, NAD83 and WGS84 should be thought of as overlapping datums. ETRF 89 and WGS 84 use the same corrections to geodetic co-ordinates (translation, scale, orientation to the celestial pole, and orientation to the celestial zero meridian). The ellipsoid parameters for both systems are based on the equatorial radius and three physical constants for the GRS80. In computing the flattening, NIMA did not use the defining GRS80 constants as given, but derived one constant from another, see [3.5.7], Datum Differences (3.29).

This resulted in two values for the reciprocal of the flattening, derived to 16 significant figures:

- GRS80: $f^{1} =$ 298.25722 21008 827 (as used by NGS)
- WGS84: $f^{1} =$ 298.25722 3563.

The action to take when confronted with two sets of co-ordinates for a single point is up to the user. If neither position determination contains a blunder, then the differences of co-ordinates should be relative small. The measured distance could differ from the value computed from the co-ordinates by a *one metre or more* due to differences in adjustment and survey methodologies. For example, a report by Ollikainen gives information about transformation between fixed stations from the EUREF89 epoch 1989.0 to the EUREF89 epoch 1992.0 in the Finnish first-order geodetic network. In the northern hemisphere one could come across the New International (GRS80) and WGS84.
(Ashkenazi, 1991, 1993; Ollikainen, 1995; Poutanen, 1995; Schwartz, 1989).

3.7.8 Blunder Detection

Families of survey stations with co-ordinates in two or more datums (or epochs) do exist in most countries. An oversight will take place when the *unaware user* is contradicted with sets of co-ordinates for survey points. However, the differences of position co-ordinates could be *relatively* small.

The distances measured by EDM instruments could differ from the values computed from the co-ordinates by *ten* metres or *more* due to differences in the datums of an area. In Northern Africa (Tunisia) a user may find:

- Stations based on Clarke 1880IGN ____________________ Datum CARTHAGE
- Stations based on Clarke 1880IGN ____________________ Datum VOIROL
- Stations based on Clarke 1880DoD ____________________ UTM Grid.

The positions calculated could differ by 22" or *more* in latitude and longitude due to differences in the datum determination (IGN, 1963).

3.7.9 In Summary

Some definitions under consideration in European Standard, CEN/TC 287 Geographic Information (Harsson, 1996).

Astronomic Co-ordinates (Φ, Λ, H) or (Φ, Λ)

Astronomic latitude and astronomic longitude of a given point, with or without orthometric height.

Astronomic Latitude (Φ)

Angle from the equatorial plane to the direction of gravity through the given point, northwards treated as positive.

Astronomic Longitude (Λ)

Angle from the Zero Meridian to the celestial meridian plane of the given point, an astronomic concept beyond the scope of this standard.

Geodetic Ellipsoid

Flattened ellipsoid of rotation, usually chosen to fit the geoid as closely as possible, either locally or globally.

Geodetic Latitude (φ)

Angle from the equatorial plane to the direction of the perpendicular to the ellipsoid through the given point, northwards treated as positive.

Geodetic Longitude (λ)

Angle from the Zero Meridian Plane to the meridian plane of the given point, eastwards treated as positive.

Geographic Co-ordinates

Geographic latitude and geographic longitude, with or without height.

Geographic Latitude

A generic term for geodetic or astronomic latitude, but more often geodetic.

Geographic Longitude

Generic term for geodetic or astronomic longitude, but more often geodetic.

Geoid

Equipotential surface of the Earth's gravity field which most closely approximates mean sea level.

Greenwich Meridian Plane

Meridian plane passing through Greenwich, England, widely used as the *Zero Meridian plane*.

> *Note - this is actually only a half-plane, on the European side of the polar axis.*

Latitude (Φ, φ)

Angle from the equatorial plane to the considered direction at the given point, northwards treated as positive. The considered direction is the perpendicular to the reference surface through the given point; see astronomic latitude, geodetic latitude and geographic latitude.

Longitude (Λ, λ)

Angle from the *Zero Meridian plane* to the meridian plane of the given point, eastwards treated as positive. See *astronomic longitude*, *geodetic longitude* and *geographic longitude*.

Map Projection

Mathematical mapping of a geodetic ellipsoid, or part of a geodetic ellipsoid, to a plane.

4. Geodetic Arc Distances

There are at least three distinct geodetic curves in the field of geodesy. Each furnishes different distances and different azimuths for any one reference ellipsoid:

- the great elliptic arc
- the normal sections
- the geodesic.

The *Great Elliptic Arc* does not represent the shortest connection between two points, but it may be of more significance than the geodesic for inertial or ballistic missile computations.

The *Normal Section* is significant for EDM instrument distances, but it may be of more significance than the geodesic for inertial or ballistic missile computations. Both arcs are discussed (ACIC, 1957).

The *Geodesic* is considered the principal path in that it represents the shortest distance in a unique way between any two given points on the ellipsoidal earth (Leick, 1990). See Figure 15: The geodesic and the reference ellipsoid.

Computations of distance and azimuth, whatever the type of curve involved, are directly related to:

- formulae utilised
- quality of P_1, P_2
- reference datum
- size and shape of the reference ellipsoid.

Arc distance P_1 P_2 and also the size of angles P_n P_1 P_2, P_n P_2 P_1 (Figure 15) depend upon the characteristics and behaviour of the surface curve joining the end points.

4.1 The Great Elliptic Arc

A "great elliptic" is a curved line on the surface of the reference ellipsoid which connects two points on that surface and lies in a plane which contains the centre of the reference ellipsoid. It is similar to a geodesic, except that it is always a plane curve. It is similar to a normal section except that only one great elliptic exists between any two points. The meridians and the equator are special cases which may be considered to be great elliptic arcs as well as geodesics or normal sections.

4.2 The Normal Section

A "normal section" is a curved line on the surface of the reference ellipsoid which connects two points on that surface and lies in a plane which contains the normal at the one point and passes through the other point. It is similar to a geodesic, except that it is always a plane curve. It is different from a geodesic in that two normal sections exist between any two points, except in the cases of the meridians and the equator (Bowring, 1971).

4.3 Geodesics

Clairaut found that for any point along a geodesic on *a surface of revolution* the distance from the axis of rotation times the sine of the azimuth ($\sin \alpha$) remains constant.

The concern is to obtain accurate values of the shortest, smooth but mathematically complicated geodesic line *S* between two given points and of the azimuths of the geodesic. Two fundamental geodetic calculations on a reference ellipsoid are generally known as the direct and inverse problems (Bomford, 1977; Hopfner, 1949; Leick, 1990; Pfeifer, 1984; Urmajew, 1955).

The azimuths are conventionally calculated clockwise from True North. Generally, it is not a plane curve except when it follows a meridian or the equatorial circle. Any segment of a meridian or the equator is a geodesic.

Likewise, the S-shaped curved line between points P_1 (φ_1, λ_1) and P_2 (φ_2, λ_2) is a geodesic because it is the unique line on the mathematically defined reference ellipsoid, that represents the shortest distance and contains the normal at each point.

To solve the Direct and the Inverse problem on the *surface of a reference ellipsoid*, there are several sets of formulae. These are usually referred to by the name of their originator, e.g. (Robbins, 1950, 1962). If 0.01 m or 0".0001 accuracy is desired most of these methods can be used for distances not greater than *150* km, beyond which results might be erroneous.
Some can be used up to *400* km but only a very few from *400* km to *20 000* km, the latter being the maximum possible distance between two points on the Earth's surface.

4.3.1 Geodesics up to 20 000 km by Kivioja's Method

In theory, the computer offers a choice of integration methods in the solution of the Direct and Inverse problems, which has been extensively documented (Kivioja, 1971; Jank, 1980; Murphy, 1981).

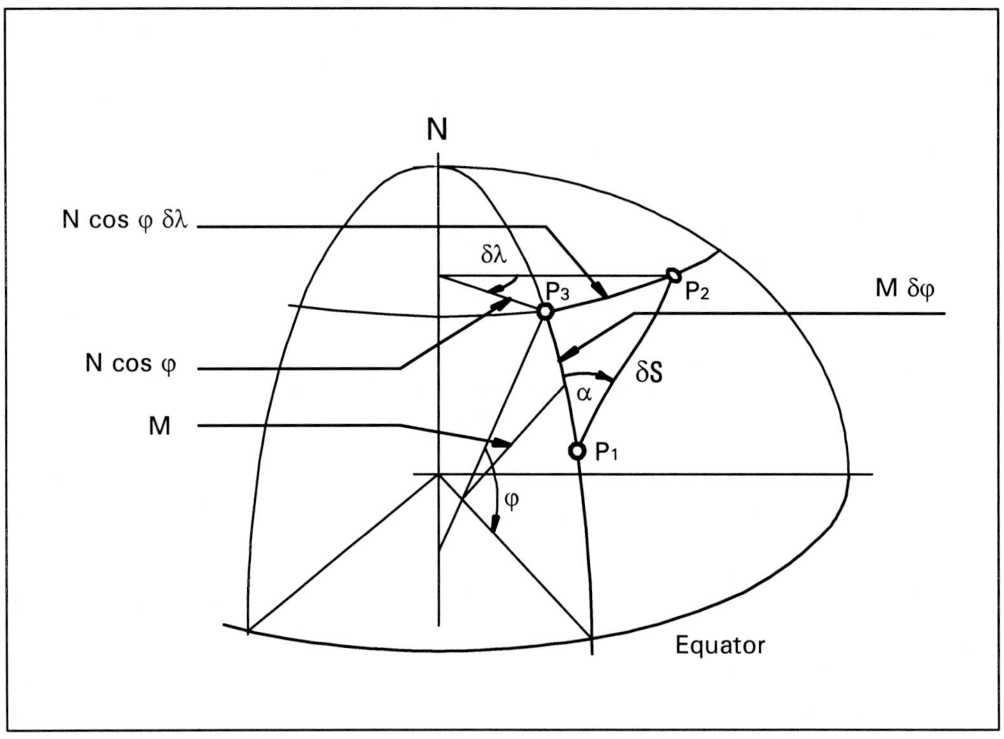

Figure 13: Geodetic triangle

Integrating the differential equations directly, the practical implementation of such a method on a computer works briefly as follows:

> If a geodesic, *length S*, is divided into *n* equal elements length δS, then, providing δS is small, the elemental triangle $P_1P_2P_3$ (Figure 13) may be considered plane. Also, the sides P_1P_2, P_2P_3, P_1P_3 can be determined. Therefore, the problem can be reliably solved with excellent accuracy by adding together the lengths of small elements, δS, of the length of the Geodetic Line, S.

To solve *the direct problem* by the algorithm, distance S is divided into n equal line elements (Figure 14), each of length δS, so that:

S = n δS

The line elements are kept in correct azimuths by estimating and correcting Clairaut's Equation:

$$C = N_1 \cos \varphi_1 \sin \alpha_1 = N_2 \cos \varphi_2 \sin \alpha_2, \text{ etc.} \tag{4.01}$$

for the Geodetic Line to force each line element to lie on the geodetic line, in which:

φ = geodetic latitude
N = radius of curvature in Prime Vertical
α = bearing of the geodetic line element

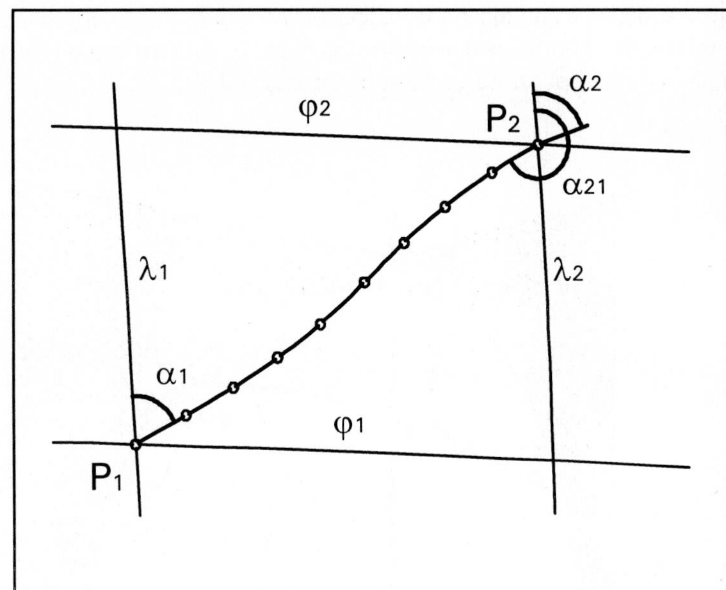

Figure 14: Geodesic divided into dS line elements

The inverse problem calculates between points P_1 and P_2 approximations for the distance S, an azimuth α and, starting from one initial point P_1, computes the co-ordinates of the second position P_2 by the same algorithm used in the direct problem. The computed co-ordinates will not generally coincide with the given values, but the differences between the computed and the known points are used to correct the initial approximations for the distance and azimuth. The program computes the latitude ($\delta\varphi_i$) and longitude ($\delta\lambda_i$) increments for each equal line element and cumulatively adds them to the latitude and longitude of the previous point. Therefore, this solution by integration is called the "point-by-point method".

As in numerical integrations in general, the accuracy of the result can be increased by decreasing the length of the geodetic line elements δS. Computing time of the program depends upon the number of integrations of the geodetic line. If *millimetre accuracy* is desired, δS should be between 100 to 200 m, and for centimetre accuracy, δS should be between 1 000 or 2 000 m but should not exceed 4 000 m (Kivioja, 1971).

Obviously, this method can be used to check any other classical method. The point-by-point method has an additional advantage of calculating intermediate geodetical positions (Figure 14). Unfortunately, the mathematical developments of the geodesic on the ellipsoid and its use on the computer are fairly complex. Because the solutions are fully documented in the literature, this section contains only the basic concept and computer algorithms BDG00000.BAS, GBD00000.BAS found in [11.7.3; 11.7.4], respectively.

4.3.2 Direct Problem

Solution of the Direct Problem

The Direct problem involves computation of the latitude, longitude, and back-azimuth of a point P_2 whose distance and azimuth from given initial point P_1 are known.

The geodetic line (geodesic) runs through points P_1 and P_2. The length of the geodetic line between points P_1 and P_2 is S. Therefore *S* runs in azimuth α_1 at P_1 and in azimuth α_2 at P_2. The back azimuth at point P_2 is obtained by adding 180° to α_2, or generally $\alpha_{2\text{-}1} = \alpha_2 \pm 180°$. Figure 14 shows the meridians λ_1 and λ_2 through P_1 and P_2, and φ_1 and φ_2 are parallels of latitude through P_1 and P_2.

The program computes the latitude and longitude increments $\delta\varphi$ and $\delta\lambda$ for each geodetic line element δS in a repetitive manner. Over the first δS, the increments are $\delta\varphi_1$ and $\delta\lambda_1$ over the second δS, the increments are $\delta\varphi_2$ and $\delta\lambda_2$; and so on, until over the last δs_n, the increments are $\delta\varphi_n$ and $\delta\lambda_n$.

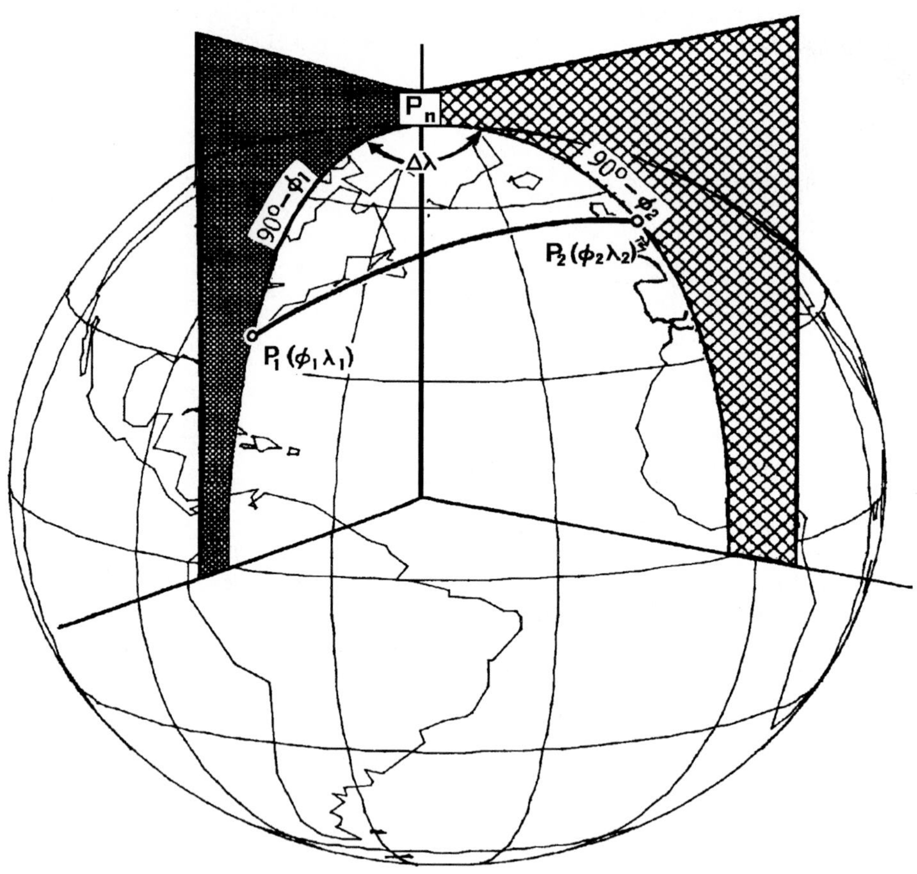

Figure 15: The geodesic and the reference ellipsoid

Starting with latitude φ_1, and longitude λ_1, at point P_1 after the first δS, the latitude will be $\varphi_1+\delta\varphi_1$, and the longitude will be $\lambda_1+\delta\lambda_1$; after the second δS, the latitude will be $\varphi_1+\delta\varphi_1+\delta\varphi_2$, and the longitude will be $\lambda_1+\delta\lambda_1+\delta\lambda_2$; and so on. The computation proceeds until all n elements δS are covered. To obtain these increments, the radii of curvature are needed for the reference ellipsoid (Kivioja, 1971; Jank, 1980).

The equation of the reference ellipsoid with semi-axes a and b is:

$X^2 / a^2 + Y^2 / a^2 + Z^2 / b^2 = 1$ (4.02)

$M = c / V^3$, radius of curvature in the Meridian, (4.03)

in which:

$c = a^2 / b$

$V = (1+e'^2 \cos^2 \varphi)^{1/2}$ (φ=geodetic latitude)

$e'^2 = (a^2 - b^2) / b^2$

$N = c / V$, the radius of curvature in the Prime Vertical (4.04)

Figure 13: Geodetic triangle shows:

$\delta S \cos \alpha = M \delta\varphi$, or $\delta\varphi = \delta S \cos \alpha / M$ (4.05)

$\delta S \sin \alpha = N \cos \varphi \, \delta\lambda$, or $\delta\lambda = \delta S \sin \alpha / (N \cos \varphi)$ (4.06)

in which δS is any curve element in azimuth α on the surface of the ellipsoid, $M \delta\varphi$ is the corresponding element of meridian between parallels φ and $\varphi+\delta\varphi$, and $N \cos \varphi \, \delta\lambda$ is the corresponding element of the parallel circle between meridians λ and $\lambda+\delta\lambda$. Clairaut's equation (4.01) for a geodetic line is used to control the azimuth of δs:

$N \cos \varphi \sin \alpha = C$ (Clairaut's Constant) (4.07)

C is constant for one geodetic line element. It can have any value between 0 m and the Equatorial Radius a. The equations (4.05, 4.06, 4.07) are required to compute a geodetic line. The shorter δS, the more accurate are the results (Kivioja, 1971).

The following examples are calculated by program BDG00000.BAS, according to Kivioja's method.

Numerical Examples

Given:	φ_1,	λ_1,	α_{1-2},	$S_{1-2} = S$
Output:	φ_2,	λ_2,	$\alpha_{2-1} = \alpha_2 \pm 180°$	

The following examples are calculated using varying number of δS line elements.

4.3.2.1 Geodetic Line - Direct Problem - 1320 km

Given: Ellipsoid Bessel 1841 a=**6 377 397.155** f^{-1}=**299.15281 285** See (J/E/K, 1959), [4.3.3.2]

Latitude φ_1:	**45° 00' 00" N**	True Bearing - α_{1-2}:	**29° 03' 15".458713**
Longitude λ_1:	**10° 00' 00" E**	True Distance - S:	**1 320 284.368 37**
n:	**10 - 250 000**		
Output:	Latitude - φ_2	Longitude - λ_2	True Bearing - α_{2-1}

n	δS_m	Latitude - φ_2	Longitude - λ_2	True Bearing - α_{2-1}
10 [22]	132028.436837	55° 00' 00".61619	19° 59' 57".49669	216° 45' 08".05494
100	13202.843683	55° 00' 00".00615	19° 59' 59".97525	216° 45' 07".40575
1 000	1320.284368	55° 00' 00".00016	19° 59' 59".99952	216° 45' 07".39937
10 000	132.028437	55° 00' 00".00010	19° 59' 59".99976	216° 45' 07".39931
100 000	13.202844	55° 00' 00".00010	19° 59' 59".99976	216° 45' 07".39931
250 000	5.281137	55° 00' 00".00010	19° 59' 59".99976	216° 45' 07".39931

[22] n of *δS-line elements*

4.3.2.2 Geodetic Line - Direct Problem - 65 km

Given:	Ellipsoid International 1924	a=**6 378 388**	**f^{-1}=297** [4.3.2.3]
Latitude φ_1:	**46° 55' 09".9100 N**	True Bearing - α_{1-2}:	**119° 17' 21".9200**
Longitude λ_1:	**7° 26' 40".4700 E**	True Distance - S:	**64 865.007**
Output: Latitude - φ_2	Longitude - λ_2	True Bearing - α_{2-1}	and intermediate points

n	Latitude - φ_2	Longitude - λ_2	True Bearing - α_{2-1}
1 000 |[23]	46° 37' 53".68297 N	8° 10' 59".93073 E	299° 49' 39".82352

If desired, the *point-by-point* method computes the co-ordinates for *intermediate geodetic positions.*

Decr.- φ n	Latitude - φ_i	Decr.- λ n	Longitude - λ_i
φ_{901}	46° 53' 27".06	λ_{901}	7° 31' 07".69
φ_{900}	46° 53' 26".03	λ_{900}	7° 31' 10".36
φ_{899}	46° 53' 25".00	λ_{899}	7° 31' 13".03
φ_{898}	46° 53' 23".76	λ_{898}	7° 31' 15".70
φ_{897}	46° 53' 22".46	λ_{897}	7° 31' 18".37
↓	↓	↓	↓
φ_{818}	46° 52' 01".57	λ_{818}	7° 34' 49".27
φ_{817}	46° 52' 00".53	λ_{817}	7° 34' 51".93
φ_{816}	46° 51' 59".50	λ_{816}	7° 34' 54".60
φ_{815}	46° 51' 58".47	λ_{815}	7° 34' 57".27
φ_{814}	46° 51' 57".44	λ_{814}	7° 34' 59".94
↓	↓	↓	↓
φ_{604}	46° 48' 20".59	λ_{604}	7° 44' 19".66
φ_{603}	46° 48' 19".55	λ_{603}	7° 44' 22".33
φ_{602}	***46° 48' 18".52***	λ_{602}	***7° 44' 24".99***
φ_{601}	46° 48' 17".48	λ_{601}	7° 44' 27".65
φ_{600}	46° 48' 16".45	λ_{600}	7° 44' 30".31
↓	↓	↓	↓
φ_{246}	46° 42' 09".14	λ_{246}	8° 00' 10".98
φ_{245}	46° 42' 08".10	λ_{245}	8° 00' 13".63
φ_{244}	46° 42' 07".06	λ_{244}	8° 00' 16".28
φ_{243}	46° 42' 06".02	λ_{243}	8° 00' 18".93
φ_{242}	46° 42' 04".98	λ_{242}	8° 00' 21".59
↓	↓	↓	↓
φ_7	46° 37' 59".95	λ_7	8° 10' 44".05
φ_6	46° 37' 58".90	λ_6	8° 10' 46".70
φ_5	46° 37' 57".86	λ_5	8° 10' 49".34
φ_4	46° 37' 56".81	λ_4	8° 10' 51".99
φ_3	46° 37' 55".77	λ_3	8° 10' 54".64
φ_2	46° 37' 54".72	λ_2	8° 10' 57".28
φ_1	46° 37' 53".68	λ_1	8° 10' 59".93

[23] n of *δS-line elements*

4.3.2.3 Geodetic Line - Direct Problem - 65 km

[4.3.2.2; 4.3.3.1] |[24]

Given:	Ellipsoid International 1924		a=**6 378 388**	f^{-1}=**297**
Latitude	φ_1:	**46° 55' 09".9100 N**	True Bearing - $\alpha_{1\text{-}2}$:	**119° 17' 21".9200**
Longitude	λ_1:	**7° 26' 40".4700 E**	True Distance - S:	**64 865.007**
Output:	Latitude - φ_2	Longitude - λ_2	True Bearing - $\alpha_{2\text{-}1}$	and intermediate points

n	Latitude - φ_2	Longitude - λ_2	True Bearing - $\alpha_{2\text{-}1}$
10 000 |[25]	46° 37' 53".68297 N	8° 10' 59".93073 E	299° 49' 39".82352

If desired, the *point-by-point method* computes the co-ordinates for *intermediate geodetic positions*

Decr.- φ n	Latitude - φ_i	Decr.- λ n	Longitude - λ_i
φ_{6011}	**46° 48' 18".52**	λ_{6011}	**7° 44' 24".99**
φ_{6010}	46° 48' 18".41	λ_{6010}	7° 44' 25".25
φ_{6009}	46° 48' 18".31	λ_{6009}	7° 44' 25".52
φ_{6008}	46° 48' 18".21	λ_{6008}	7° 44' 25".79
φ_{6007}	46° 48' 18".10	λ_{6007}	7° 44' 26".05
↓	↓	↓	↓
φ_{3646}	47° 00' 32".81	λ_{3646}	8° 06' 21".00
φ_{3645}	47° 00' 32".86	λ_{3645}	8° 06' 21".37
φ_{3644}	47° 00' 32".91	λ_{3644}	8° 06' 21".75
φ_{3643}	47° 00' 32".96	λ_{3643}	8° 06' 22".12
↓	↓	↓	↓
φ_6	46° 37' 54".20	λ_6	8° 10' 58".60
φ_5	46° 37' 54".10	λ_5	8° 10' 58".87
φ_4	46° 37' 53".99	λ_4	8° 10' 59".13
φ_3	46° 37' 53".89	λ_3	8° 10' 59".40
φ_2	46° 37' 53".78	λ_2	8° 10' 59".66
φ_1	46° 37' 53".68	λ_1	8° 10' 59".93

[24] See (Baeschlin, 1948): pp 146-151

[25] n of δS-line elements

4.3.2.4 Geodetic Line - Direct Problem - 15 000 km

Given:	Ellipsoid International 1924	a=**6 378 388**	f^{-1}=**297** (J/E/K, 1959)
Latitude φ_1:	**50° 00' 00" N**	True Bearing - $\alpha_{1\text{-}2}$:	**140° 00' 00".00000**
Longitude λ_1:	**10° 00' 00" E**	True Distance - S:	**15 000 000.000**
n :	**1000 - 750 000**		
Output:	Latitude - φ_2	Longitude - λ_2	True Bearing - $\alpha_{2\text{-}1}$

n	Latitude - φ_2	Longitude - λ_2	True Bearing - $\alpha_{2\text{-}1}$
1 000[26]	62° 57' 03".181655 S	105° 05' 38".406919 E	294° 46' 41".578009
10 000	62° 57' 03".203645 S	105° 05' 38".300738 E	294° 46' 41".484846
100 000	62° 57' 03".203865 S	105° 05' 38".299675 E	294° 46' 41".483913
750 000	62° 57' 03".203867 S	105° 05' 38".299664 E	294° 46' 41".483904

4.3.3 Inverse Problem

Solution of Inverse Problem

The Inverse problem calculates the distance and azimuths between two geodetic positions, P_1 and P_2.

Given:	φ_1,	λ_1,	φ_2,	λ_2
Output:	$S_{1\text{-}2} = S$	$\alpha_{1\text{-}2}$	$\alpha_{2\text{-}1} = \alpha_2 \pm 180°$	

Using Clairaut's equation for the Geodetic Line, the Inverse problems can be solved. The point-by-point method "stakes out" the Geodetic Line. It computes the co-ordinates for each intermediate geodetical position (Kivioja, 1971).

Important:

In order to solve this problem it is advisable that the position of Station P_1 is situated south of Station P_2. Furthermore, the Line P_1-P_2 should not run exactly East-West or exactly North-South. In case this may occur during data processing, the computer will issue an error message. See [2.2], Error Messages, or the BASIC-Handbook, *which explains the various error conditions. Some calculations in the Western or Southern Hemisphere can be mirrored and calculated in the N/E Hemisphere (Figure 16).*

Using equations (4.05, 4.06, 4.07), it is possible to estimate an approximate azimuth and an increment at P_1. Thus, the distance S and the azimuth α between points P_1 (φ_1, λ_1) and P_2 (φ_2, λ_2) are calculated.
Use $\delta\varphi = (\varphi_2 - \varphi_1) / n$ if the $\alpha < 45°$ or use $\delta\lambda = (\lambda_2 - \lambda_1) / n$ if the azimuth $\alpha > 45°$, or use both.
If $\delta\varphi$ is used, compute M, N, cos φ , and α for latitude $\varphi + \frac{1}{2}\,\delta\varphi$, then δs from (4.05), and then $\delta\lambda$ from (4.06), and the azimuth α from (4.07). Do this *n* times accumulating the increments of distance and longitude.

A computed longitude difference will of course not correspond with $\lambda_2 - \lambda_1$, because none of the approximate azimuths will guide the geodesic exactly through P_2, but this geodetic line will cut the parallel of φ_2 either East or West from P_2. Using these disagreements, the estimated azimuth can be substantially improved. If desired, the computer can show the latitudes and longitudes at the last set of increments.

[26] n of *δS-line elements*

Kivioja's paper gives some hints, e.g. integration could also utilise the differential elements $\delta\varphi$, $\delta\lambda$, or $\delta\alpha$, rather than δS, using rearrangements of equations (Kivioja, 1971; Jank, 1980).

The following examples are calculated by program GBD00000.BAS, according to Kivioja's method.

4.3.3.1 Geodetic Line - Inverse Problem - 65 km

Given:	Ellipsoid International 1924	a=**6 378 388**	f^{-1}=**297** [4.3.2.2; 4.3.2.3]
Latitude φ_1:	**46° 37' 53".6830 N**	Latitude - φ_2:	**46° 55' 09".9100 N**
Longitude λ_1:	**8° 10' 59".9308 E**	Longitude - λ_2:	**7° 26' 40".4700 E**
Output: True Distance - S	True Bearing - α_{1-2}	True Bearing - α_{2-1}	and intermediate points

n	Geodesic - S_m	True Bearing - α_{1-2}	True Bearing - α_{2-1}
10 000 [27]	64 865.00777	299° 49' 39".81880	119° 17' 21".91522

4.3.3.2 Geodetic Line - Inverse Problem - 1320 km

Given:	Ellipsoid Bessel 1841	a=**6 377 397.155**	f^{-1}=**299.15281 285** [4.3.2.1]
Latitude φ_1:	**45° 00' 00" N**	Latitude φ_2:	**55° 00' 00" N**
Longitude λ_1:	**10° 00' 00" E**	Longitude λ_2:	**20° 00' 00" E**
n:	**10 - 250 000**		
Output:	True Distance - S	True Bearing - α_{1-2}	True Bearing - α_{2-1}

n	δS_m	Geodesic - S_m	True Bearing - α_{1-2}	True Bearing - α_{2-1}
10 [27]	132029.572567	1320295.725668	29° 03' 22".80827	216° 45' 17".27885
100	13202.844791	1320284.479108	29° 03' 15".53074	216° 45' 07".49602
1 000	1320.284369	1320284.369475	29° 03' 15".45943	216° 45' 07".40016
5 000	264.056874	1320284.368415	29° 03' 15".45874	216° 45' 07".39924
10 000	132.028437	1320284.368381	29° 03' 15".45872	216° 45' 07".39921
25 000	52.811375	1320284.368372	29° 03' 15".45871	216° 45' 07".39920
100 000	13.202844	1320284.368370	29° 03' 15".45871	216° 45' 07".39920
250 000	5.281137	1320284.368370	29° 03' 15".45871	216° 45' 07".39920

[27] n of *δS-line elements*

4.3.3.3 Geodetic Line - Inverse Problem - 81 km

Given:	Ellipsoid International 1924	a=**6 378 388**	**f^{-1}=297** [28]
Latitude φ_1:	**46° 55' 09".9100 N**	Latitude φ_2:	**47° 03' 31".7965 N**
Longitude λ_1:	**7° 26' 40".4700 E**	Longitude λ_2:	**8° 29' 09".9324 E**
Output: Latitude - φ_2	Longitude - λ_2	True Bearing - $\alpha_{2\text{-}1}$	and intermediate points

n	Geodesic S_m	True Bearing - $\alpha_{1\text{-}2}$	True Bearing - $\alpha_{2\text{-}1}$
10 000 [29]	80 734.46365	78° 33' 05".13058	259° 18' 46".86569

Decr.- φ n	Latitude - φ_i	Decr.- λ n	Longitude - λ_i
φ_{10000}	46° 55' 09".96	λ_{10000}	7° 26' 40".84
φ_{9999}	46° 55' 10".01	λ_{9999}	7° 26' 41".21
φ_{9998}	46° 55' 10".06	λ_{9998}	7° 26' 41".59
↓	↓	↓	↓
φ_{9399}	46° 55' 41".08	λ_{9399}	7° 30' 25".63
φ_{9398}	46° 55' 41".13	λ_{9398}	7° 30' 26".00
φ_{9397}	46° 55' 41".19	λ_{9397}	7° 30' 26".37
↓	↓	↓	↓
φ_{7669}	46° 57' 09".99	λ_{7669}	7° 41' 13".09
φ_{7668}	46° 57' 10".04	λ_{7668}	7° 41' 13".46
φ_{7667}	46° 57' 10".10	λ_{7667}	7° 41' 13".84
↓	↓	↓	↓
φ_{5629}	46° 58' 53".53	λ_{5629}	7° 53' 57".33
φ_{5628}	46° 58' 53".58	λ_{5628}	7° 53' 57".70
φ_{5627}	46° 58' 53".63	λ_{5627}	7° 53' 58".08
↓	↓	↓	↓
φ_{4830}	46° 59' 33".69	λ_{4830}	7° 58' 56".88
φ_{4829}	46° 59' 33".74	λ_{4829}	7° 58' 57".25
φ_{4828}	46° 59' 33".79	λ_{4828}	7° 58' 57".63
↓	↓	↓	↓
φ_{2659}	47° 01' 21".72	λ_{2659}	8° 12' 31".43
φ_{2658}	47° 01' 21".77	λ_{2658}	8° 12' 31".80
φ_{2657}	47° 01' 21".82	λ_{2657}	8° 12' 32".18
↓	↓	↓	↓
φ_{1958}	47° 01' 56".26	λ_{1958}	8° 16' 54".63
φ_{1957}	47° 01' 56".31	λ_{1957}	8° 16' 55".01
φ_{1956}	47° 01' 56".36	λ_{1956}	8° 16' 55".38
↓	↓	↓	↓
φ_{769}	47° 02' 54".46	λ_{769}	8° 24' 21".28
φ_{768}	47° 02' 54".51	λ_{768}	8° 24' 21".66
φ_{767}	47° 02' 54".56	λ_{767}	8° 24' 22".04

[28] See (Baeschlin, 1948): pp 146-151

[29] n of δS-line elements

4.4 Calculation of the Arc of the Meridian

The arc of the meridian is complicated to evaluate because the meridional radius of curvature varies continuously with latitude. To solve the direct question on the arc of the meridian of a reference ellipsoid, the method starts with the differential formulae and proceeds by series development.

Determining the arc G_m measured from the equator to geodetic latitude $B_i°$. Latitude $B_i°$ is the angle PME $=\varphi$P between the major axis of the ellipsoid and the normal to the tangent plane at an arbitrary point on the surface of the ellipsoid measured at the point of intersection of the normal with the equatorial plane (Figure 9).

Considering any point along the arc as an infinitely small part of a curve, corresponding to an infinitely small change in latitude, the length of the whole curve G_m is defined from the equator to a point in latitude $B_0°$.

It is now necessary to determine, first, the length of a very short arc at the equator, and then accumulate the lengths of all the small elements which form the whole arc G_m. Since the limits of the arc have already been

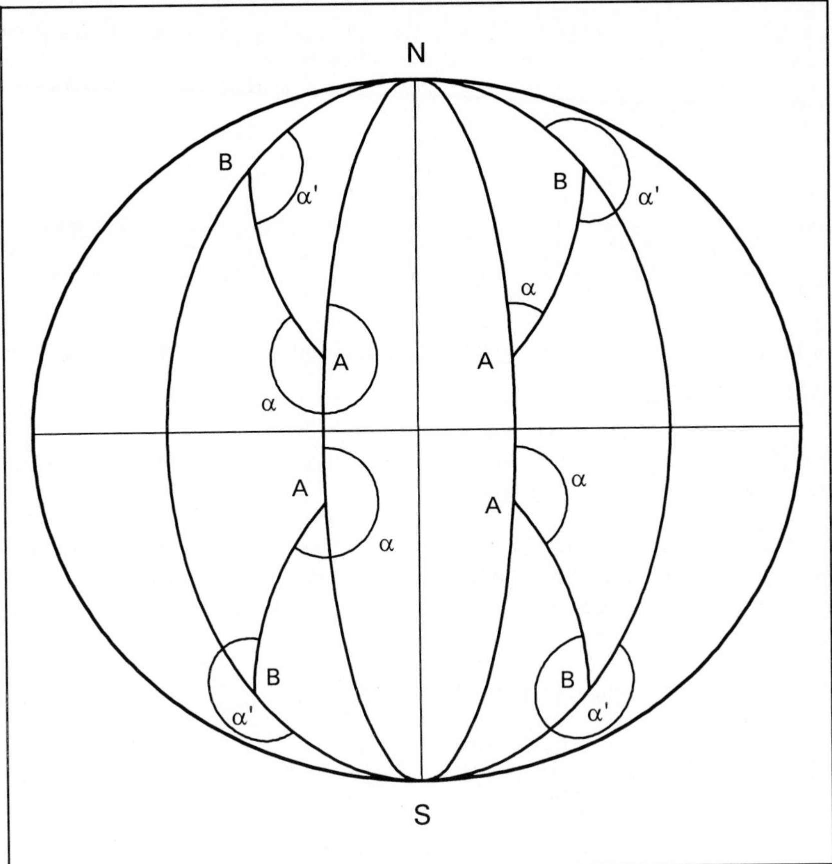

Figure 16: Different positions of the geodesic A - B

specified, the arc distance on the ellipsoid, G_m, may be written as the integral:

$$\delta G_m = M \Delta B = c\,(1+e'^2 \cos^2 B)^{-3/2}\, \delta B \text{ (J/E/K, 1959),} \tag{4.08}$$

in which:

c = polar radius of curvature

e'^2 = second eccentricity squared

M = ρ = radius of curvature in the meridian (4.03).

Obviously these parameters are closely related.

After integration of this expression, the equation is simplified to a form suitable for calculation as shown by (Helmert, 1880).

4.4.1 Direct Computation

$$G_m = c\,[\,E_0\,B + E_2 \sin(2B) + E_4 \sin(4B) + E_6 \sin(6B) + E_8 \sin(8B)\,] \tag{4.09}$$

in which the coefficients E_0, E_2, E_4, E_6, E_8 are expressed in terms of e' as given by (J/E/K, 1959):

$$\begin{aligned}
E_0 &= 1 - \frac{3}{4}e'^2 + \frac{45}{64}e'^4 - \frac{175}{256}e'^6 + \frac{11025}{16384}e'^8 - \frac{43659}{65536}e'^{10}\\
E_2 &= -\frac{3}{4}e'^2 + \frac{15}{16}e'^4 - \frac{525}{512}e'^6 + \frac{2205}{2048}e'^8 - \frac{72765}{65536}e'^{10}\\
E_4 &= +\frac{15}{64}e'^4 - \frac{105}{256}e'^6 + \frac{2205}{4096}e'^8 - \frac{10395}{16384}e'^{10}\\
E_6 &= -\frac{35}{512}e'^6 + \frac{315}{2048}e'^8 - \frac{31185}{131072}e'^{10}\\
E_8 &= +\frac{315}{16384}e'^8 - \frac{3465}{65536}e'^{10}
\end{aligned} \tag{4.10}$$

See Table 6, in which a and b are the semi-major axis and semi-minor axis of the ellipsoid used.

4.4.2 Inverse Computation

The arc of the meridian from the equator B_0 to B_i is G_m as before. Most computing algorithms calculate the inverse solution by a time-consuming iterative process.

Rather than use the coefficients E_2, E_4, E_6, E_8 as derived for this equation, better consistency is obtained by reversing the equation used in the direct computation. This was done by Krack in 1982.
An explanation of the inversion can be found in a report described by (Krack, 1982). He explains in the paper a way for an accurate solution.

In principle, formulae (4.09) are divided by c E_0. This results in:

$$\frac{G}{c\,E_0} = B + \frac{E_2}{E_0}\sin(2B) + \frac{E_4}{E_0}\sin(4B) + \frac{E_6}{E_0}\sin(6B) + \frac{E_8}{E_0}\sin(8B) \tag{4.11}$$

or respectively the abbreviated formulae for g and e_2, e_4, e_6, e_8 in:

$$g = B + e_2 \sin(2B) + e_4 \sin(4B) + e_6 \sin(6B) + e_8 \sin(8B) \tag{4.11a}$$

By reversing (4.11a) Krack finally gets the equation:

$$B^R = g + f_2 \sin(2g) + f_4 \sin(4g) + f_6 \sin(6g) + f_8 \sin(8g) \tag{4.12}$$

in which the coefficients f_2, f_4, f_6, f_8 are expressed in terms of e' as:

$$f_2 = +\frac{3}{8}e'^2 - \frac{3}{16}e'^4 + \frac{213}{2048}e'^6 - \frac{255}{4096}e'^8 + \frac{166479}{655360}e'^{10}$$

$$f_4 = +\frac{21}{256}e'^4 - \frac{21}{256}e'^6 + \frac{533}{8192}e'^8 - \frac{152083}{327680}e'^{10} \tag{4.13}$$

$$f_6 = +\frac{151}{6144}e'^6 - \frac{3171}{86016}e'^8 + \frac{2767911}{9175040}e'^{10}$$

$$f_8 = +\frac{38395}{4587520}e'^8 - \frac{427277}{4587520}e'^{10}$$

The coefficients calculated in (4.10) and (4.13) are substituted in formulae (4.09) and (4.12) respectively.

Direct and inverse problems can be solved with good accuracy using the formulae mentioned above. It must be stressed that computing the various trigonometric functions for B^R, $2B^R$, $4B^R$, $6B^R$, $8B^R$ and $2g^R$, $4g^R$, $6g^R$, $8g^R$ are time consuming.

A higher computational speed may be achieved by the equations converted to a nested form and the use of a single trigonometric function in both direct and inverse computations. The equations (4.09, 4.10, 4.12, 4.13) are modified so that the *only trigonometric function* appearing in them is a *cosine (x).*

This is done by substituting the appropriate trigonometric identities from the following reduction table (Table 15), as given by (Adams, 1949), in the formulae (4.09, 4.12). New constants for the calculation of *direct* and *inverse* constants are obtained. These constants in terms of e' and *one* cosine function allow calculation of the arc G_m and latitude B° directly. See [6.1.1] (6.04-6.09, 6.22-6.23).

A higher computational speed may be achieved by the use *n* in substitution for e'^2, see the equations described by (Krack, 1983), (3.19-3.23).

$\sin 2\alpha$	=				$2 \sin\alpha \cos\alpha$
$\sin 4\alpha$	=			$8 \sin\alpha \cos^3\alpha$	$- 4 \sin\alpha \cos\alpha$
$\sin 6\alpha$	=		$32 \sin\alpha \cos^5\alpha$	$- 32 \sin\alpha \cos^3\alpha$	$+6 \sin\alpha \cos\alpha$
$\sin 8\alpha$	=	$128 \sin\alpha \cos^7\alpha$	$- 192 \sin\alpha \cos^5\alpha$	$+80 \sin\alpha \cos^3\alpha$	$- 8 \sin\alpha \cos\alpha$

Table 15: Reduction table

5. Conformal Projections in General

A projection is a systematic representation of a portion of the earth's curved surface upon a plane. The projections have been devised with a geodetical curvilinear lattice of latitude and longitude. The co-ordinate systems are based on projections and are determined by the orientation of a particular area of the earth's curved surface. A grid is a system of co-ordinates on the earth's surface expressed in linear units: X, Y or Eastings, Northings (Maling, 1992).

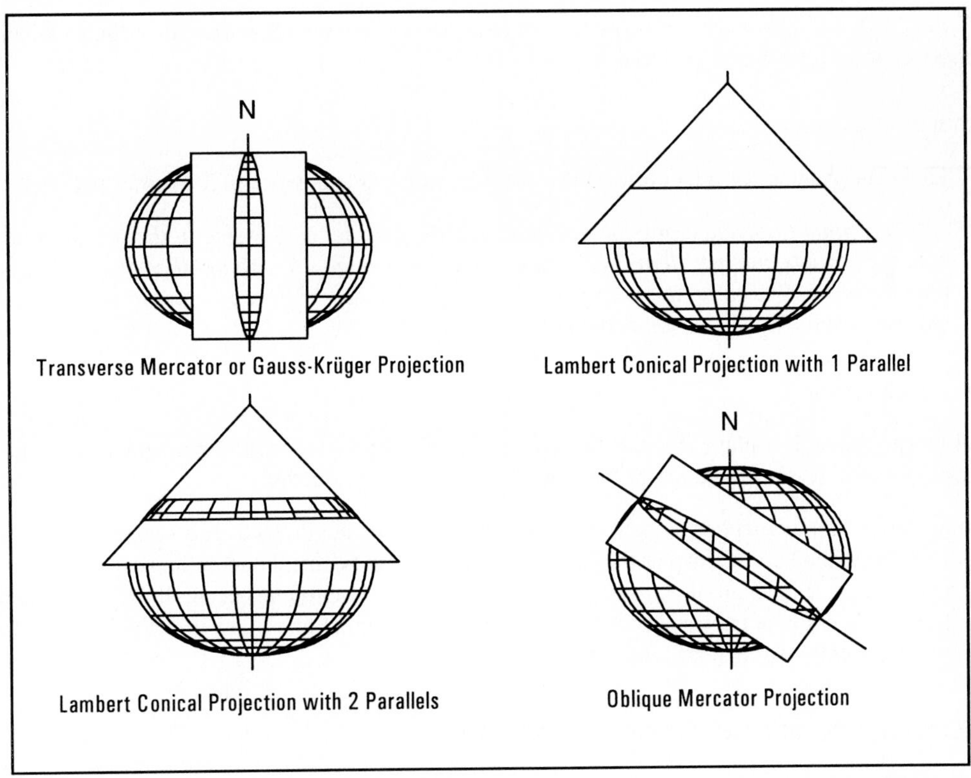

Figure 17: Tangency or secancy of the projection surfaces

The mathematical function defines a relationship of co-ordinates between the ellipsoid and a sphere, between that sphere and a plane, or both, between the ellipsoid and a plane.

A full treatment of map projections is beyond the scope of the book. For more information the interested reader should refer to (Sacks, 1950; Hotine, 1946, 1947) - Nos. 65, 66 being regarded as "classics" - (Hotine, 1969; Grafarend / Engels, 1995e; Grafarend, 1995a, 1995d, 1996c).

5.1 Scope and Terminology

The name *Transverse Mercator projection, or Gauss-Krüger (GK) in Europe,* see [6], derives from the fact that this projection is the transverse case of the *normal Mercator projection* (Grafarend, 1996c). Transverse Mercator co-ordinates are best suited to an area or country which has a large extent in a north-south direction and a relatively small extent in an East - West direction. It is the most commonly used conformal projection system in the world.

For mapping an area of considerable extent in longitude, a geometrical system - the *Lambert Conformal Conical projection* based on the idea of an east-west centre line - can be used. In such a system the differences of latitude to be covered will be comparatively small, see [7].

If the Transverse Mercator projection is considered to be unsuitable because of zone changes every 6° in longitude, an oblique aspect projection must be preferred. The cylinder, with a strip of zero distortion, may be rotated into a predefined azimuth passing through an arbitrary central point of the area, thus creating an Oblique Mercator projection.

The *Oblique Mercator conformal projection,* see [8], is a skew form of projection. The skew projection is suited to a country which is orientated at some azimuth intermediate between 0° and 90°, e.g. the countries of Brunei Darussalam, Madagascar, Malaysia and Switzerland are on oblique Mercator projections.

Although the projection character and formulae are based on a skew meridian, the co-ordinates obtained are finally *rotated* on to a North-South grid axis.

Conformal or Orthomorphic

Arthur R. Hinks, in (Adams, 1921), defines orthomorphic, which is another term for conformal, as follows:

> "*If at any point the scale along the meridian and the parallel is the same in the two directions and the parallels and meridians of the map are at right angles to one another, then the shape of any very small area on the map is the same as the shape of the corresponding small area upon the earth. The projection is then called orthomorphic*".

Tangency or Secancy

The local tangent plane is a plane that has been *developed* from another regular mathematical surface, such as the *cylinder* of the Mercator projection or the *cone* of the Lambert projection.

The plotting surface may be either tangent or secant to the reference ellipsoid. If a secant surface is used, one true length line is defined on the map at the line of secancy. For the conical and cylindrical projections, the line of tangency is a single true length line. In the secant case, two true length lines occur at the two lines of secancy. Note that secancy is not a geometrical property, but only a conceptual one in a projection such as the Lambert Conformal Conical with two Standard Parallels (Maling, 1992).

5.2 Conversions and Transformations

In converting co-ordinates from one positional reference system to the co-ordinates of that position represented in any other reference system, the co-ordinate conversions involve co-ordinate calculations and sometimes datum transformations (Floyd, 1985).

Reference systems can be categorised into three broad groups:

- geodetic co-ordinates in terms of a curvilinear lattice of latitudes and longitudes
- plane co-ordinates in terms of a rectilinear lattice of X and Y, or Eastings and Northings
- latitudes, longitudes, and height (space co-ordinates) in terms of a Cartesian system of X, Y, and Z.

Algorithms are provided for any geodetic datum for the following projections (Figure 17):

- Transverse Mercator ______________________ GK000000.BAS
- Oblique Mercator ______________________ OM000000.BAS
- Lambert Conformal Conical ______________________ LCC00000.BAS.

Calculations apply specifically to the process of co-ordinate conversions and datum transformation (Maling, 1992).

Changes can be listed as follows:

- change in datum ______ TRM00000.BAS
- change in epoch, see (Ollikainen, 1995) ______ TRM00000.BAS
- change in projection ______ see projection example [10.2.4].

Co-ordinate conversions

- Converting geodetic co-ordinates to plane co-ordinates referenced to the same ellipsoid, datum and epoch or vice versa
- Plane co-ordinates on one projection to plane co-ordinates on another projection, both projections referenced to the same ellipsoid, datum and epoch, for instance due to change in scale factor, origin, orientation of grid axes.

Datum Transformation

- Transformation of geodetic co-ordinates based on one ellipsoid, datum and epoch to geodetic or plane co-ordinates based on another ellipsoid, datum and epoch
- Transformation of plane co-ordinates on one projection based on one ellipsoid, datum and epoch to plane co-ordinates on the same type of projection referenced to another ellipsoid, datum and epoch
- Transformation of plane co-ordinates on one projection based on one ellipsoid, datum and epoch to plane co-ordinates on a different type of projection referenced to another ellipsoid, datum and epoch.

Plane Co-ordinates and Origins

Two types of plane co-ordinates and three plane co-ordinate origins have the following significance:

- *True origin* is the fundamental origin of the projection that defines and orients the projection surface
- *True* co-ordinates are those reckoned from the true origin of the projection at a scale given by the projection parameters
- Normally the *grid origin* is located by design to the west and south of the region of interest. Resulting grid co-ordinates are always positive-valued
- *Grid* co-ordinates are at the same scale and referenced to an origin situated more conveniently for a particular area of interest
- *False Eastings* and *False Northings* are X, Y constants used in the translation of co-ordinates from a true origin to a false origin assigned to the true origin of the grid system.

α	= geodetic azimuth ______	reckoned from north
T	= projected geodetic azimuth	
t	= grid azimuth ______	reckoned from north
γ	= convergence angle	
δ	= (t - T) ______	arc-to-chord correction

5.3 Symbology

Convergence is an angular difference between geodetic north and grid north. Geodesic azimuths are referred to meridians: the projected geodetic azimuth or geodetic north. Grid azimuths are referred to north-south grid lines. Defined another way, the convergence angle is the angle between a north-south grid line and the meridian as represented on the plane grid.

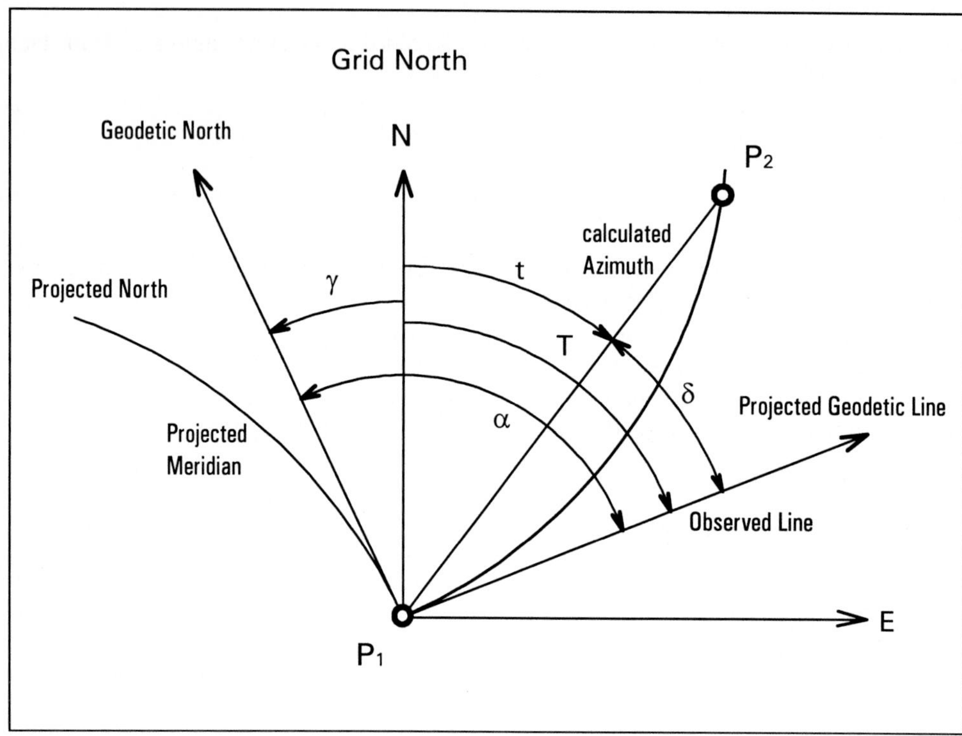

Figure 18: Azimuth and convergence

A geodetic azimuth is symbolised as "α". A convergence angle is symbolised as "γ " (Stem, 1989a; Bomford, 1977).

In the Northern Hemisphere, when a point is east of the Central Meridian, grid north is east of true north: γ is positive. When a point is west of the Central Meridian, grid north is west of true north: γ is negative. In the Southern Hemisphere the reverse condition is true.

Grid Azimuth and Projected Geodetic Azimuth

The projection of the geodetic azimuth α onto the grid is not the grid azimuth, but the *projected geodetic azimuth* symbolised as "T". Convergence γ is defined as the difference between a geodetic and a projected geodetic azimuth. Hence by definition, $\alpha = T + \gamma$, and the sign of γ should be applied accordingly. The value of γ varies with latitude according to $\gamma = \delta\lambda \sin \varphi$. The angle obtained from two projected geodetic azimuths is a true representation of an observed angle.

When an azimuth is computed from two plane co-ordinate pairs, the resulting quantity is the grid azimuth symbolised as "t". The angular difference between the projected geodetic azimuth T and grid azimuth t is a calculable quantity symbolised as "δ", or more often identified as the (t - T) correction. The relationship between these azimuths is shown in Figure 18. For sign convention it is defined as $\delta = (t - T)$ and is also identified as the *arc-to-chord* correction, which should always be considered (Stem, 1989a).

Figure 19 and Figure 20 illustrate *the Projected Geodetic lines T*, the bold lines in both figures - which always bow away from the Central Line - and *the grid lines t.* Both figures illustrate the (t - T) correction in the projections of a traverse.

Given the definition of α and δ, we obtain:

$$t = \alpha - \gamma + \delta \tag{5.01}$$

The arc-to-chord correction δ or (t - T) is normally determined for a pair of points on the ellipsoid, such as P_1 and P_2, whose grid co-ordinates are known. There are two corrections: one to be applied at P_1 for the line P_{1-2} and another one to be applied at P_2 for the line P_{2-1} (Stem, 1989a).

The Elementary Concept of Scale and True Distances

The scale factor is the ratio of infinitesimal linear distance in any direction at a point on the plane grid to the corresponding true distance on the ellipsoid. Thus, the grid scale factor is the measure of the linear distortion that has been mathematically imposed on ellipsoid distances so they may be projected onto a plane. The grid scale factor - symbolised by *k* - is constant at a point, regardless of the azimuth, when conformal projections are used.

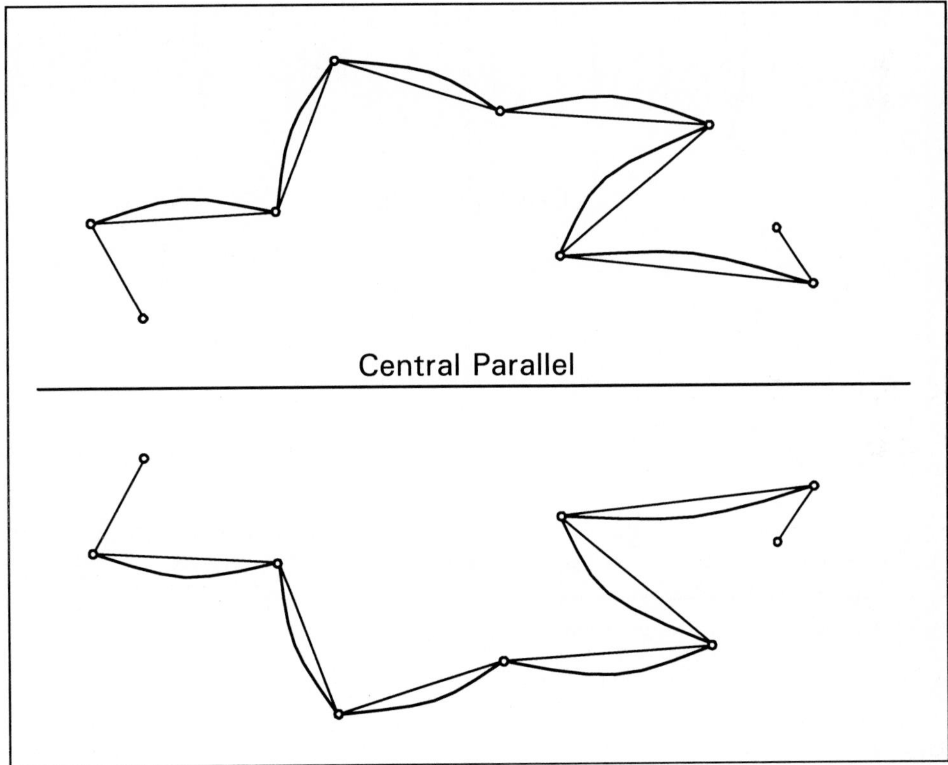

Figure 19: Lambert projection - projected geodetic vs. grid angles

In order to obtain the True distance, S, from the grid distance, D, derived from grid co-ordinates, or alternatively, to convert a true distance to a grid distance (for plotting on the map or projection) it is necessary to calculate the Scale Factor and apply it in the correct sense:

$$D = k\ S, \text{ or}$$
$$S = D / k \tag{5.02}$$

The scale factor changes along a line so slowly that for most purposes it may be taken as constant within any 10 km square and equal to the mean of the value at the centre at the square considered.

To apply a correction to an ellipsoidal distance or plane grid distance, it is necessary to use the values of all points making up the line. A more efficient method is to apply Simpson's Rule of numerical analysis to compute a line scale-factor:

$$k = \frac{1}{6}\left(k_1 + 4\, k_m + k_2 \right) \tag{5.03}$$

See the example of e.g. [6.2.3], in which k_1 and k_2 are the scale factors at the end of each line and k_m is the scale factor at the midpoint of the line (Bomford, 1977; Maling, 1992; Stem, 1989a).

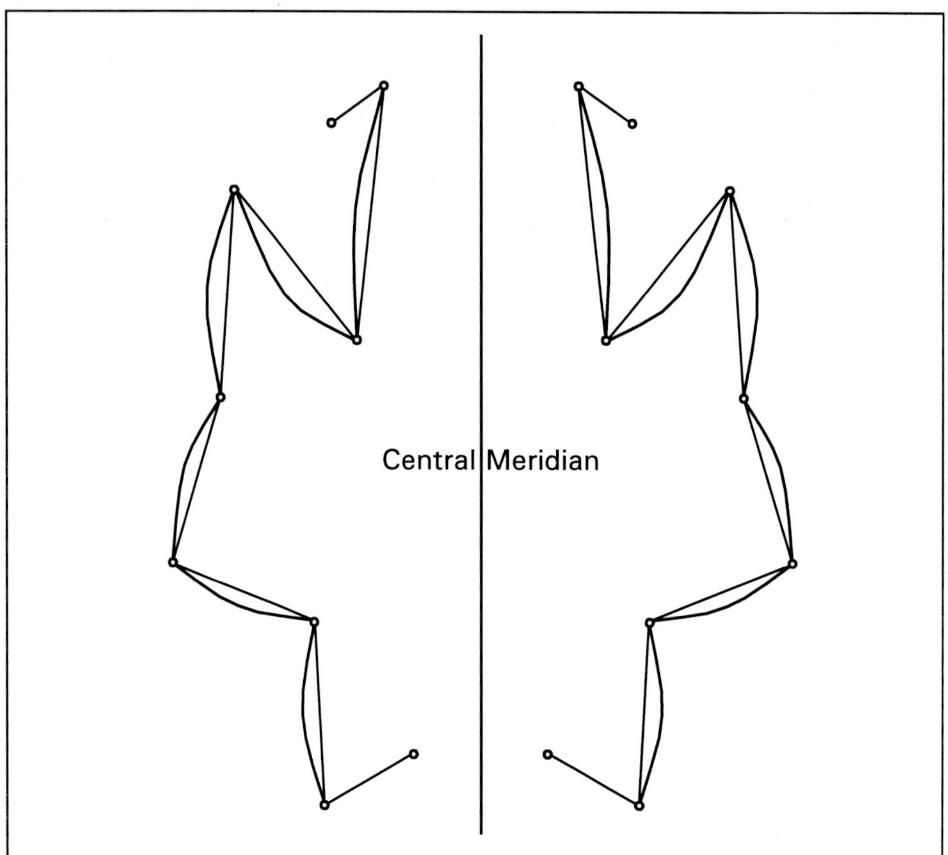

Figure 20: Transverse Mercator vs. grid angles

Interesting illustrations on the subject can be found in (Arrighi, 1994).

5.4 Practice

At least 12 significant figures are adequate for the desired accuracy in projection zone widths that may be encountered. Therefore, double precision variables will be required on most computers. This book contains very efficient algorithms and program listings for co-ordinate conversion (0.1 mm accuracy or better) and co-ordinate transformations (Vincenty, 1985, 1986; Stem, 1989; Floyd, 1985).

6. Gauss-Krüger Projection

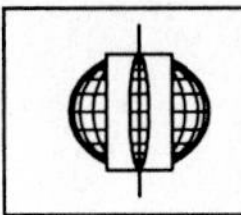

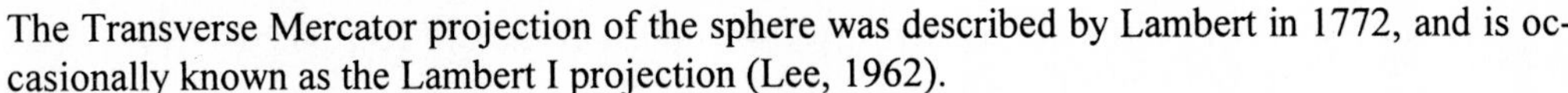

The Transverse Mercator projection of the sphere was described by Lambert in 1772, and is occasionally known as the Lambert I projection (Lee, 1962).

6.1 Transverse Mercator Projection System

The Transverse Mercator conformal projection, or *Gauss-Krüger (GK) projection* in continental Europe - the projection first employed by the geodesists *Gauss and Schreiber* - is used because of its longer dimension in a north-south direction. The abbreviation TM is used for *Thematic Mapper* (Landsat), and UTM *(Universal Transverse Mercator)* grid is a derivative of the general GK projection (Burton, 1996). See [6.3]: pp 111.

Origin and Central Meridian

The true origin of this projection is at the intersection of the *Central Meridian (CM)* with the *Equator,* and a grid origin may be established to the south and west of that area.

False Origin

The definition of the north-south location of the grid origin: if a *False Northing (Y)* is specified rather than a *Latitude* of *False Origin*, the False Origin is located at the intersection of the Central Meridian with the Equator. Then the False Northing, or Y constant of the False Origin, would be negative-valued in the northern hemisphere, moving the grid origin northwards.
In the southern hemisphere the False Northing would be *positive-valued.* The *False Easting* is the *X* constant assigned to the False Origin. It would be *positive* in either hemisphere.

Region

The equations for the Gauss-Krüger projection become unstable near the poles. The Gauss-Krüger projection should be limited to a region bounded by a maximum latitude and a longitudinal distance from the central meridian, which will depend on the purpose of the projection.

6.1.1 Gauss-Krüger or Transverse Mercator Mapping Equations

Symbols and Definitions

All angles are expressed in *radians.* See program GK000000.BAS [11.7.5]

a	semi-major axis of the ellipsoid
b	semi-minor axis of the ellipsoid
f	flattening of the ellipsoid
k_0	grid scale factor assigned to the central meridian
φ_0	parallel of geodetic latitude grid origin
λ_0	Central Meridian (CM)
E_0	false easting (constant assigned to the CM)
N_0	false northing (constant assigned to the latitude of grid origin)
φ	parallel of geodetic latitude, positive north
λ	meridian of geodetic longitude, positive east
E	easting co-ordinate on the projection
N	northing co-ordinate on the projection
γ	meridian convergence
δ_{1-2}	arc-to-chord correction (t - T), for a line from P_1 to P_2
k	point grid scale factor
k_{1-2}	line grid scale factor (P_1–P_2)

ω	rectifying latitude	
S	meridional distance	
S_0	meridional distance from the equator to φ_0, multiplied by the CM scale factor	
ΔN	$N_2 - N_1$ - difference in northings	(6.01)
ΔE	$E_2 - E_1$ - difference in eastings	(6.02)
E'	$E - E_0$	(6.03)
e^2	first eccentricity squared	
e'^2	second eccentricity squared	
n	second flattening	
R	radius of curvature in the Prime Vertical	
r_0	geometric mean radius of curvature scaled to the grid	
r	radius of the rectifying sphere	
t	tan φ, see formulae direct and inverse computation	
t	grid azimuth, see formulae (t - T)	
η^2	$e'^2 \cos^2 \varphi$	

Constants for Meridional Arc

Compute constants for meridional arc as given below, or use (3.03 - 3.08):

$$c = a / (1 - e^2)^{½} \tag{6.04}$$

$$r = a (1+n^2 / 4) / (1+n), \text{ see (2.03)} \tag{6.05}$$

$$\begin{aligned}
U_0 &= c [((((- 86625 / 8 e'^2+11025) / 64 e'^2 - 175) / 4 e'^2+45) / 16 e'^2 - 3) / 4 e'^2] \\
U_2 &= c [(((- 17325 / 4 e'^2+3675) / 256 e'^2 -175 / 12) e'^2+15) / 32 e'^4] \\
U_4 &= c [- 1493 / 2+735 e'^2] / 2048 e'^6 \\
U_6 &= c [(- 3465 / 4 e'^2+315) / 1024 e'^8]
\end{aligned} \tag{6.06}$$

$$\begin{aligned}
V_0 &= ((((16384 e'^2 - 11025) / 64 e'^2+175) / 4 e'^2 - 45) / 16 e'^2+3) / 4 e'^2 \\
V_2 &= (((- 20464721 / 120 e'^2+19413) / 8 e'^2-1477) / 32 e'^2+21) / 32 e'^4 \\
V_4 &= ((4737141 / 28 e'^2 - 17121) / 32 e'^2+151) / 192 e'^6 \\
V_6 &= (- 427277 / 35 e'^2+1097) / 1024 e'^8
\end{aligned} \tag{6.07}$$

Meridional Arc formula

$$\omega_0 = \varphi_0 + \sin \varphi_0 \cos \varphi_0 (U_0 +U_2 \cos^2 \varphi_0 +U_4 \cos^4 \varphi_0 +U_6 \cos^6 \varphi_0) \tag{6.08}$$

$$S_0 = k_0 \omega_0 r \tag{6.09}$$

Direct Computation

Input: geodetic co-ordinates of a point P (φ, λ)
Output: grid co-ordinates of a point P (E, N), convergence angle (γ), scale factor (k)

$$L = (\lambda - \lambda_0) \cos \varphi \tag{6.10}$$

$$\omega = \varphi + \sin \varphi \cos \varphi (U_0 + U_2 \cos^2 \varphi + U_4 \cos^4 \varphi + U_6 \cos^6 \varphi) \tag{6.11}$$

$$S = k_0 \omega r \tag{6.12}$$

$$R = k_0 a / (1 - e^2 \sin^2 \varphi)^{½} \tag{6.13}$$

$$\begin{aligned}
A_1 &= - R \\
A_3 &= 1 / 6 (1 - t^2 + \eta^2) \\
A_5 &= 1 / 120 (5 - 18 t^2 + t^4 + \eta^2 (14 - 58 t^2)] \\
A_7 &= 1 / 5040 (61 - 479 t^2 + 179 t^4 - t^6)
\end{aligned} \tag{6.14}$$

$$
\begin{aligned}
A_2 &= \tfrac{1}{2} R\, t \\
A_4 &= 1/12\, [\, 5 - t^2 + \eta^2 (\, 9 + 4\eta^2 \,) \,] \\
A_6 &= 1/360\, [\, 61 - 58\, t^2 + t^4 + \eta^2 (\, 270 - 330 t^2 \,) \,]
\end{aligned} \qquad (6.15)
$$

$$E = E_0 + A_1 L\, [\, 1 + L^2 (\, A_3 + L^2 (\, A_5 + A_7 L^2 \,)\,)\,] \qquad (6.16)$$

$$N = S - S_0 + N_0 + A_2 L^2 (\, 1 + L^2 (\, A_4 + A_6 L^2 \,)\,] \qquad (6.17)$$

$$
\begin{aligned}
C_1 &= - t \\
C_3 &= 1/3\, (\, 1 + 3\eta^2 + 2\eta^4 \,) \\
C_5 &= 1/15\, (\, 2 - t^2 \,)
\end{aligned} \qquad (6.18)
$$

$$
\begin{aligned}
F_2 &= \tfrac{1}{2} (\, 1 + \eta^2 \,) \\
F_4 &= 1/12\, [\, 5 - 4\, t^2 + \eta^2 (\, 9 - 24\, t^2 \,) \,]
\end{aligned} \qquad (6.19)
$$

$$\gamma = C_1 L\, [\, 1 + L^2 (\, C_3 + C_5 L^2 \,) \,] \qquad (6.20)$$

$$k = k_0\, [\, 1 + F_2 L^2 (\, 1 + F_4 L^2 \,) \,] \qquad (6.21)$$

Inverse Computation

Input: grid co-ordinates of a point P (E, N)

Output: geodetic co-ordinates P (φ, λ) convergence angle γ, grid scale factor k

$$\omega = (\, N - N_0 + S_0 \,) / (\, k_0\, r \,) \qquad (6.22)$$

$$\varphi_f = \omega + (\, \sin\omega \cos\omega \,)\, (\, V_0 + V_2 \cos^2\omega + V_4 \cos^4\omega + V_6 \cos^6\omega \,) \qquad (6.23)$$

$$R_f = k_0\, a / (\, 1 - e^2 \sin^2\varphi_f \,)^{1/2} \qquad (6.24)$$

$$Q = E' / R_f, \text{ in which } E' = E - E_0 \qquad (6.25)$$

$$
\begin{aligned}
B_2 &= - \tfrac{1}{2}\, t_f\, (\, 1 + \eta_f^2 \,) \\
B_4 &= - 1/12\, [\, 5 + 3\, t_f^2 + \eta_f^2 (\, 1 - 9\, t_f^2 \,) - 4\, \eta_f^4 \,] \\
B_6 &= 1/360\, [\, 61 + 90\, t_f^2 + 45\, t_f^4 + \eta_f^2 (\, 46 - 252\, t_f^2 - 90\, t_f^4 \,) \,]
\end{aligned} \qquad (6.26)
$$

$$
\begin{aligned}
B_3 &= - 1/6\, (\, 1 + 2\, t_f^2 + \eta_f^2 \,) \\
B_5 &= 1/120\, [\, 5 + 28\, t_f^2 + 24\, t_f^4 + \eta_f^2 (\, 6 + 8\, t_f^2 \,) \,] \\
B_7 &= - 1/5040\, (\, 61 + 662\, t_f^2 + 1320\, t_f^4 + 720\, t_f^6 \,)
\end{aligned} \qquad (6.27)
$$

$$\varphi = \varphi_f + B_2 Q^2\, [\, 1 + Q^2 (\, B_4 + B_6 Q^2 \,) \,] \qquad (6.28)$$

$$L = Q\, [\, 1 + Q^2 (\, B_3 + Q^2 (\, B_5 + B_7 Q^2 \,)\,)\,] \qquad (6.29)$$

$$\lambda = \lambda_0 - L / \cos\varphi_f \qquad (6.30)$$

$$
\begin{aligned}
D_1 &= t_f \\
D_3 &= - 1/3\, (\, 1 + t_f^2 - \eta_f^2 - 2\, \eta_f^4 \,) \\
D_5 &= 1/15\, (\, 2 + 5\, t_f^2 + 3\, t_f^4 \,)
\end{aligned} \qquad (6.31)
$$

$$
\begin{aligned}
G_2 &= \tfrac{1}{2} (\, 1 + \eta_f^2 \,) \\
G_4 &= 1/12\, (\, 1 + 5\, \eta_f^2 \,)
\end{aligned} \qquad (6.32)
$$

$$\gamma = D_1 Q\, (\, 1 + Q^2 (\, D_3 + D_5 Q^2 \,)\,) \qquad (6.33)$$

$$k = k_0\, (\, 1 + G_2 Q^2 (\, 1 + G_4 Q^2 \,)\,) \qquad (6.34)$$

Arc-to-Chord Correction δ = (t - T)

Grid azimuth *(t)*, geodetic azimuth *(α)*, convergence angle *(γ)*, and arc-to-chord correction *(δ)* at any given point are related as follows: $t = \alpha - \gamma + \delta$; see (5.01).

Input: P_1 (E_1, N_1), and P_2 (E_2, N_2)
Output: δ_{1-2}

$$N_m = \tfrac{1}{2} (N_1 + N_2) \quad (6.35)$$

$$\omega = (N_m - N_0 + S_0) / (k_0 r) \quad (6.36)$$

$$\varphi_f = \omega + V_0 \sin \cos \omega \quad (6.37)$$

$$F = (1 - e^2 \sin^2 \varphi_f) (1 + \eta_f^2) / (k_0 a)^2 \quad (6.38)$$

$$E_3 = 2 E'_1 + E'_2 \quad (6.39)$$

$$\delta_{1-2} = - 1 / 6 \Delta N E_3 F (1 - (1 / 27) E_3^2 F) \quad (6.40)$$

Note

The computer subroutines were written in the most suitably economical form. Therefore, these basic equations may deviate from the subroutine listings in this book.

The equations in this section are found in (Krack, 1982; Hooijberg, 1996), and (Floyd, 1985; Stem, 1989a) based on (Meade, 1987; Vincenty, 1984a).

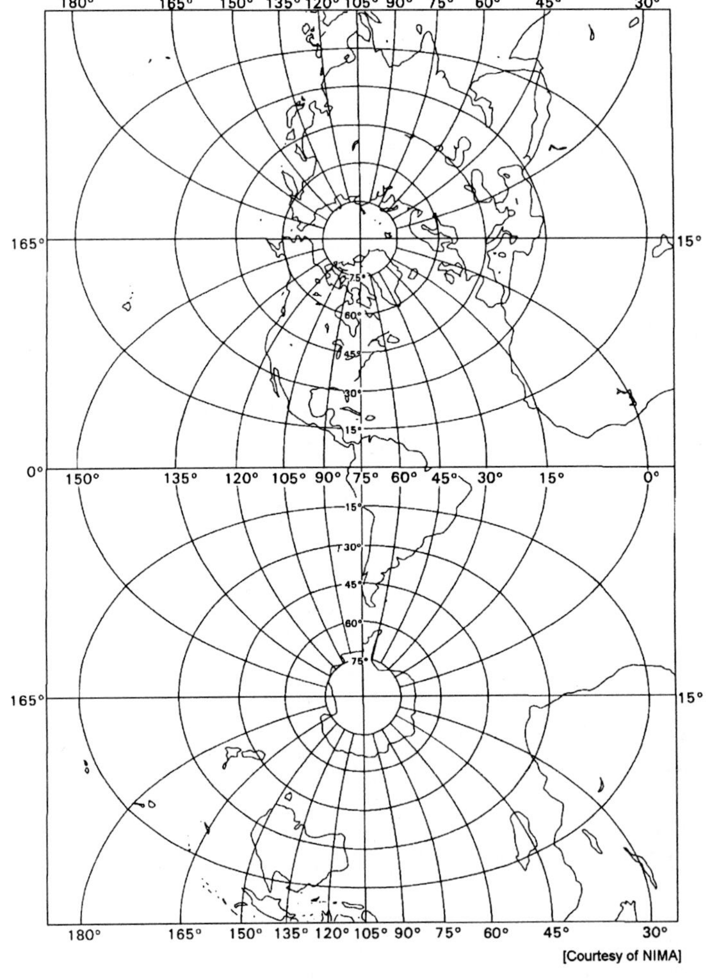

Figure 21: Transverse Mercator projection

6.2 Gauss-Krüger Projection Applications

6.2.1 Reference and GK-projection Systems of Australia

Australia is a vast continent of over 9 000 000 square kilometres, with a population of a little more than 17 million, the overwhelming majority of which live in a few cities on the coastal margins. Developing and maintaining a continent-wide geodetic control network represents an enormous technical challenge. A further challenge has been to carry out this task using cost-effective technologies in 1966 (Rizos, 1993).

The enormity and importance of the undertaking to survey an entire continent in little more than a decade is rightly referred to as *"... one of the survey wonders of the world"*. The resulting Australian Geodetic Datum of 1966 was the first datum constructed from traversing observations rather than triangulations. AGD66 will be superseded by the Australian Geodetic Datum of 1994 (AGD94), to be implemented by *the year 2000*.

See also [6.3.2], UTM of Australia (Rüeger, 1994).

New South Wales Integrated Survey Grid

In the State of New South Wales all surveys other than the national surveys are based on the New South Wales Integrated Survey Grid (ISG). See (ISG, 1972) Tables.

Defining parameters of the Australian National Spheroid (ANS), based on IAU65, are: semi-major axis, a=6 378 160 metres, and reciprocal flattening f^{-1}=298.25 exactly.

The Integrated Survey Grid System has the following specifications:

- Projection type — Transverse Mercator (Gauss-Krüger Type)
- Reference Ellipsoid — ANS 1966
- Zones — 2° wide
- Longitude of Origin — Central Meridian of each zone
- Latitude of Origin — 0° ______ the Equator
- Unit — metre
- False Northing — 5 000 000 m ___ South of the True Origin
- False Easting — 300 000 m ___ West of the True Origin
- Scale Factor at the Central Meridian — 0.99994
- Latitude Limits of System — North ______ 28° S; South ______ 39° S
- Zone numbering — Each 6° zone of the Australian Map Grid (AMG) is sub-divided into three ISG zones: 1 = west, 2 = centre and 3 = east
- Zone identification — number consists of an AMG- and an ISG-zone number separated by an oblique
- Limits of overlap — 0° 15'.

The distinctive numbers zone identification

The first part indicates the corresponding AMG zone number, the second part indicates the number of the subdivision from 1 to 3, the numbers increasing eastwards.

Zone number	54/2	54/3	55/1	55/2	55/3	56/1	56/2
CM	141°	143°	145°	147°	149°	151°	153°

Gauss-Krüger Projection - Integrated Survey Grid of NSW - Australia

Zone:	*ISG Zone 55/2*		
Reference Ellipsoid:	ANS 1966	False Easting:	**300 000 m**
Semi-Major axis:	**6 378 160**	False Northing:	**5 000 000 m**
Reciprocal Flattening:	**298.25**	Parallel of Origin φ_o:	**0° 00' 00" S**
Scale Factor:	**0.99994**	C.M. of zone λ_o:	**147° 00' 00" E**

Conversion of AGD66 Geographicals to ISG-Grid

The Direct calculation is to convert Geodetic Co-ordinates into GK-Planar Co-ordinates:

Input:	Latitude	Longitude	Output:	Easting	Northing	Convergence	Scale Factor

Latitude:	**28° 45' 01".2592 S**	Easting=	422 145.5146
Longitude:	**148° 15' 02".1012 E**	Northing=	1 817 938.9749
Convergence=	- 0° 36' 05".7527	Scale Factor=	1.000 124 053 567

Conversion of ISG-Grid to AGD66 Geographicals

The Inverse process is to convert GK-Planar Co-ordinates into Geodetic Co-ordinates:

Input:	Latitude	Longitude	Output:	Easting	Northing	Convergence	Scale Factor

Easting:	**422 145.5146**	Latitude=	28° 45' 01".2592 S
Northing:	**1 817 938.9749**	Longitude=	148° 15' 02".1012 E
Convergence=	- 0° 36' 05".7527	Scale Factor=	1.000 124 053 568

6.2.2 Reference and GK-projection Systems of the People's Republic of China

In China the first official datum, the *Beijing State Reference System of 1954* (BCS1954), was introduced in 1954, based upon Krassovsky 1940.

The datum *Xi'an State Reference System of 1980* (XCS1980), based upon IAG75, superseded BCS1954 in the year 1978.

However, BCS1954 is still in use for *nautical charts* (Zhu, 1986; Xiang, 1988).

The Xi'an State Reference Grid System of 1980 has the following specifications:

Projection	Transverse Mercator (Gauss-Krüger Type)	
Zones	1½°, 3° and 6° wide	
Reference Ellipsoid	IAG75, see specification Table 8 |[30]	
Longitude of Origin	Central Meridian of each zone	
Latitude of Origin	0° ______________	the Equator
Unit	metre	
False Northing	0 m ____________	Northern Hemisphere
	10 000 000 m ____	Southern Hemisphere
False Easting	500 000 m	
Scale Factor at the Central Meridian	1.0000 (exactly), creates a line of zero distortion	

1. The specifications for the *Xi'an State Reference System* is a uniform cover of the world in zones which correspond to the 6° longitudinal units, starting at the Greenwich Meridian with the distinctive *"Baumgart"* number 1 on the zone from 0°E to 6°E, and increasing eastwards to prefixed number 60 on the zone from 6°W to 0°E The zone number is prefixed to the false easting in most cases, i.e. the false easting for the GK Zone 20 is 20 500 000 meters.

2. The prefixed numbers can be found as follows for a 6° zone:

   - The n is found by the formula: $n =(\varphi_o + 3) / 6$ (6.41)
   - The φ_o is found by the formula: $\varphi_o = 6 n - 3$ (6.42)

3. These prefixed numbers n' can be found as follows for a 3° zone:

   - The n' is found by the formula: $n' = 2 n - 1$ (6.43)
   - The φ_o is found by the formula: $\varphi_o = 3 n'$ (6.44)

4. Change of zone hazard.
   The natural value of the co-ordinate Easting Y= - 200.2493 receives a distinctive "Baumgart" number, $+20 \times 10^6$ (= number for 20th [6°] zone) +500 000 (= False Easting), which gives Y= **20** 499 799.7507.

[30] Krassovsky 1940, for offshore and nautical charts only

Transverse Mercator Projection of the People's Republic of China

Zone Parameters	6° width		
Reference Ellipsoid:	IAG75	False Easting:	**500 000 m**
Semi-Major axis:	**6 378 140**	False Northing:	**0 m**
Recipr. Flattening:	**298.257 222 0**	Parallel of Origin φ_o:	**0°00' 00" N**
Scale Factor:	**1.0000**	C.M. of zone λ_o:	**117° 00' 00" E**

Conversion of Xi'an80 Geographicals to Transverse Mercator Grid

The Direct calculation is to convert Geodetic Co-ordinates into GK-Planar Co-ordinates:

Input:	Latitude	Longitude	Output:	Easting	Northing	Convergence	Scale Factor

Latitude:	**50° 26' 51".4848 N**	Easting Y=	499 799.7507
Longitude:	**116° 59' 49".8505 E**	Northing X=	5 590 641.5892
Convergence=	- 0° 00' 07".8257	Scale Factor=	1.000 000 000 492

Conversion of Transverse Mercator Grid to Xi'an80 Geographicals

The Inverse process is to convert GK-Planar Co-ordinates into Geodetic Co-ordinates:

Input:	Latitude	Longitude	Output:	Easting	Northing	Convergence	Scale Factor

Easting Y:	**499 799.7507**	Latitude=	50° 26' 51".4848 N
Northing X:	**5 590 641.5892**	Longitude=	116° 59' 49".8505 E
Convergence=	- 0° 00' 07".8257	Scale Factor=	1.000 000 000 492

6.2.3 Reference and GK-projection Systems of the CIS

In 1928 the USSR (now CIS) |[31] decided to adopt a Soviet National Geodetic System comprising the Gauss-Krüger projection in zones of 3° and 6° width.

In 1942 the entire Soviet National Geodetic System was transformed from the Bessel ellipsoid to the "Pulkovo System 1942 of Survey Co-ordinates" (CK42), based on the *Krassovsky ellipsoid.* This was performed prior to the advent of digital computers and all the computations and data assimilations were done manually by the Central Scientific Research Institute for Geodesy, Air Survey and Cartography (CNIIGAik). A large-scale, totally comprehensive adjustment of all horizontal control data within the USSR represented an impossible task (Bogdanov, 1995; Dupuy, 1954; Izotov, 1959).
After 1960, the System 42 was introduced to the signatories of *the Warsaw Pact Agreement,* the Allied Countries in Eastern Europe, with the year of realisation, e.g. System 42/83 in the Eastern States of the FRG (Ihde, 1991, 1995).

The Soviet Unified Vertical Datum is "the Baltic System of Heights", the Kronstadt Datum being near St. Petersburg.

The Soviet Unified Reference Grid System has the following specifications:

- Projection: Transverse Mercator (Gauss-Krüger Type)
- Zones 3° and 6° wide
- Reference Ellipsoids Krassovsky 1940 (National Surveys)
  SGS90 (used by GLONASS)
- Longitude of Origin Central Meridian of each zone
- Latitude of Origin 0° ____________ the Equator
- Unit metre
- False Northing 0 m ____________ Northern Hemisphere
  10 000 000 m ____ Southern Hemisphere
- False Easting 500 000 m
- Scale Factor at the Central Meridian 1.0000 (exactly), creates a line of zero distortion.

Further specifications for this Gauss-Krüger Grid are:

1. The specifications for the CK42 / Transverse Mercator is a uniform cover of the world in zones which correspond to the 6° longitudinal units, starting at the Greenwich Meridian, rather than the 180° meridian, with a distinctive *"Baumgart"* number 1 on the zone from 0°E to 6°E, and increasing eastwards to prefixed number 60 on the zone from 6°W to 0°E. In other words, the UTM and CK zones differ by 30.
2. The Central Meridians, e.g. the 21° meridian, starting with on the zone from 18°E to 24°E, and proceeding eastwards on the zone from 54°E to 60°E, receive the prefixed numbers 4, 5, 6, ... ,10. Zone numbers are prefixed to the false easting in most cases, i.e. the false easting for the CK Zone 7 is **7** 500 000 m.
3. These prefixed numbers n' can be found as follows for a 3° zone:
   - The n' is found by the formula: $n' = \varphi_o / 3$ (6.45)
   - The φ_o is found by the formula: $\varphi_o = 3\,n'$ (6.46)
4. The prefixed numbers can be found as follows for a 6° zone:
   - The n is found by the formula: $n = (\varphi_o + 3) / 6$ (6.47)
   - The φ_o is found by the formula: $\varphi_o = 6\,n - 3$ (6.48)

[31] The Commonwealth of Independent States, formerly the USSR

Transverse Mercator of the Commonwealth of Independent States

Zone Parameters	3° width (Hristov, 1959)		
Reference Ellipsoid:	Krassovsky 1940	False Easting:	**500 000 m**
Semi-Major axis:	**6 378 245**	False Northing:	**0 m**
Recipr. Flattening:	**298.3**	Parallel of Origin φ_o:	**0° 00' 00" N**
Scale Factor:	**1.0000**	C.M. of zone λ_o:	**24° 00' 00" E**

Conversion of Pulkovo42 Geographicals to CK42-Grid

The Direct calculation is to convert Geodetic Co-ordinates into GK-Planar Co-ordinates:

Input: Latitude Longitude		Output: Easting Northing Convergence	Scale Factor
Latitude:	**42° 24' 17".6282 N**	Easting=	537 127.9991
Longitude:	**24° 27' 03".5764 E**	Northing=	4 696 793.1726
Convergence=	+0° 18' 14".8964	Scale Factor=	1.000 016 952 766
Latitude:	**42° 19' 53".2714 N**	Easting=	589 184.1732
Longitude:	**25° 04' 55".3915 E**	Northing=	4 689 104.7299
Convergence=	+0° 43' 43".4014	Scale Factor=	1.000 097 819 629

Conversion of CK42-Grid to Pulkovo42 Geographicals

The Inverse process is to convert GK-Planar Co-ordinates into Geodetic Co-ordinates:

Input: Latitude Longitude		Output: Easting Northing Convergence	Scale Factor
Easting:	**537127.9991**	Latitude=	42° 24' 17".6282 N
Northing:	**4696793.1726**	Longitude=	24° 27' 03".5764 E
Convergence=	+0° 18' 14".8964	Scale Factor=	1.000 016 952 766
Easting:	**589184.1732**	Latitude=	42° 19' 53".2714 N
Northing:	**4689104.7299**	Longitude=	25° 04' 55".3915 E
Convergence=	+0° 43' 43".4014	Scale Factor=	1.000 097 819 628

Ellipsoidal and Grid Calculation (manual)

Using GBD00000.BAS gives:

True Distance $S_{1\text{-}2}$:	*52 618.1566*	
True Azimuth $\alpha_{1\text{-}2}$:	*98° 42' 19".4791*	
True Azimuth $\alpha_{2\text{-}1}$:	*279° 07' 50".4463*	
Grid Distance $D_{1\text{-}2}$:	52 620.8838	
True Distance $S_{calc\ 1\text{-}2}$:	52 618.1594	
See (5.03), using		
Scale Factor $k_{1\text{-}2}$:	1.000 051 776 925	

Grid bearing	$t_{1\text{-}2}$	98° 24' 05".6449
+ convergence	γ_1	+0° 18' 14".8964
- (t - T)	$\delta_{1\text{-}2}$	+1".0625
True Azimuth	$\alpha_{1\text{-}2}$:	98° 42' 19".4788
Grid bearing	$t_{2\text{-}1}$	278° 24' 05".6449
+ convergence	γ_2	+0° 43' 43".4014
- (t - T)	$\delta_{2\text{-}1}$	- 1".4009
True Azimuth	$\alpha_{2\text{-}1}$:	279° 07' 50".4472

6.2.4 Reference and GK-projection Systems of the Federal Republic of Germany

The use of the Gauss-Krüger projection as a national system was introduced to Germany in 1922.

The Western part of the FRG uses System 1922, ellipsoid Bessel 1841, Das Deutsche Hauptdreiecksnetz (DHDN), Rauenberg Datum, with projection zones of 3° width.

Vertical datum: NN (Normal Null = NAP, Amsterdam Datum 1684 - 0.013 m).

The Eastern part of the FRG uses System 42/83, Krassovsky 1940, Einheitliches Astronomisch-Geodätisches Netz (EAGN) and Staatliches Trigonometrisches Netz 1st Order (STN 1.0), Datum Pulkovo 1942, with projection zones of 6° width.

Vertical datum: HN (Höhen Null = Kronstadt Datum).

The Gauss-Krüger Grid System has the following specifications:	
• Projection:	Gauss-Krüger (Transverse Mercator Type)
• Reference Ellipsoid (W)	Co-ordinate System 1922 Bessel for FRG - West
Zones:	3° wide. See (Krüger, 1919).
• Reference Ellipsoid (E)	Co-ordinate System 42/83 Krassovsky for FRG - East
Zones	6° wide
• Longitude of Origin	Central Meridian of each zone
• Latitude of Origin	0° ______________ the Equator
• Unit	metre [32]
• False Northing	0 m
• False Easting	500 000 m
• Scale Factor at the Central Meridian	1.0000 (exactly), creates a line of zero distortion.

Further specifications for this Gauss-Krüger Grid are:

1. the northern axes of the FRG-West coincides with the meridians at 3°, 6°, 9°, 12°, 15°, etc.
2. the northern axes of the FRG-East coincides with the meridians at 3°, 9°, 15°, 21°, etc., see CK42, [6.2.3]
3. The Eastings receive a prefixed *"Baumgart" number.* These distinctive numbers n' can be found for a 3° zone as follows:
   - The n' is found by the formula: $n' = \varphi_o / 3$ (6.49)
   - The φ_o is found by the formula: $\varphi_o = 3\ n'$ (6.50)
4. The prefixed numbers can be found for a 6° zone as follows:
   - The n is found by the formula: $n = (\varphi_o + 3) / 6$ (6.51)
   - The φ_o is found by the formula: $\varphi_o = 6\ n - 3$ (6.52)
5. The Zone Number is prefixed to the false easting, i.e. the natural value of the co-ordinate FRG-W Easting Y = 83 038.4725 receives a distinctive "Baumgart" number, $+4*10^{6}$ (=number for 12°) +500 000 (=false easting) and gives Y = **4 5**83 038.4725. Consequently, the false easting for the GK Zone 4 is **4 5**00 000 metres.
6. Change of Zone Hazard. Erroneous calculations are easily detected due to the large differences between *Co-ordinate System 1922* and *Co-ordinate System 1942 / 83* of 13-15 m in the Eastings, and of 600-800 m in the Northings.
7. A difference between the Kronstadt Datum and Amsterdam Datum in the order of 0.15 m must be considered.

[32] See (Strasser, 1957; J/E/K, 1959), International metre since 1944

For further reading, see (Groten, 1974; Göhler, 1991; Ihde, 1991, 1993, 1995; J/E/K, 1956, 1959; Ledersteger, 1954; Strauss, 1991; Waalewijn, 1986, 1987; Wittke, 1949; Wolf, 1987).

Gauss-Krüger Projection - the Federal Republic of Germany

Zone Parameters for:	3° width - Zone 4		
Reference Ellipsoid (W):	Bessel 1841	False Easting:	**500 000 m**
Semi-Major axis:	**6 377 397.155**	False Northing:	**0 m**
Recipr. Flattening:	**299.15281 285**	Parallel of Origin φ_o:	**0° 00' 00" N**
Scale Factor:	**1.0000**	C.M. of zone λ_o:	**12° 00' 00" E**

Conversion of Rauenberg1922 Geographicals to Transverse Mercator Grid

The Direct calculation is to convert Geodetic Co-ordinates into GK-Planar Co-ordinates:

Input: Latitude Longitude		Output: Easting Northing Convergence	Scale Factor
Latitude:	**49° 06' 48".5214 N**	Easting=	583 038.4725
Longitude:	**13° 08' 15".2808 E**	Northing=	5 442 314.5539
Convergence=	+ 0° 51' 36".2387	Scale Factor=	1.000 084 690 758
Latitude:	**49° 04' 24".0502 N**	Easting=	529 236.5163
Longitude:	**12° 24' 00".7092 E**	Northing=	5 437 305.9385
Convergence=	+ 0° 18' 08".5338	Scale Factor=	1.000 010 498 510
Latitude:	**48° 40' 10".5910 N**	Easting=	558 734.5829
Longitude:	**12° 47' 51".1207 E**	Northing=	5 392 642.1667
Convergence=	+ 0° 35' 56".0258	Scale Factor=	1.000 042 374 730

Conversion of Transverse Mercator Grid to Rauenberg1922 Geographicals

The Inverse process is to convert GK-Planar Co-ordinates into Geodetic Co-ordinates:

Input: Latitude Longitude		Output: Easting Northing Convergence	Scale Factor
Easting:	**583 038.4725**	Latitude=	49° 06' 48".5214 N
Northing:	**5 442 314.5539**	Longitude=	13° 08' 15".2808 E
Convergence=	+ 0° 51' 36".2387	Scale Factor=	1.000 084 690 758
Easting:	**529 236.5163**	Latitude=	49° 04' 24".0502 N
Northing:	**5 437 305.9385**	Longitude=	12° 24' 00".7092 E
Convergence=	+ 0° 18' 08".5338	Scale Factor=	1.000 010 498 510
Easting:	**558 734.5829**	Latitude=	48° 40' 10".5910 N
Northing:	**5 392 642.1667**	Longitude=	12° 47' 51".1207 E
Convergence=	+ 0° 35' 56".0258	Scale Factor=	1.000 042 374 730

6.2.5 Reference and GK-projection Systems of Great Britain

In 1852 an ellipsoid was selected by the National Survey: Airy 1830, and it was calculated by Colonel Alexander R. Clarke. OSGB36 is the system used for National mapping in Great Britain (Macdonald, 1991).

In 1963, the re-triangulation of Great Britain has been computed on the same Airy ellipsoid, resulting in the Transverse Mercator of Great Britain.

A national GPS network was designed to be controlled by OSGB(SN)77, and its successor OSGB(SN)80 - *the Scientific Network* for precise surveying and geodesy - as this was the best set of co-ordinates available (Ashkenazi, 1986a). However, the introduction of a new European datum (ETRF89, with OSGRS80) has changed the intentions (OS, 1995b).

The breakthrough of the *Channel Tunnel* has enabled Ordnance Survey and IGN to investigate the possibilities of an improved vertical connection between Britain and France, through the tunnel. It should greatly improve the geodetical vertical links with continental Europe using UELN (*Unified European Levelling Network)* (Alberda, 1960, 1963; Bordley / Calvert, 1985; Waalewijn, 1986, 1987).

The Transverse Mercator Grid Systems of Great Britain have the following specifications:

- Projection — Transverse Mercator of Great Britain
- Reference Ellipsoid — Airy (National projection grid only)
OSGRS80 (associated with GRS80)
- Latitude of True Origin — 49° North ______ of the Equator
- Longitude of True Origin — 2° West ______ of Greenwich
- Unit — metre
- False Northing — 100 000 m ______ North of True Origin
- False Easting — 400 000 m ______ West of True Origin
- Scale Factor at the central meridian — 0.99960 12717
- The latitude and longitude of the false origin, used by the formulae:
  1. Latitude of False Origin — 49° 46' ______ North (approx.)
  2. Longitude of False Origin — 7° 33' ______ West (approx.)
- True Northing — N= ______ Y+100 000
- True Easting — E= ______ X - 400 000.

In order to keep all co-ordinates within the map system positive, 400 000 m are added to all Eastings since these would otherwise be negative for points west of 2°W, also, 100 000 m are subtracted from all Northings. In the United Kingdom the longitudinal difference is *11° width approximately.* It follows that there are some linear distortions present e.g. in the Western Isles (Maling, 1992).

Transverse Mercator Projection of Great Britain

Zone Parameters	*Gt. Britain, excl. N. Ireland* (OS, 1950, 1995a).		
Reference Ellipsoid:	Airy 1830	False Easting:	**400 000 m**
Semi-Major axis:	**6 377 563.396**	False Northing:	**- 100 000 m**
Reciprocal Flattening:	**299.32496 459**	Parallel of Origin φ_o:	**49° 00' 00" N**
Scale Factor:	**0.99960 12717**	Central Meridian λ_o:	**2° 00' 00" W**

Conversion of OSGB36 Geographicals to Transverse Mercator Grid

The Direct calculation is to convert Geodetic Co-ordinates into GK-Planar Co-ordinates:

Input:	Latitude	Longitude	Output:	Easting	Northing	Convergence	Scale Factor

Sta. Framingham (OS, 1950, 1995a) |[33].

Latitude	φ_1:	**52° 34' 26".8915 N**	Easting	E_1=	626 238.2477
Longitude	λ_1:	**1° 20' 21".1080 E**	Northing	N_1=	302 646.4119
Convergence	γ_1=	+ 2° 39' 10".4691	Scale Factor	k_1=	1.000 229 694 988
Sta. Caister Water Tower					
Latitude	φ_2:	**52° 39' 27".2531 N**	Easting	E_2=	651 409.9029
Longitude	λ_2:	**1° 43' 04".5177 E**	Northing	N_2=	313 177.2703
Convergence	γ_2=	+ 2° 57' 26".5561	Scale Factor	k_2=	1.000 377 316 229
Intermediate point					
Latitude	φ_m:	**52° 36' 57".6300 N**	Easting	E_m=	638 824.0761
Longitude	λ_m:	**1° 31' 42".1863 E**	Northing	N_m=	307 911.8536
			Scale Factor	k_m=	1.000 301 560 388

Conversion of Transverse Mercator Grid to OSGB36 Geographicals

The Inverse process is to convert GK-Planar Co-ordinates into Geodetic Co-ordinates:

Input:	Latitude	Longitude	Output:	Easting	Northing	Convergence	Scale Factor

Easting	E_1:	**626 238.2477**	Latitude	φ_1=	52° 34' 26".8915 N
Northing	N_1:	**302 646.4119**	Longitude	λ_1=	1° 20' 21".1080 E
Convergence	γ_1=	+ 2° 39' 10".4692	Scale Factor	k_1=	1.000 229 694 581
Easting	E_2:	**651 409.9029**	Latitude	φ_2=	52° 39' 27".2531 N
Northing	N_2:	**313 177.2703**	Longitude	λ_2=	1° 43' 04".5177 E
Convergence	γ_2=	+ 2° 57' 26".5561	Scale Factor	k_2=	1.000 377 315 457
Easting	E_m:	**638 824.0761**	Latitude	φ_m=	52° 36' 57".6300 N
Northing	N_m:	**307 911.8536**	Longitude	λ_m=	1° 31' 42".1863 E
			Scale Factor	k_m=	1.000 301 559 823

Ellipsoidal and Grid Calculation (manual)

Using GBD00000.BAS gives:			Grid bearing	t_{1-2}	67° 17' 50".7461
			+ convergence	γ_1	+ 2° 39' 10".4691
True Distance	S_{1-2}:	*27 277.4892*	- (t - T)	δ_{1-2}	- 6".2563
True Azimuth	α_{1-2}:	*69° 57' 07".4747*	True Azimuth	α_{1-2}:	69° 57' 07".4715
True Azimuth	α_{2-1}:	*250° 15' 10".8182*			
			Grid bearing	t_{2-1}	247° 17' 50".7461
Grid Distance	D_{1-2}:	27 285.7326	+ convergence	γ_2	+2° 57' 26".5561
True Distance	$S_{calc\ 1-2}$:	27 277.4891	- (t - T)	δ_{2-1}	+6.4798
See (5.03), using			True Azimuth	α_{2-1}:	250° 15' 10".8224
Scale Factor	k_{1-2}:	1.000 302 208 795			

[33] Data from (OS, 1950: pp 17-30; OS, 1995a: pp 15-23)

Other Datums in Use

Great Britain's primary triangulation station co-ordinates may be obtained from Ordnance Survey based on the following datums:

- OSGB36 ____________ Geographical co-ordinates
- OSGB(SN)70 ____________ Geographical co-ordinates
- OSGB(SN)80 ____________ Geographical and Cartesian co-ordinates
- ED50 ____________ Geographical co-ordinates
- ED87 ____________ Geographical and Cartesian co-ordinates
- WGS84 ____________ Geographical and Cartesian co-ordinates
- OSGRS80 ____________ Geographical and Cartesian co-ordinates

(Bordley / Calvert, 1985; OS, 1995b).

Reference Ellipsoid Airy 1841

a	= 20 923 713 (*unique* in feet of Bar O_1)
b	= 20 853 810 (*unique* in feet of Bar O_1)
a - b	= 69 903
a	= 6 377 563.396|35 34083 m
b	= 6 356 256.909|58 9071 m
1 / f	= 299.32496|45937 94257 75717
f	= 0.00334 08506 41566 34|149
e^2	= 0.00667 05400 00123 42|876
n	= 0.00167 32203 10356 83|948
feet of Bar O_1	= 0.30480 07491

Table 16: Derived ellipsoidal parameters of Airy's figure of the earth

6.2.6 Reference and GK-projection Systems of Ireland

History of Mapping

In 1952 the primary triangulation network of N. Ireland was reobserved and computed on the Airy ellipsoid.

In this adjustment four stations of the old "Principal Triangulation of Great Britain and Ireland " were held fixed. Production of a new series of large scale plans in Northern Ireland, based on the "1952 Datum" was started.

In 1964, a *new primary triangulation of Ireland* was observed by the Ordnance Survey, Dublin, and as this network was tied to the N. Ireland primary, consideration was given to a new adjustment which would incorporate both primary networks (OSI, 1995; OSNI, 1994).

Scale checks by electronic distance measurement (Tellurometer) revealed a systematic scale error of +35 ppm in the N. Ireland primary network. Correcting such a scale error in the new adjustment without sacrificing the required close correspondence with the 1952 positions was impossible without distorting the network. The only other choice available was to reduce the dimensions of Airy ellipsoid by 35 ppm. Geodetic co-ordinates would then remain the same, but lengths calculated on the new ellipsoid from these geodetic co-ordinates would be in much closer agreement with EDM measurements.

Replacing the previous grid scale factor of unity by a grid scale factor of 1.000035 was necessary so that a reduction in size of the original Airy ellipsoid should not affect the existing relationship between geodeticals and the rectangular co-ordinates of the Irish National Grid. The new ellipsoid was labelled *Airy Modified.*

Ireland(1965) Mapping Adjustment

The "Ireland(1965) Datum" was adopted to give a best mean fit of the "1965" adjustment to the "1952" positions of all ten stations in N. Ireland, and a mean scale in agreement with measured lengths available.

The (1965) adjustment was made in a single block, holding fixed the datum position of "Slieve Donard" and a position of "Cuilcagh" computed from an adopted length of 123 426.130 m for the line joining the two stations. To control scale in the adjustment, a length of 69 377.529 m for the line "Knockascagh - Carrigfadda" derived from a separate adjustment of a fully measured quadrilateral was taken as an observation of *weight 6* relative to weight unity for the angle observations.

The "Ireland(1965) Datum" is defined by:

National datum:	Ireland(1965)		
Reference Ellipsoid: Airy Modified	Semi-major axis, a	=	6 377 340.189 m
	e^2	=	0.006 670 540 152
Fundamental Station:	Slieve Donard		
	Latitude	=	54° 10' 48".2675 N
	Longitude	=	5° 55' 11".8675 W
Azimuth to "Cuilcagh":	Azimuth	=	271° 50' 12".276

After adjustment, root mean square (RMS) correction to an observed angle was ± 0.96 seconds. RMS scale change over 19 lines was ± 7 ppm.

A comparison of the 1965 adjustment values with the 1952 values for the ten N. Ireland stations showed RMS changes of ± 0.25 m E and ± 0.23 m N. The maximum shift at any station was 0.57 m.

Laplace observations and scale checks. Subsequent Laplace observations in 1966 of a line in the North and one in the South indicated a difference of about 0.5 seconds of arc between the two lines, thus very little internal distortion. The mean rotation required by the 1965 adjustment to satisfy Laplace conditions was +2.27 seconds. No geoidal section observations have been carried out to determine geoid/ellipsoid separations.

Subsequent *Geodimeter 8* EDM work in 1969 showed that the overall scale of the 1965 adjustment was too large by 5 ppm. All factors considered, it was decided not to carry out a further readjustment of the *Ireland (1965) Datum.* Ordnance Survey, Dublin, adopted the *Ireland(1965) Datum* for its large-scale mapping program.

In the *Ireland(1975) mapping adjustment*, which was designed to provide a common co-ordinate system for the whole of Ireland for mapping purposes, the 1952 co-ordinates of *all N. Ireland primary stations* were held completely fixed, and in the South *three primaries* were held fixed in areas where large-scale mapping had commenced. This adjustment was labelled "1975 Mapping Adjustment". As it is based on 1952 and 1965 co-ordinates of the primaries, the datum is termed as the 1965 Datum and is derived as above (Cory, 1995).

Ordnance Survey, Dublin and Ordnance Survey, Belfast adopted the new 1975 Mapping Adjustment for its large-scale mapping program.

Other Datums in Use

The European Datum of 1950 (ED50) was first defined by the US Army Map Service. It was computed and implemented by German geodesists under leadership of Erwin Gigas at the "Bamberger Institut für Erdmessung", now IfAG, in 1945-1947.

The NATO-Forces adopted the *US Army UTM-grid* for Western Europe as the common European Datum of 1950 on the International Reference Ellipsoid (Burton, 1996). See [6.3.1] - Footnotes About the UTM, and Table 17 |[34].

Connections between Great Britain and France were observed in 1951 and 1963 which allowed the introduction of ED50 to the mainland of Great Britain. As the Retriangulation of Great Britain 1935-1950 had included connections to Ireland, the Irish primary stations were included in the OSGB(SN) Scientific Network adjustments of 1970 and 1980 and the European Datum adjustment of 1987 (ED87).

The Ireland primary triangulation station co-ordinates (Northern part) may be obtained from Ordnance Survey based on the following datums:

- Ireland(1975)__________________ Geographical co-ordinates
- OSGB(SN)70 __________________ Geographical co-ordinates
- OSGB(SN)80 __________________ Geographical and Cartesian co-ordinates
- ED50 ________________________ Geographical co-ordinates
- ED87 ________________________ Geographical and Cartesian co-ordinates
- WGS84 ______________________ Geographical and Cartesian co-ordinates
- IRENET95____________________ Geographical and Cartesian co-ordinates.

Vertical Datum of Ireland

Clarendon Dock. The primary levelling of Northern Ireland was carried out during the years 1952-1958. The datum to which this levelling and later secondary and tertiary networks are referred is Mean Sea Level (MSL) Belfast which was derived from readings of the mareograph (tide gauge) at Clarendon Dock, Belfast, over the

[34] Note: this is not the definition for *UTM grid,* as given by some textbooks on Geodesy

period 1951-1956. All height information depicted on the large-scale maps is shown in metres above MSL Belfast.

Poolbeg Lighthouse. Low water mark of the spring tide on the April 8, 1837 at Poolbeg Lighthouse. Initially this datum was fixed for County Dublin and it was subsequently adopted as the national datum in 1842. Heights on earlier map series are with respect to this datum and were given in feet.

Malin Head 1970. MSL at the tide gauge at Malin Head, County Donegal, was adopted as the *national datum* in 1970 from readings taken over the period between January 1960 and December 1969.

All heights on recent mapping using the Ireland(1975) are in international metres above Malin Head. MSL Malin Head was selected for a recent map series covering the whole island of Ireland but its difference from MSL Belfast is insignificant for mapping at small scales. The approximate difference between the datums is given as Poolbeg Datum + 2.5 m = Malin Head Datum.

Mapping

OS of Ireland decided to perform digital data entry and now compiles all its new mapping data in a coded digital format. This material is stored in a complex database which in future will form the basis of Geographical Information Systems (GIS).

Transverse Mercator Projection of Ireland

The Irish Transverse Mercator Grid System has the following specifications:

- Projection — Transverse Mercator of Ireland
- Reference Ellipsoid — Airy Modified
- Latitude of True Origin — 53° 30' 00" ______ North of the Equator
- Longitude of True Origin — 8° 00' 00" ______ West of Greenwich
- Unit — metre
- False Northing — 250 000 m ______ South of True Origin
- False Easting — 200 000 m ______ West of True Origin
- Scale Factor at the central meridian — 1.000 035
- The latitude and longitude of the false origin, used by the formulae:
  1. Latitude of False Origin — 51° 13' ______ North (approx.)
  2. Longitude of False Origin — 10° 52' ______ West (approx.)
- True Northing — N= ______ Y - 250 000
- True Easting — E= ______ X - 200 000.

Note

In order to keep all co-ordinates within the map system positive, 200 000 m are added to all Eastings since these would otherwise be negative for points west of 8°W, also, 250 000 m are added to all Northings. South of the False Origin add 1 000 000 m to the Northing.

Transverse Mercator Projection of Ireland

Zone Parameters	*Whole country* (Codd, 1995).		
Reference Ellipsoid:	Airy Modified	False Easting:	**200 000 m**
Semi-Major axis:	**6 377 340.189**	False Northing:	**250 000 m**
Reciprocal Flattening:	**299.32496 459**	Parallel of Origin φ_o:	**53° 30' 00" N**
Scale Factor:	**1.000 035**	Central Meridian λ_o:	**8° 00' 00" W**

Conversion of Ireland1965 Geographicals to Transverse Mercator Grid

The Direct calculation is to convert Geodetic Co-ordinates into GK-Planar Co-ordinates:

Input: Latitude Longitude			Output: Easting Northing Convergence		Scale Factor
Sta. Howth					
Latitude	φ_1:	**53° 22' 23".1566 N**	Easting	E_1=	328 546.3442
Longitude	λ_1:	**6° 04' 06".0065 W**	Northing	N_1=	237 617.1863
Convergence	γ_1=	+ 1° 33' 01".5983	Scale Factor	k_1=	1.000 237 756 774
Sta. Slieve Donard					
Latitude	φ_2:	**54° 10' 48".2675 N**	Easting	E_2=	335 788.0111
Longitude	λ_2:	**5° 55' 11".8675 W**	Northing	N_2=	327 685.8494
Convergence	γ_2=	+ 1° 41' 12".7496	Scale Factor	k_2=	1.000 261 204 965
Intermediate point					
Latitude	φ_m:	**53° 46' 35".8440 N**	Easting	E_m=	332 170.3172
Longitude	λ_m:	**5° 59' 41".5055 W**	Northing	N_m=	282 650.9963
			Scale Factor	k_m=	1.000 249 331 183

Conversion of Transverse Mercator Grid to Ireland1965 Geographicals

The Inverse process is to convert GK-Planar Co-ordinates into Geodetic Co-ordinates:

Input: Latitude Longitude			Output: Easting Northing Convergence		Scale Factor
Easting	E_1:	**328 546.3442**	Latitude	φ_1=	53° 22' 23".1566 N
Northing	N_1:	**237 617.1863**	Longitude	λ_1=	6° 04' 06".0065 W
Convergence	γ_1=	+1° 33' 01".5983	Scale Factor	k_1=	1.000 237 756 760
Easting	E_2:	**335 788.0111**	Latitude	φ_2=	54° 10' 48".2675 N
Northing	N_2:	**327 685.8494**	Longitude	λ_2=	5° 55' 11".8675 W
Convergence	γ_2=	+ 1° 41' 12".7496	Scale Factor	k_2=	1.000 261 204 943

Transverse Mercator Projection of Ireland

Ellipsoidal and Grid Calculation (manual)

Using GBD00000.BAS gives:

True Distance	$S_{1\text{-}2}$:	*90 336.7870*
True Azimuth	$\alpha_{1\text{-}2}$:	*6° 09' 19".8725*
True Azimuth	$\alpha_{2\text{-}1}$:	*186° 16' 30".7833*
Grid Distance	$D_{1\text{-}2}$:	90 359.3150
True Distance	$S_{\text{calc } 1\text{-}2}$:	90 336.7867
See (5.03), using		
Scale Factor	$k_{1\text{-}2}$:	1.000 249 381 078

Grid bearing	$t_{1\text{-}2}$	4° 35' 48".4283
+ convergence	γ_1	+1° 33' 01".5983
- (t - T)	$\delta_{1\text{-}2}$	- 29".8445
True Azimuth	$\alpha_{1\text{-}2}$:	6° 09' 19". 8711
Grid bearing	$t_{2\text{-}1}$	184° 35' 48".4283
+ convergence	γ_2	+1° 41' 12".7496
- (t - T)	$\delta_{2\text{-}1}$	+30".3944
True Azimuth	$\alpha_{2\text{-}1}$:	186° 16' 30".7835

6.2.7 Reference and GK-projection Systems of Italy

The Italian National System is based on the Gauss projection system (Transverse Mercator grid).

The Italian Geodetic Network, datum IGM1940 (Hayford reference ellipsoid oriented to Rome - M.- Mario, Astronomical Definition 1940 - Sistema Nazionale di Riferimento). See [9.1.3.2; 9.1.3.4].

There is also an Italian geodetic network datum IGM1983, see [9.1.3.3], based upon the recomputation using Laplace azimuths and EDM of the last twenty years, and the Sistema Geodetico Catastali, see [9.1.3.5].

The Gauss-Boaga Grid System has the following specifications:

- Projection — Gauss-Boaga IGM1940 (Gauss-Krüger Type)
- Reference Ellipsoid — International
- Latitude of True Origin — 0° ________ the Equator
- Longitude of True Origin — 9° East ________ of Greenwich (Zone 1) or
  15° East ________ of Greenwich (Zone 2)
- Unit — metre
- False Northing — 0 m
- False Easting — 1 500 000 m _____ (Zone 1) or
  2 520 000 m _____ (Zone 2)
- Scale Factor at the central meridian — 0.9996.

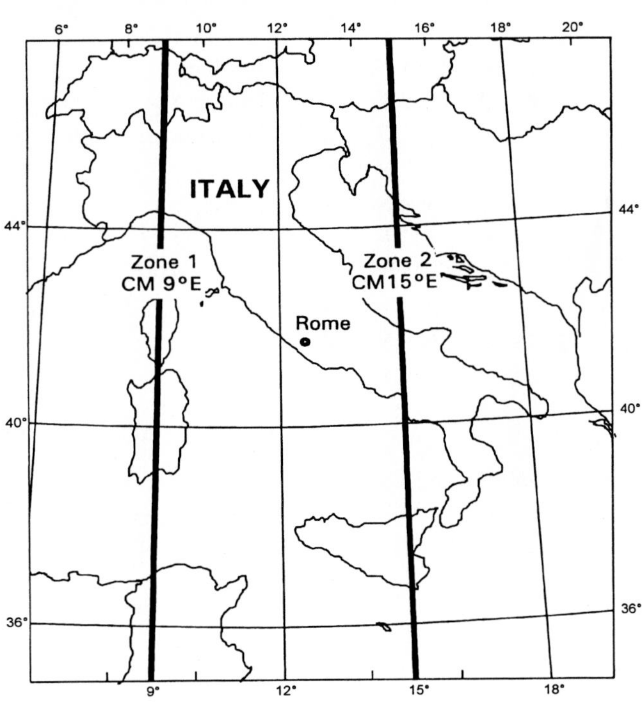

Figure 22: Transverse Mercator Zone System of Italy

In order to keep all co-ordinates within the map system positive, 1 500 000 m and 2 520 000 m are added to all Eastings of Zone 1 and Zone 2, respectively.

Reference and Projection Systems of Italy

Zone Parameters	*Italy (West), Zone 1*		
Reference Ellipsoid:	International 1924	False Easting:	**1 500 000 m**
Semi-Major axis:	**6 378 388**	False Northing:	**0 m**
Reciprocal Flattening:	**297**	Parallel of Origin φ_o:	**0° 00' 00" N**
Scale Factor:	**0.9996**	Central Meridian λ_o:	**9° 00' 00" E**

Conversion of IGM1940 Geographicals to Gauss-Boaga Grid

The Direct calculation is to convert Geodetic Co-ordinates into GK-Planar Co-ordinates:

Input:	Latitude	Longitude	Output:	Easting	Northing	Convergence	Scale Factor

Sta. 122901

Latitude	φ_1:	**43° 07' 37".2498 N**	Easting	E_1=	1 748 583.3314
Longitude	λ_1:	**12° 03' 21".0772 E**	Northing	N_1=	4 779 539.7803
Convergence	γ_1=	+ 2° 05' 24".3748	Scale Factor	k_1=	1.000 360 175 786

Sta. 122902

Latitude	φ_2:	**43° 10' 56".3950 N**	Easting	E_2=	1 745 880.7939
Longitude	λ_2:	**12° 01' 31".3119 E**	Northing	N_2=	4 785 593.6080
Convergence	γ_2=	+ 2° 04' 16".8915	Scale Factor	k_2=	1.000 343 725 137

Sta. 122903

Latitude	φ_3:	**43° 10' 58".2214 N**	Easting	E_3=	1 754 840.9775
Longitude	λ_3:	**12° 08' 08".2725 E**	Northing	N_3=	4 785 980.0205
Convergence	γ_3=	+ 2° 08' 49".0357	Scale Factor	k_3=	1.000 398 924 047

Sta. 122904

Latitude	φ_4:	**43° 08' 33".9638 N**	Easting	E_4=	1 756 952.5211
Longitude	λ_4:	**12° 09' 34".3542 E**	Northing	N_4=	4 781 602.5743
Convergence	γ_4=	+ 2° 09' 42".2394	Scale Factor	k_4=	1.000 412 227 239

Sta. 122905

Latitude	φ_5:	**43° 05' 22".5540 N**	Easting	E_5=	1 756 826.3169
Longitude	λ_5:	**12° 09' 18".9196 E**	Northing	N_5=	4 775 684.1496
Convergence	γ_5=	+ 2° 09' 23".9764	Scale Factor	k_5=	1.000 411 439 977

Sta. 122906

Latitude	φ_6:	**43° 05' 00".5858 N**	Easting	E_6=	1 752 591.4774
Longitude	λ_6:	**12° 06' 10".5236 E**	Northing	N_6=	4 774 847.3132
Convergence	γ_6=	+ 2° 07' 14".1998	Scale Factor	k_6=	1.000 384 898 587

See datum transformations [9.1.3.4; 9.1.3.5].

Reference and Projection Systems of Italy

Conversion of Gauss-Boaga Grid to IGM1940 Geographicals

The Inverse process is to convert GK-Planar Co-ordinates into Geodetic Co-ordinates:

Input:	Latitude	Longitude	Output:	Easting	Northing	Convergence	Scale Factor

Sta. 122901						
	Easting	E_1:	**1 748 583.3314**	Latitude	φ_1=	43° 07' 37".2498 N
	Northing	N_1:	**4 779 539.7803**	Longitude	λ_1=	12° 03' 21".0772 E
	Convergence	γ_1=	+2° 05' 24".3748	Scale Factor	k_1=	1.000 360 175 495
Sta. 122902						
	Easting	E_2:	**1 745 880.7939**	Latitude	φ_2=	43° 10' 56".3950 N
	Northing	N_2:	**4 785 593.6080**	Longitude	λ_2=	12° 01' 31".3119 E
	Convergence	γ_2=	+2° 04' 16".8915	Scale Factor	k_2=	1.000 343 724 862
Sta. 122903						
	Easting	E_3:	**1 754 840.9775**	Latitude	φ_3=	43° 10' 58".2214 N
	Northing	N_3:	**4 785 980.0205**	Longitude	λ_3=	12° 08' 08".2725 E
	Convergence	γ_3=	+ 2° 08' 49".0357	Scale Factor	k_3=	1.000 398 923 707
Sta. 122904						
	Easting	E_4:	**1 756 952.5211**	Latitude	φ_4=	43° 08' 33".9638 N
	Northing	N_4:	**4 781 602.5743**	Longitude	λ_4=	12° 09' 34".3542 E
	Convergence	γ_4=	+ 2° 09' 42".2394	Scale Factor	k_4=	1.000 412 227 239
Sta. 122905						
	Easting	E_5:	**1 756 826.3169**	Latitude	φ_5=	43° 05' 22".5540 N
	Northing	N_5:	**4 775 684.1496**	Longitude	λ_5=	12° 09' 18".9196 E
	Convergence	γ_5=	+ 2° 09' 23".9765	Scale Factor	k_5=	1.000 411 439 626
Sta. 122906						
	Easting	E_6:	**1 752 591.4774**	Latitude	φ_6=	43° 05' 00".5858 N
	Northing	N_6:	**4 774 847.3132**	Longitude	λ_6=	12° 06' 10".5236 E
	Convergence	γ_6=	+ 2° 07' 14".1998	Scale Factor	k_6=	1.000 384 898 269

6.2.8 Reference and GK-projection Systems of Norway

The National Reference and projection Systems of Norway is based on the Gauss-Krüger (Transverse Mercator) grid system.

The Norwegian geodetic network datum NGO1948, is based upon a calculation using 133 First Order trigonometrical stations in Southern Norway and Laplace azimuths, on the Bessel Modified reference ellipsoid - oriented to Oslo Observatory λ=10° 43' 22".5 E, definition 1948 (Figure 24: Zones of projection in the system NGO1948).

There is also a Norwegian geodetic network, datum ED50 [6.3.6], and in the near future ETRF89, based on geodetic network datum GRS80 (Harsson, 1995, 1996).

The NGO1948 grid system has the following specifications:

- Projection — NGO1948 (Gauss-Krüger Type)
- Reference Ellipsoid — Bessel Modified
- Latitude of True Origin — 58° N
- Longitude of True Origin CM λ_{o1} — 4° 40' 00" West of Oslo — Zone I
  Longitude of True Origin CM λ_{o2} — 2° 20' 00" West of Oslo — Zone II
  Longitude of True Origin CM λ_{o3} — 0° 00' 00" East of Oslo — Zone III
  Longitude of True Origin CM λ_{o4} — 2° 30' 00" East of Oslo — Zone IV
  Longitude of True Origin CM λ_{o5} — 6° 10' 00" East of Oslo — Zone V
  Longitude of True Origin CM λ_{o6} — 10° 10' 00" East of Oslo — Zone VI
  Longitude of True Origin CM λ_{o7} — 14° 10' 00" East of Oslo — Zone VII
  Longitude of True Origin CM λ_{o8} — 18° 20' 00" East of Oslo — Zone VIII
- Unit — metre
- False Northing — 0 m
- False Easting — 0 m
- Scale Factor at the central meridian — 1.0000 (exactly).

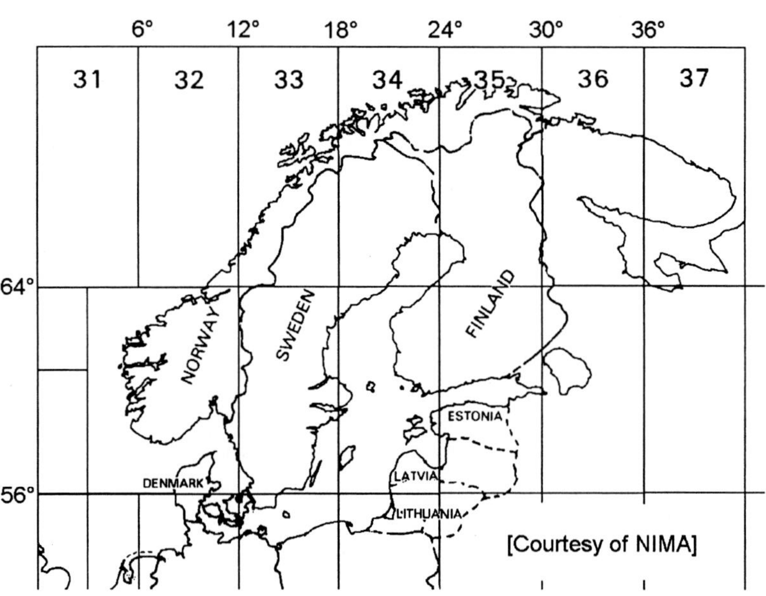

Figure 23: UTM zones of northern Europe

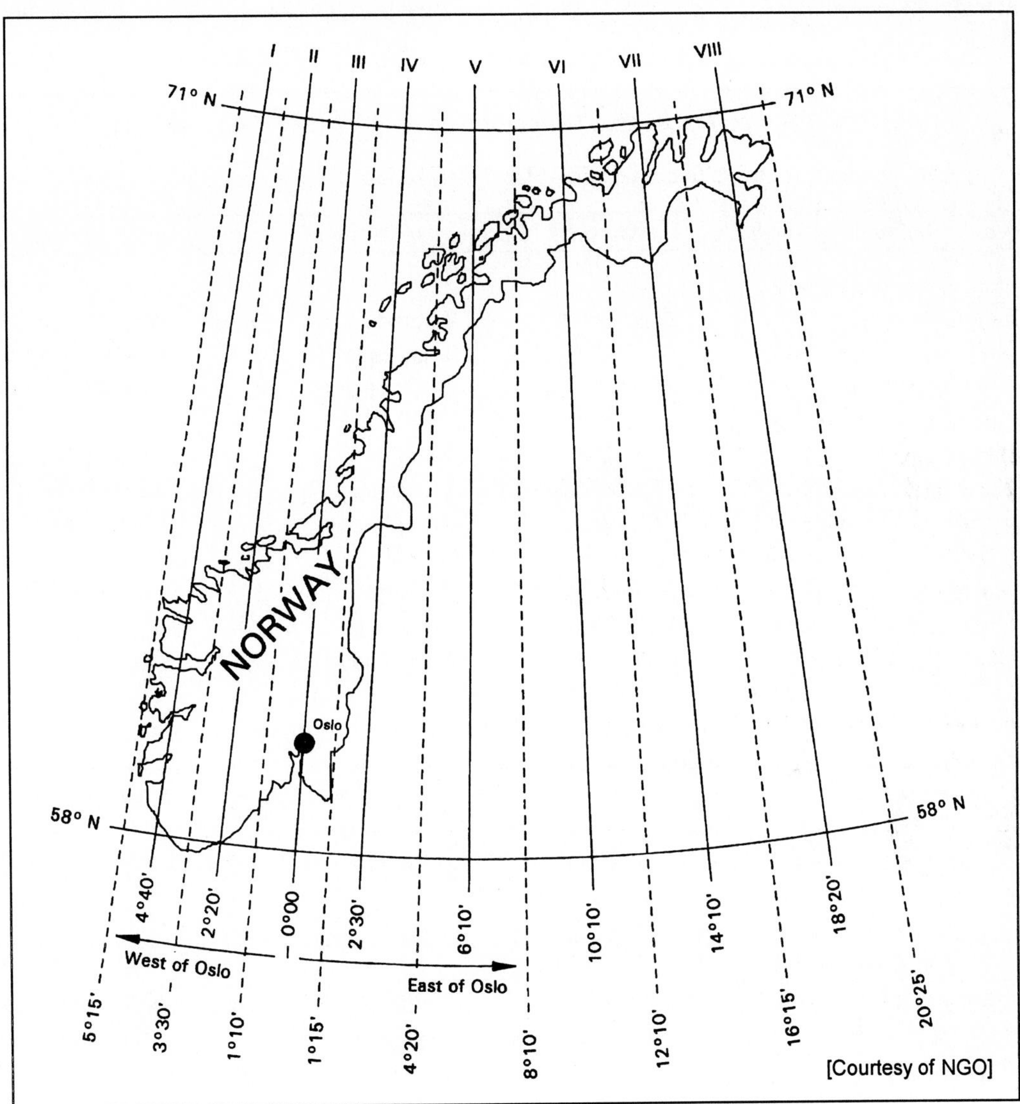

Figure 24: Zones of projection in the system NGO1948

NGO1948 - Reference and GK-projection Systems of Norway

Zone Parameters	Norway, see [6.2.8]		
Reference Ellipsoid:	Bessel Modified	False Easting:	**0 m**
Semi-Major axis:	**6 377 492.0176**	False Northing:	**0 m**
Reciprocal Flattening:	**299.1528**	Parallel of Origin φ_o:	**58° 00' 00" N**
Scale Factor:	**1**	Central Meridian λ_o:	see [6.2.8]

Conversion of Transverse Mercator Grid to NGO1948 Geographicals

The Inverse process is to convert GK-Planar Co-ordinates into Geodetic Co-ordinates:

Input:	Latitude	Longitude	Output:	Easting	Northing	Convergence	Scale Factor
Sta. Eigeberg		Zone I					
	Easting	E_1:	**- 26 917.603**	Latitude		φ_1=	58° 51' 00".4665 N
	Northing	N_1:	**94 775.788**	Longitude		λ_1=	5° 07' 58".9019 W
	Convergence	γ_1=	- 0° 23' 56".8416	Scale Factor		k_1=	1.000 008 879 640
Sta. Vigra		Zone I					
	Easting	E_2:	**1 324.544**	Latitude		φ_2=	62° 33' 11".7688 N
	Northing	N_2:	**507 257.766**	Longitude		λ_2=	4° 38' 27".3025 W
	Convergence	γ_2=	+0° 01' 22".2635	Scale Factor		k_2=	1.000 000 021 485
Sta. SK Tårn		Zone III					
	Easting	E_3:	**- 26 057.276**	Latitude		φ_3=	60° 08' 34".0483 N
	Northing	N_3:	**238 766.889**	Longitude		λ_3=	0° 28' 08".5887 W
	Convergence	γ_3=	- 0° 24' 24".4684	Scale Factor		k_3=	1.000 008 318 895
Sta. Vassfjell		Zone III					
	Easting	E_4:	**- 18 129.156**	Latitude		φ_4=	63° 15' 41".7049 N
	Northing	N_4:	**586 249.214**	Longitude		λ_4=	0°21' 39".7572 W
	Convergence	γ_4=	- 0° 19' 20".7772	Scale Factor		k_4=	1.000 004 024 369
Sta. Bodø		Zone IV					
	Easting	E_5:	**51 163.144**	Latitude		φ_5=	67° 17' 01".1104 N
	Northing	N_5:	**1 035 084.479**	Longitude		λ_5=	3° 41' 13".0665 E
	Convergence	γ_5=	+ 1° 05' 41".6800	Scale Factor		k_5=	1.000 032 029 353
Sta. Tromsø		Zone V					
	Easting	E_6:	**79 846.04**	Latitude		φ_6=	69° 39' 44".1832 N
	Northing	N_6:	**1 301 229.43**	Longitude		λ_6=	8° 13' 29".6930 E
	Convergence	γ_6=	+ 1° 55' 48".1367	Scale Factor		k_6=	1.000 077 979 082
Sta. Domen		Zone VIII					
	Easting	E_7:	**74 640.941**	Latitude		φ_7=	70° 19' 59".3979 N
	Northing	N_7:	**1 375 941.444**	Longitude		λ_7=	20° 19' 12".8774 E
	Convergence	γ_7=	+ 1° 52' 15".9244	Scale Factor		k_7=	1.000 068 136 692

6.2.9 Reference and SPC Systems of the USA

In the United States of America the State Plane Co-ordinate System (SPCS27) was introduced in 1933. The individual projections forming the State Co-ordinate Systems 1927 were based upon the Clarke 1866 ellipsoid: North American Datum of 1927 (NAD27) (Schwartz, 1989).

Referring to the redefinition of the North American Datum from NAD27 into North American Datum of 1983 (NAD83), which includes the change from the Clarke 1866 ellipsoid to the GRS80 ellipsoid, gave the National Geodetic Survey the possibility of considering changing the projection system to provide the best projection within the boundaries of a particular State (Osterhold, 1993). See also [3.5.7].

Each State has its own system of co-ordinates. Several States have more than one Transverse Mercator or Lambert zone, and a few States have a combination of both Lambert and Transverse Mercator (Dracup, 1994).

There is one State, Alaska, in which all three systems are employed: a small part of Alaska makes use of the *Hotine Oblique Mercator projection.*

Units

The following States use *international feet and metres:*
- Arizona, Michigan, Montana, Oregon, South Carolina and Utah.

The following States use *US survey feet and metres:*
- California, Colorado, Connecticut, Indiana, Maryland, Nebraska, North Carolina, Texas and Wyoming
- All other States *only metres*.

The changes in co-ordinates due to the readjustment of the NAD83. Changes in latitude and longitude from NAD27 to NAD83 are estimated (Vincenty, 1976a):

- φ = -24 to +40 m and λ= -40 to +100 m within the ConUS |[35], and
- φ = -40 to -120 m and λ= +110 to +140 m in Alaska.

The US State Plane Co-ordinate System has the following specifications:

- Projection: Transverse Mercator
Lambert Conformal Conic, see [7.2.4]
Oblique Mercator, see [8.2.2]
- Reference Ellipsoid: Clarke 1866 _____ NAD27
GRS80 _________ NAD83
- Longitude of Origin: Central Meridian, see *specifications of a State*
- Latitude of Origin: See *specifications of a State*
- Unit: See *specifications of a State*
- False Northing: See *specifications of a State*
- False Easting: See *specifications of a State*
- Scale Factor at the central meridian: 0.9999 (*general case, see State specification*)
- Latitude Limits of SPCS: State Boundaries.

The specifications for the SPCS27 are given in e.g. (Mitchell, 1945) and for the SPCS83 in e.g. NOAA Technical Manual Memorandum NOS NGS 5 (Stem, 1989a).

---

[35] Contiguous States of North America

6.2.9.1 SPCS27 - Transverse Mercator - State Alaska Zone 6 - USA

Zone Parameters:	*SPCS GK Alaska Zone 6* |[36]		
Reference Ellipsoid:	Clarke 1866 *(NAD27)*	False Easting:	**500 000 feet**
Semi-Major axis:	**20 925 832.16 ft** |[37]	False Northing:	**0 feet**
Recipr. Flattening:	**294.978 698 2**	Parallel of Origin φ_o	**54° 00' 00" N**
Scale Factor:	**0.9999**	C.M. of zone λ_o:	**158° 00' 00" W**

Conversion of NAD27 Geographicals to Transverse Mercator Grid

The Direct calculation is to convert Geodetic Co-ordinates into GK-Planar Co-ordinates:

Input:	Latitude	Longitude	Output:	Easting	Northing	Convergence	Scale Factor

Latitude:	**71° 00' 00" N**	Easting=	857 636.1684 feet
Longitude:	**155° 00' 00" W**	Northing=	6 224 356.3177 feet
Convergence=	+ 2° 50' 12".5915	Scale Factor=	1.000 045 283 883

Conversion of Transverse Mercator Grid to NAD27 Geographicals

The Inverse process is to convert GK-Planar Co-ordinates into Geodetic Co-ordinates:

Input:	Latitude	Longitude	Output:	Easting	Northing	Convergence	Scale Factor

Easting:	**857 636.1684 feet**	Latitude=	71° 00' 00" N
Northing:	**6 224 356.3177 feet**	Longitude=	155° 00' 00" W
Convergence=	+ 2° 50' 12".5915	Scale Factor=	1.000 045 283 882

[36] (Claire, 1968)

[37] US survey feet

6.2.9.2 SPCS27 - Transverse Mercator - The Hawaiian Islands - USA

The plane co-ordinate system adopted for the Hawaiian Islands consists of five zones, Zone 1, 2, 3, 4, and 5 for Hawaii, Maui (Maui, Lanai, Molokai and Kahoolawe), Oahu, Kauai and Niinau respectively, each of which is based on the Transverse Mercator projection. In the zones 1 to 4 the scale along the central meridian in each zone is reduced, in order to minimise the scale error throughout the zone.

Zone 5 is so narrow that no reduction in scale is made along its central meridian. An example on NAD27 for State Hawaii Zone 1 (NOAA-C&GS, 1954):

SPCS27 - Transverse Mercator - State Hawaii - Zone 1

Zone Parameters:	*SPCS GK Hawaii, Zone 1*		
Reference Ellipsoid:	Clarke 1866 *(NAD27)*	False Easting:	**500 000 feet**
Semi-Major axis:	**20 925 832.16 ft** |[38]	False Northing:	**0 feet**
Recipr. Flattening:	**294.978 698 2**	Parallel of Origin φ_o	**18° 50' 00" N**
Scale Factor:	**0.9999667**	C.M. of zone λ_o:	**155° 30' 00" W**

Conversion of NAD27 Geographicals to Transverse Mercator Grid

The Direct calculation is to convert Geodetic Co-ordinates into GK-Planar Co-ordinates:

Input: Latitude Longitude		Output: Easting Northing Convergence	Scale Factor
Latitude:	**19° 37' 23".477 N**	Easting=	332 050.9332 feet
Longitude:	**155° 59' 16".911 W**	Northing=	287 068.3516 feet
Convergence=	- 0° 09' 50".0412	Scale Factor=	0.999 999 078 952
Latitude:	**19° 31' 24".578 N**	Easting=	568 270.0632 feet
Longitude:	**155° 18' 06".262 W**	Northing=	250 663.2492 feet
Convergence=	+ 0° 03' 58".5274	Scale Factor=	0.999 972 050 233

Conversion of Transverse Mercator Grid to NAD27 Geographicals

The Inverse process is to convert GK-Planar Co-ordinates into Geodetic Co-ordinates:

Input: Latitude Longitude		Output: Easting Northing Convergence	Scale Factor
Easting:	**332 050.9332 feet**	Latitude=	19° 37' 23".4770 N
Northing:	**287 068.3516 feet**	Longitude=	155° 59' 16".9110 W
Convergence=	- 0° 09' 50".0412	Scale Factor=	0.999 999 078 952
Easting:	**568 270.0632 feet**	Latitude=	19° 31' 24".5780 N
Northing:	**250 663.2492 feet**	Longitude=	155° 18' 06".2620 W
Convergence=	+0° 03' 58".5274	Scale Factor=	0.999 972 050 233

[38] US survey feet

6.2.9.3 SPCS83 - Transverse Mercator - State New Jersey - (New York - East) - USA

Zone Parameters for: *SPCS GK New Jersey* (Greenfield, 1992).			
Reference Ellipsoid:	GRS80	False Easting:	**150 000 m**
Semi-Major axis:	**6 378 137 m**	False Northing:	**0 m**
Recipr. Flattening:	**298.25722 21008 827**	Parallel of Origin φ_o:	**38° 50' 00" N**
Scale Factor:	**0.9999**	Central Meridian λ_o:	**74° 30' 00" W**

Conversion of NAD83 Geographicals to Transverse Mercator Grid

The Direct calculation is to convert Geodetic Co-ordinates into GK-Planar Co-ordinates:

Input:	Latitude	Longitude	Output:	Easting	Northing	Convergence	Scale Factor

Latitude:	**38° 52' 34".5376 N**	Easting=	114 614.7312
Longitude:	**74° 54' 28".1235 W**	Northing=	4 844.0181
Convergence=	- 0° 15' 21".4634	Scale Factor=	0.999 915 413 288

Conversion of Transverse Mercator Grid to NAD83 Geographicals

The Inverse process is to convert GK-Planar Co-ordinates into Geodetic Co-ordinates:

Input:	Latitude	Longitude	Output:	Easting	Northing	Convergence	Scale Factor

Easting:	**114 614.7312**	Latitude=	38° 52' 34".5376 N
Northing:	**4 844.0181**	Longitude=	74° 54' 28".1235 W
Convergence=	- 0° 15' 21".4634	Scale Factor=	0.999 915 413 288

UTM Reference Ellipsoids

Listing of current preferred UTM reference ellipsoids

- Australian National Spheroid__________ AGD66, AGD84
- Bessel 1841 __________ Tokyo Datum area
- Clarke 1866 ellipsoid__________ NAD27 area
- Clarke 1866 ellipsoid __________ NAD27, zones 47-55
- Clarke 1866 ellipsoid__________ NAD27, for Philippines and Mariana Islands
- Clarke 1880DoD
- Everest 1830
- GRS67 __________ SAD69 area
- GRS80 (New International)__________ NAD83 area
- International 1924 __________ ED50
- WGS72
- WGS84.

Table 17: List of preferred UTM ellipsoids

6.3 The Universal Transverse Mercator Grid System

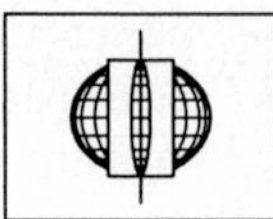

The Story of the Universal Transverse Mercator Grid

At the outbreak of World War I in August 1914, the Artillery of the Allied Forces in France had to resort to the *French Survey System* [10.1.1].

This survey system, which was based on projection of curvilinear co-ordinates, uses rectangular co-ordinates, continuous from sheet to sheet, for calculating distance and direction.

During World War II the choice of a *non-metric unit* for the *squares* on what was otherwise a metric map was made at the insistence of the British artillery, who *failed* to realise that the squares were designed as a means of location and not of measurement; nevertheless, due to the incompatibility of the British method, the French survey method was eventually adopted by the *Allied Forces*.
After the end of World War II in 1945, the *US Forces* felt that if they had taken the universal military maps with a metric grid in their hands at the beginning of war, they should have saved many difficulties (Seymour, 1980).

Consequently, the Universal Transverse Mercator (UTM) grid system was devised by the US Army Map Service (AMS) and adopted by the US Joint Chiefs of Staff as the official US Army grid in 1947. AMS became a component of the Defense Mapping Agency, now National Imagery and Mapping Agency (NIMA), in 1972 and is currently part of NIMA Bethesda.

(Ayres, 1995): *"The UTM grid system is defined for military maps by a transverse Mercator projection (Gauss-Krüger-type) for areas between latitudes 80° S and 84° N, with a scale factor of k_0 = 0.9996 at the longitude of the origin which is the central meridian of a UTM grid zone (Thus excepting the two polar caps |*[39]*).* Overlap. *The UTM grid extends to 84° 30' N and 80° 30' S, providing a 30' overlap with the UPS grid".*

The Universal Polar Stereographic (UPS) grid is applicable world-wide for the polar regions south of 80° S and north of 84° N The UPS grid extends to 83° 30' N and 79° 30' S, providing a 30' overlap with the UTM grid. The grid is not discussed in this book.

The US Corps of Engineers accomplished the task to introduce UTM as a world-wide metric mapping base from pole-to-pole for all military mapping, in 6° longitudinal zones to specifications of the UTM. The grid zones are 6° wide and numbered from 1 through 60 beginning at 180° W and proceeding eastwards to 180° E. For example, Zone 1 is 180° W to 174° W, Zone 30 is 6° W to 0°, etc. The UTM System in local use varies between regions, so the limits within which the map grid is to be used need to be defined. See Table 17: List of preferred UTM ellipsoids.

Longitude of Origin

The Central Meridian (CM) of each projection zone is: 3°, 9°, 15°, 21°, 27°, 33°, 39°, 45°, 51°, 57°, 63°, 69°, 75°, 81°, 87°, 93°, 99°, 105°, 111°, 117°, 123°, 129°, 135°, 141°, 147°, 153°, 159°, 165°, 171°, 177° East and West of Greenwich.

False Co-ordinates

Having grid quantities changing from positive to negative is inconvenient as the grid zero lines are crossed. To avoid this, UTM has a special convention for recording false values added to avoid negative grid co-ordinates within a grid zone:

[39] NIMA - DoD is contemplating revising the above polar limits for UTM for latitudes 80° S to 80° N

- the central meridian is assigned a false easting value of 500 000 m
- the equator is assigned a false northing value of 10 000 000 m in the Southern Hemisphere, and
- the equator is assigned a false northing value of 0 m in the Northern Hemisphere,

which are added to the scaled X, Y cartesian co-ordinates.

Scale Factor

The Scale Factor $k_0 = 0.9996$ is assigned to the central meridian, and creates two lines of zero distortion at a distance of about 180 km from the central meridian.

The Universal Transverse Mercator grid system has the following specifications:		
• Projection type	Transverse Mercator (Gauss-Krüger Type)	
• Reference Ellipsoid	see [6.3.1], (Table 17)	
• Zones	6° wide	
• Longitude of Origin	Central Meridian of each zone	
• Latitude of Origin	0°	the Equator
• Unit	metre	
• False Northing	0 m 10 000 000 m	Northern Hemisphere Southern Hemisphere
• False Easting	500 000 m	
• Scale Factor at the central meridian	0.9996 (exactly)	
• Latitude Limits of System	North South	84° N 80° S
• Starting zone, numbering	No. 1, zone from 180° W to 174° W, increasing eastwards	
• Ending zone number	No. 60, zone from 174° E to 180° E	
• Limits of zones	by meridians whose longitudes are multiples of 6° West or East of Greenwich.	

Table 18: UTM specifications

6.3.1 Footnotes About the UTM Grid

The reader should note some *inconsistency* in the description of the Universal Transverse Mercator grid (UTM) in literature concerning the source of the UTM grid system, properties of the UTM, and its relationship to the World Geodetic System 1984 (WGS84) (Fister, 1980).

(Burton, 1996):

- "Because the UTM grid and the transverse Mercator projection are dependent, the projection is sometimes called a "UTM projection" when used concerning the UTM grid
- A "UTM projection" is a transverse Mercator projection with the imposed scale factor and origin parameters of the UTM grid system. Other grid systems for certain parts of the world, such as shown in [11.2] are based on the transverse Mercator projection but their origins and scale factors are different. Those systems are being progressively replaced by the UTM-grid,

with the intent to cover all the military mapping of the world with a universal metric grid system eventually

- The projection is computed using the defining parameters of an ellipsoid associated with the datum of the latitude and longitude and geographic co-ordinates as input. Thus, the UTM grid is computed from the projection cartesian co-ordinates without regard to the datum. See [3.4], Geodetic Reference Datum

- There is no dependent relationship between the UTM grid system, a particular datum and an ellipsoid. The grid system is universal because it is defined world-wide and is independent of the datum

- If grid values for the same geographic co-ordinate are computed on, e.g. the WGS84 ellipsoid and the Hayford ellipsoid, the difference in values will reflect the difference in the two ellipsoids

- However, when detail or features are positioned on the two grids relative to their geographic co-ordinates on the WGS84 Datum and their geographic co-ordinates on the European Datum 1950, the difference reflected in grid values will be differences in the two datums. Both grids are on the UTM grid system but common features have different geographic co-ordinates and consequently different grid co-ordinates because they are on different reference systems

- To update the DoD NIMA specifications, only the reference ellipsoid requires changing, because each zone consists of an identical Universal Transverse Mercator (GK) grid system

- In conclusion, a UTM grid is generated by transforming a geographic co-ordinate - geodetic latitude (φ) and longitude (λ) - into a 2-D rectangular co-ordinate (easting and northing) using the transverse Mercator projection equations. The input is scale factor, central meridian, and ellipsoid parameters".

In geodetic applications and for topographical mapping and/or charting of 1:1 000 000 scales or larger, the geodetic latitude and longitude must be associated with a (complete) geodetic datum.

The UTM specifications, i.e. the defining constants, appear in many manuals of the US DoD NIMA Topographic Center (SDHG), Washington DC, USA, originator of the system, e.g. (DA, 1958) |[40].

Global Datum

The European Datum of 1950 (ED50) and its later versions such as *ED79, ED87, EUREF89, and ETRF89* do not cover the whole world. WGS84 is the *only global datum* in use at this time. See for features of the Baltic area: (Vermeer, 1995); for ED87: (Poder, 1989).

[40] Now obsolete and replaced by the NIMA (DMA) TM 8358 series, which states current authorative guidance for the use of grids and reference systems

Note about Optimising UTM

For an interesting feature about optimising UTM by mathematical geodesy, the interested reader may consult the report of Grafarend about the optimisation of the global transverse Mercator projection:

> *"The fundamental equations which govern conformal mapping - the Korn-Lichtenstein partial differential equations subject to a system of integrability conditions of Laplace-Beltrami type - are investigated in terms of real analysis.*
>
> *It is started with a system of isometric parameters, which cover the surface of a biaxial ellipsoid and an equidistant mapping of the central meridian by a dilatation factor, which is optimised over the zone controlled by the Central Meridian.*
>
> *The determination of the dilatation factor by means of various global deformation measures - such as Cauchy-Green deformation tensor, Euler-Lagrange deformation tensor, simultaneous diagonalisation of a pair of symmetric matrices, and extension of the Tissue deformation portrait as an optimisation problem - is one objective.*
>
> *Investigation of the boundary value problem, the principal deformations and the calculation of the unknown, optimal, dilatation factor is performed to arrive at a minimum of total area distortion"*
>
> (Grafarend, 1995a).

6.3.2 UTM Grid Reference System of Australia

Introduction to Australia

For reference and projection systems of Australia, see also (ISG, 1972: [6.2.1]; NMC, 1986).

Between 1858 and 1966, geodetic surveys in Australia were computed on either a State or regional basis using no fewer than four different ellipsoids and as many as twenty co-ordinate origins. Some of the larger States employed two or more origins simultaneously and various values were adopted for the imperial system units of length then in use.

In April 1965, the National Mapping Council (NMC) adopted *the Australian National Spheroid (ANS)* and recommended that all geodetic surveys in Australia and Papua New Guinea were re-computed and adjusted on the newly defined *Australian Geodetic Datum.* The Australian Geodetic Datum was proclaimed on October 6, 1966.

With the introduction of the Australian Geodetic Datum of 1966 (AGD66) occurring during a period of gradual change (from the imperial system with measurements in yards) to the *metric system,* the opportunity was taken to change to the Universal Transverse Mercator Grid with measurements in metres and was called the *Australian Map Grid* (AMG).

Positioning

In Australia, research in geodesy is undertaken through University establishments. Activities in support of the national geodetic infrastructure are co-ordinated through the State and Federal government agencies by the Intergovernmental Committee on Surveying and Mapping (ICSM). This committee has established a working level Geodesy Group, chaired by AUSLIG, with representatives from State and Federal agencies. The Geodesy Group plans and implements approved national projects (Rüeger, 1994).

Conventional Networks

The labour intensive, traditional methods of geodetic surveys have basically ceased with the advent of GPS. No major field work activities have been undertaken except in some areas of new urban development, particularly in Western Australia and Queensland.

6.3.2.1 Australian Geodetic Datum

Geodetic surveys in Australia are computed on the ANS, based on IAU65, for which the defining parameters are semi-major axis, a=6 378 160 metres, and reciprocal flattening, f^{-1}=298.25 (exactly).

The Australian Geodetic Datum of 1966 (AGD66) Reference Meridian Plane of zero longitude is defined as being parallel to the Bureau International de l'Heure (BIH) mean meridian plane near Greenwich. This in turn gives a value of 149° 00' 18".885 E for the plane contained by the vertical through *the Mount Stromlo Photo Zenith Tube and the CIO.*

The position of the centre of the ellipsoid is defined by the co-ordinates of:

Johnston Geodetic Station:

Geodetic Latitude	25° 56' 54".5515 S
Geodetic Longitude	133° 12' 30".0771 E
Ellipsoidal Height	571.2 metres

The co-ordinates for Johnston Geodetic Station were derived from astronomical observations at 275 stations on the geodetic survey distributed all over Australia. The ellipsoidal height was adopted to be 571.2 metres, which is equal to the height of the station above the geoid as computed by trigonometrical levelling in 1965.

Since 1966, there have been several readjustments of the *National Geodetic Survey.* The latest adjustment, *Geodetic Model of Australia 1982 (GMA82),* has used the most recent observations available, including Satellite Doppler observations, Satellite Laser Ranging (SLR) and Very Long Baseline Interferometry (VLBI).

Recognising the need to convert to a geocentric geodetic datum, the Council resolved that the GMA82 adjustment would be adopted as the first step in the conversion process. The GMA82 adjustment maintained the Australian Geodetic Datum as originally defined.

6.3.2.2 Geocentric Datum of Australia

In 1988, the ICSM resolved to adopt an Earth-Centred, Earth-Fixed (ECEF) orthogonal co-ordinate system to

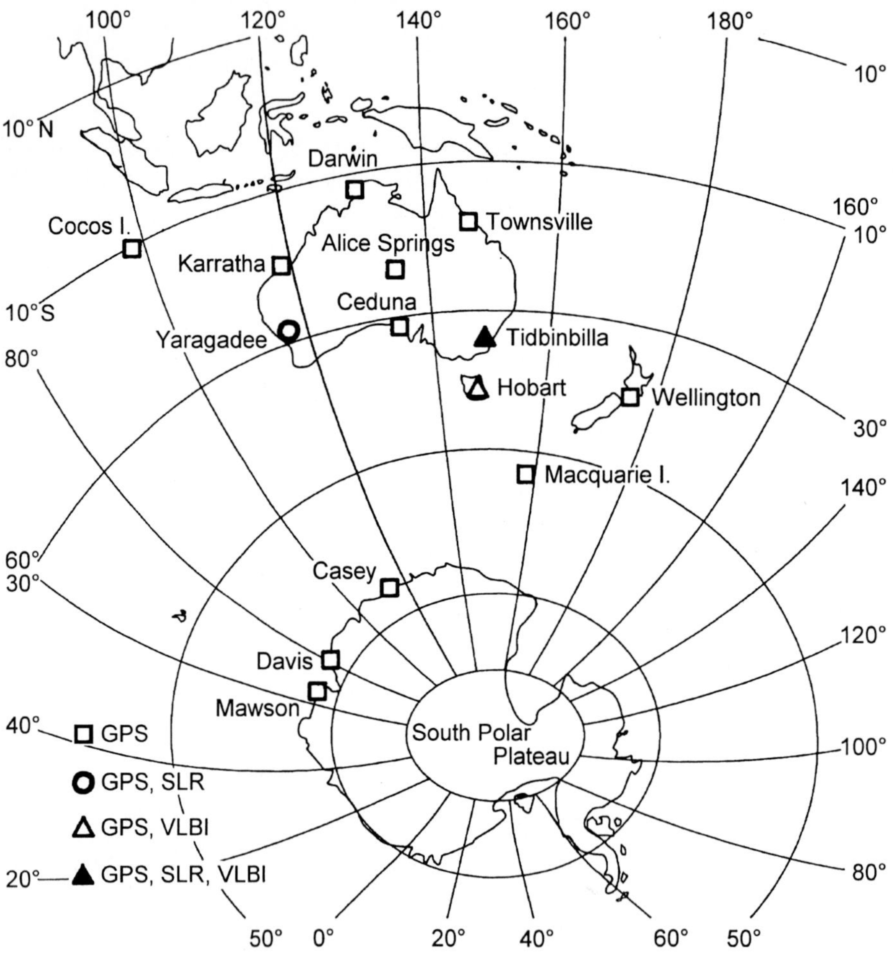

Figure 25: Australian AFN and ARGN sites

be implemented by *the year 2000,* replacing the regional Australian Geodetic Datum of 1966 (AGD66).

The Geocentric Datum of Australia (GDA94) is realised through the estimated positions of the Australian Fiducial Network and the Australian National Network. The GRS80 ellipsoid has been adopted to provide geodetic curvilinear co-ordinates, for which the defining parameters are semi-major axis, a=6 378 137 m, and reciprocal flattening, f^{-1} = 298.25722 21008 827 (AGS, 1995).

Australian Fiducial Network

The Australian Fiducial Network (AFN) consists of eight permanent, continuously operating, GPS receivers on the Australian mainland, including Tasmania. The network was initially observed during the International GPS Service for GeoDynamics (IGS) epoch '92 campaign, July-August, 1992. The AFN, in conjunction with six additional sites beyond the Australian mainland, forms the Australian Regional GPS Network (ARGN). Data from all the ARGN sites are being used for a range of scientific, geodynamic and other projects including integrity monitoring and legal traceability (Figure 25).

The AFN (and hence GDA94) station co-ordinates are based on the ITRF92 at the epoch 1994.0 and are estimated to have a precision of about 2-4 parts in 10^9.

Australian National Network

The Australian National Network (ANN), consisting of seventy-eight GPS campaign points spaced at approximately 500 km intervals across Australia, was observed in the period between 1992 and 1994. The network has an estimated co-ordinate precision of one part in 10^8. The major uncertainties in a position stem from the fact that the stations are not, in general, permanent monuments, but ground marks.

State / Territory GPS Networks

Western Australia will observe their State network during the period 1996-1997, but most other States and Territories have observed an approximately 100 km GPS network, which has been designed to integrate with the ANN and AFN. The GPS baselines have precision of about 0.5 - 1.0 ppm.

The State/Territory networks are connected to the existing conventional Australian Geodetic Datum (AGD66 / AGD84). A geodetic readjustment will be undertaken in 1995, using the AFN and ANN co-ordinates to constrain the terrestrial data and State / Territory GPS networks in terms of GDA94.

6.3.2.3 Australian Map Grid

Co-ordinates on the Australian Map Grid (AMG) are derived from a Transverse Mercator projection of latitudes and longitudes on the Australian Geodetic Datum.

Limits of the Australian Map Grid

The Universal Transverse Mercator Grid system is of world-wide application, but the geodetic reference system in local use varies between regions, so the limits within which the Australian Map Grid is to be used need to be defined. The Australian Map Grid covers the Australian mainland, Tasmania and features close to their shores. The grid does not cover Lord Howe, Macquarie, Cocos, Christmas, Norfolk, Heard and McDonald Islands; neither does it cover the Australian Antarctic Territory [11.3].

The Australian Map Grid corresponds with the Universal Transverse Mercator Grid, as follows:

- zones are 6° wide plus overlapping belts of 80 kilometres at each grid junction
- AMG zones are Zone 49 with CM 111° E to Zone 57 with CM 159° E
- Northings are defined *by adding* 10 000 000 metres to the negative value of N (Southern Hemisphere).

The National Mapping Council has adopted the following definitions for general usage:

- Datum ______ Australian Geodetic Datum
- Ellipsoid ______ Australian National Spheroid
- 1966 Co-ordinate Set ______ AGD66 (geodetic co-ordinates)
  AMG66 (grid co-ordinates)
- 1982 Adjustment / Least Squares Solution ______ GMA82
- 1984 Adopted Co-ordinate Set ______ AGD84 (geodetic co-ordinates)
  AMG84 (grid co-ordinates)
- 1988 Adopted Co-ordinate Set ______ AGD94 (geodetic co-ordinates).

The GMA82 adjustment is a truly ellipsoidal adjustment. Therefore, any observations used in conjunction with the AGD84 co-ordinate set should first be reduced to the Australian National Spheroid using the appropriate geoid-ellipsoid separation values in terms of N = +4.9 metres at Johnston Geodetic Station.

The *test line* located within the limits of the AMG lies in Victoria and connects Tri-gonometrical stations Flinders Peak and Buninyong. This line is nearly 55 kilometres long in an azimuth of 307°, and runs across the boundary between AMG Zones 54 and 55 (Figure 26).

The numerical examples of both test lines, as computed by Kivioja's and Vincenty's Formulae are given, see the next pages.

Geodetic Line - Inverse Problem

Given:	Australian National Spheroid	a= **6 378 160**		f^{-1}= **298.25**
Latitude |[41] φ_1:	**37° 57' 09''.1288 S**	Latitude |[42] - φ_2:		**37° 39' 15''.5571 S**
Longitude λ_1:	**144° 25' 24''.7866 E**	Longitude - λ_2:		**143° 55' 30''.6330 E**
Output:	True Distance - S	True Bearing - α_{1-2}		True Bearing - α_{2-1}

n	Geodesic - S_{1-2}	True Bearing - α_{1-2}	True Bearing - α_{2-1}
1 000	54 972.1599	306° 52' 07''.3416	127° 10' 27''.0839

[41] Sta. "Flinders Peak"

[42] Sta. "Buninyong"

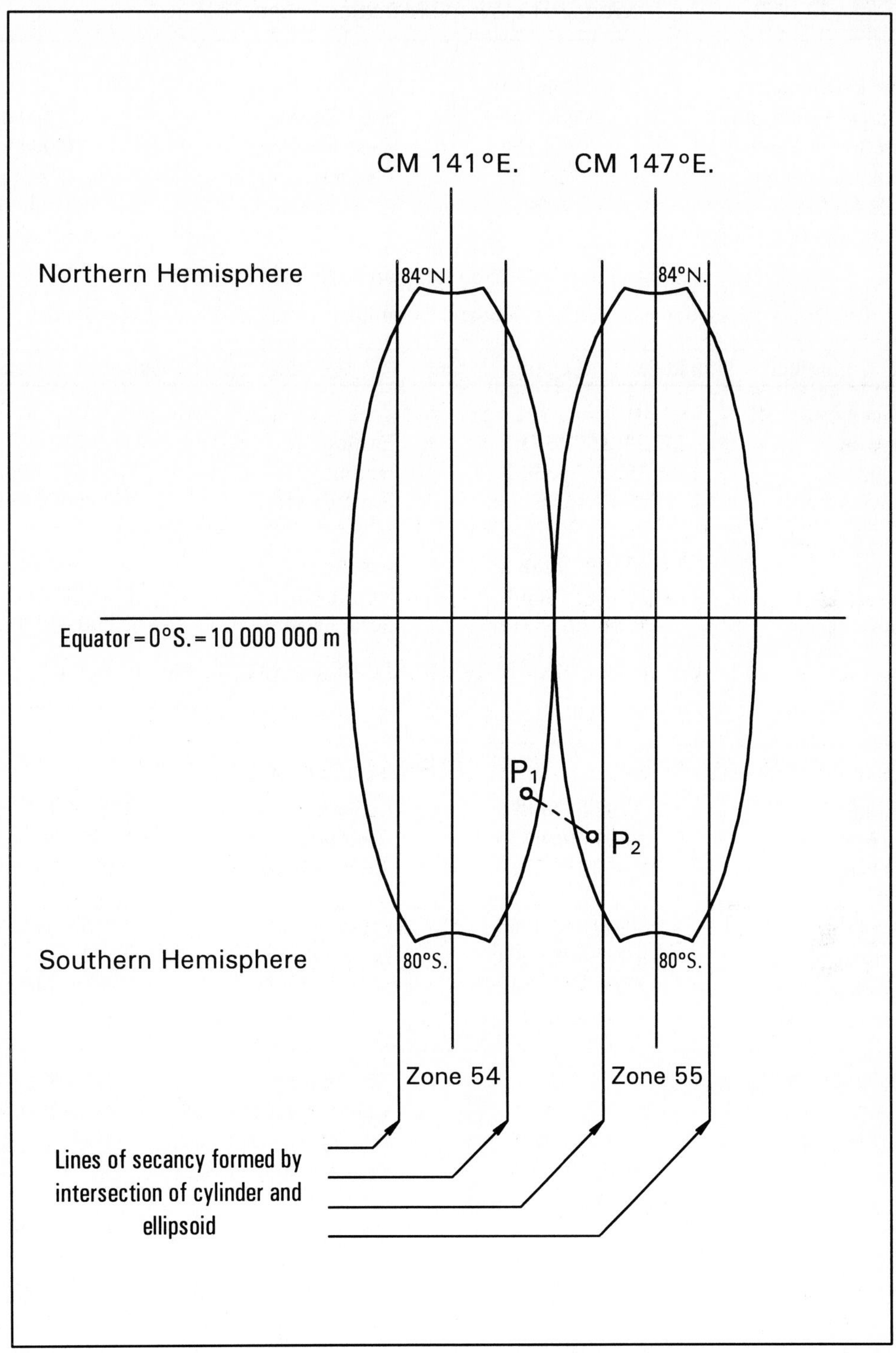

Figure 26: Conversion between UTM Zone 54 and UTM Zone 55

Universal Transverse Mercator - Australia

Zone Parameters:	Zone 54		
Reference Ellipsoid:	ANS 1966	False Easting:	**500 000 m**
Semi-Major axis:	**6 378 160**	False Northing:	**10 000 000 m**
Recipr. Flattening:	**298.25**	Parallel of Origin φ_0:	**0° 00' 00" S**
Scale Factor:	**0.9996**	C.M. of zone λ_0:	**141° 00' 00" E**

Conversion of AGD66 Geographicals to UTM-Grid

The Direct calculation is to convert Geodetic Co-ordinates into GK-Planar Co-ordinates |[43]

Input: Latitude Longitude		Output: Easting Northing Convergence Scale Factor	
Sta. Buninyong			
Latitude:	**37° 39' 15".5571 S**	Easting=	758 053.0896
Longitude:	**143° 55' 30".6330 E**	Northing=	5 828 496.9735
Convergence=	- 1° 47' 16".6717	Scale Factor=	1.000 420 299 351
Sta. Flinders Peak			
Latitude:	**37° 57' 09".1288 S**	Easting=	800 817.4065
Longitude:	**144° 25' 24".7866 E**	Northing=	5 793 905.6504
Convergence=	- 2° 06' 25".5312	Scale Factor=	1.000 714 682 440

Conversion of UTM-Grid to AGD66 Geographicals

The Inverse process is to convert GK-Planar Co-ordinates into Geodetic Co-ordinates:

Input: Latitude Longitude		Output: Easting Northing Convergence Scale Factor	
Easting:	**758 053.0896**	Latitude=	37° 39' 15".5571 S
Northing:	**5 828 496.9735**	Longitude=	143° 55' 30".6330 E
Convergence=	- 1° 47' 16".6717	Scale Factor=	1.000 420 299 203
Easting:	**800 817.4065**	Latitude=	37° 57' 09".1288 S
Northing:	**5 793 905.6504**	Longitude=	144° 25' 24".7866 E
Convergence=	- 2° 06' 25".5312	Scale Factor=	1.000 714 682 041

Ellipsoidal and Grid Calculation (manual)

Using GBD00000.BAS gives:

True Distance	$S_{1\text{-}2}$:	*54 972.1599*
True Azimuth	$\alpha_{1\text{-}2}$:	*127° 10' 27".0839*
True Azimuth	$\alpha_{2\text{-}1}$:	*306° 52' 07".3416*
Grid Distance	$D_{1\text{-}2}$:	55 003.1493
True Distance	$S_{calc\ 1\text{-}2}$:	54 972.1704
See (5.03), using		
Scale Factor	$k_{1\text{-}2}$:	1.000 563 537 799

Grid bearing	$t_{1\text{-}2}$	128° 58' 07".6874
+ convergence	γ_1	- 1° 47' 16".6717
- (t - T)	$\delta_{1\text{-}2}$	+23".9243
True Azimuth	$\alpha_{1\text{-}2}$:	127° 10' 27".0914
Grid bearing	$t_{2\text{-}1}$	308° 58' 07".6874
+ convergence	γ_2	- 2° 06' 25".5312
- (t - T)	$\delta_{2\text{-}1}$	- 25".1751
True Azimuth	$\alpha_{2\text{-}1}$:	306° 52' 07".3313

[43] (NMC, 1986: [4.6])

Universal Transverse Mercator - Australia

Zone Parameters:	Zone 55		
Reference Ellipsoid:	ANS 1966	False Easting:	**500 000 m**
Semi-Major axis:	**6 378 160**	False Northing:	**10 000 000 m**
Recipr. Flattening:	**298.25**	Parallel of Origin φ_0:	**0° 00' 00" S**
Scale Factor:	**0.9996**	C.M. of zone λ_0:	**147° 00' 00" E**

Conversion of AGD66 Geographicals to UTM-Grid

The Direct calculation is to convert Geodetic Co-ordinates into GK-Planar Co-ordinates |[44]

Input: Latitude Longitude	Output: Easting Northing Convergence		Scale Factor
Sta. Buninyong			
Latitude:	**37° 39' 15".5571 S**	Easting=	228 742.0764
Longitude:	**143° 55' 30".6330 E**	Northing=	5 828 074.2081
Convergence=	+1° 52' 46".3554	Scale Factor=	1.000 506 410 833
Sta. Flinders Peak			
Latitude:	**37° 57' 09".1288 S**	Easting=	273 629.4358
Longitude:	**144° 25' 24".7866 E**	Northing=	5 796 305.2356
Convergence=	+1° 35' 06".7564	Scale Factor=	1.000 231 177 620

Conversion of UTM-Grid to AGD66 Geographicals

The Inverse process is to convert GK-Planar Co-ordinates into Geodetic Co-ordinates:

Input: Latitude Longitude	Output: Easting Northing Convergence		Scale Factor
Easting:	**228 742.0764**	Latitude=	37° 39' 15".5571 S
Northing:	**5 828 074.2081**	Longitude=	143° 55' 30".6330 E
Convergence=	+1° 52' 46".3555	Scale Factor=	1.000 506 410 633
Easting:	**273 629.4358**	Latitude=	37° 57' 09".1288 S
Northing:	**5 796 305.2356**	Longitude=	144° 25' 24".7866 E
Convergence=	+1° 35' 06".7564	Scale Factor=	1. 000 231 177 548

Ellipsoidal and Grid Calculation (manual)

Using GBD00000.BAS gives:			Grid bearing	t_{1-2}	125° 17' 20".0532
			+ convergence	γ_1	+1° 52' 46".3554
True Distance	S_{1-2}:	*54 972.1599*	- (t - T)	δ_{1-2}	- 20".6818
True Azimuth	α_{1-2}:	*127° 10' 27".0839*	True Azimuth	α_{1-2}:	127° 10' 27".0904
True Azimuth	α_{2-1}:	*306° 52' 07".3416*			
			Grid bearing	t_{2-1}	305° 17' 20".0532
Grid Distance	D_{1-2}:	54 992.2053	+ convergence	γ_2	+ 1° 35' 06".7564
True Distance	$S_{calc\ 1-2}$:	54 972.1497	- (t - T)	δ_{2-1}	+19".4756
See (5.03), using			True Azimuth	α_{2-1}:	306° 52' 07".3340
Scale Factor	k_{1-2}:	1.000 364 832 614			

[44] (NMC, 1986: [5.4])

6.3.2.4 The Geodetic Datum for Australian Offshore Islands and External Territories

In September 1980, the National Mapping Council adopted the US DoD World Geodetic System of 1972 (WGS72) as the geodetic datum for all mapping of Australian offshore islands and External Territories lying outside the limits of the Australian Map Grid (NMC, 1986).

The National Mapping Council also decided that rectangular co-ordinates shall be obtained from the WGS72 co-ordinates by the UTM projection north of latitude 80° S and by the Polar Stereographic Grid system south of latitude 79°30' S.

The *Australian National Spheroid* and the *WGS72 ellipsoid* differ in size, shape and orientation with a resulting vector difference of approximately *195 metres* between their respective centres. Due to these dissimilarities, the vector difference between the related co-ordinate reference systems varies over the area in which the Australian National Spheroid is to be used, but is generally of the order of 200 metres.

The World Geodetic System of 1972 Datum

The *US DoD World Geodetic System of 1972 (WGS72) datum* is a geocentric co-ordinate reference system in world-wide use. The WGS72 mean earth ellipsoid is defined with semi-major axis, a=6 378 135 metres and reciprocal flattening, f^{-1}=298.26 [3.5.4].

Use of the UTM-Grid System in conjunction with the WGS72 Datum

The characteristics of the UTM-grid system, see [6.3], are identical with those already given for the AMG in [6.3.2.3]. Grid zone numbers correspond with the AMG and are numbered eastwards from Zone 1 with CM 177° West to Zone 60 with CM 177° East.

The test line located outside the limits of the AMG and in the vicinity of the Australian Territory of Norfolk Island, has one of its terminals coinciding with trigonometrical station "M" on Norfolk Island. The other terminal is a completely fictitious trigonometrical station, designated "X", lying nearly 57 kilometres in an azimuth of 69°37'50" from station "M". This line runs across the boundary between UTM-grid Zones 58 and 59.

The numerical examples of both test lines, as computed by Kivioja's and Vincenty's formulae, are given, see (NMC, 1986: [5.8]).

Geodetic Line - Direct Problem

Given:	Ellipsoid WGS72	a= **6 378 135**	f^{-1}= **298.26**
Latitude [45] φ_1:	**29° 03' 23".1530 S**	True Bearing - $\alpha_{1\text{-}2}$:	**69° 37' 49".9812**
Longitude λ_1:	**167° 57' 06".6320 E**	True Distance - S:	**56 959.8244**
Output:	Latitude - φ_2	Longitude - λ_2	True Bearing - $\alpha_{2\text{-}1}$

n	Latitude - φ_2	Longitude - λ_2	True Bearing - $\alpha_{2\text{-}1}$
1 000	28° 52' 35".1710 S	168° 29' 57".1520 E	249° 21' 55".6572

[45] Station "M"

Universal Transverse Mercator - Australia

Zone Parameters:	Zone 58		
Reference Ellipsoid:	WGS72	False Easting:	**500 000 m**
Semi-Major axis:	**6 378 135**	False Northing:	**10 000 000 m**
Recipr. Flattening:	**298.26**	Parallel of Origin φ_0:	**0° 00' 00" S**
Scale Factor:	**0.9996**	C.M. of zone λ_0:	**165° 00' 00" E**

Conversion of WGS72 Geographicals to UTM-Grid

The Direct calculation is to convert Geodetic Co-ordinates into GK-Planar Co-ordinates:

Input: Latitude Longitude Output: Easting Northing Convergence Scale Factor

Sta. "M"			
Latitude:	**29° 03' 23".1530 S**	Easting=	787 420.4874
Longitude:	**167° 57' 06".6320 E**	Northing=	6 782 165.2011
Convergence=	- 1° 26' 04".5906	Scale Factor=	1.000 619 550 391
Sta. "X"			
Latitude:	**28° 52' 35".1710 S**	Easting=	841 341.1579
Longitude:	**168° 29' 57".1520 E**	Northing=	6 800 667.2098
Convergence=	- 1° 41' 29".3410	Scale Factor=	1.001 038 122 238

Conversion of UTM-Grid to WGS72 Geographicals

The Inverse process is to convert GK-Planar Co-ordinates into Geodetic Co-ordinates:

Input: Latitude Longitude Output: Easting Northing Convergence Scale Factor

Easting:	**787 420.4874**	Latitude=	29° 03' 23".1530 S
Northing:	**6 782 165.2011**	Longitude=	167° 57' 06".6320 E
Convergence=	- 1° 26' 04".5907	Scale Factor=	1.000 619 550 564
Easting:	**841 341.1579**	Latitude=	28° 52' 35".1710 S
Northing:	**6 800 667.2098**	Longitude=	168° 29' 57".1520 E
Convergence=	- 1° 41' 29".3412	Scale Factor=	1.001 038 122 747

Ellipsoidal and Grid Calculation (manual)

Using BDG00000.BAS gives:

True Distance	S_{1-2}:	*56 959.8244*
True Azimuth	α_{1-2}:	*69° 37' 49".9812*
True Azimuth	α_{2-1}:	*249° 21' 55".6572*
Grid Distance	D_{1-2}:	57 006.6929
True Distance	$S_{calc\ 1-2}$:	56 959.8085
See (5.03), using		
Scale Factor	k_{1-2}:	1.000 823 113 481

Grid bearing	t_{1-2}	71° 03' 40".2081
+ convergence	γ_1	- 1° 26' 04".5906
- (t - T)	δ_{1-2}	- 14".3764
True Azimuth	α_{1-2}:	69° 37' 49".9939
Grid bearing	t_{2-1}	251° 03' 40".2081
+ convergence	γ_2	- 1° 41' 29".3410
- (t - T)	δ_{2-1}	+15".2211
True Azimuth	α_{2-1}:	249° 21' 55".6460

6.3.3 UTM Grid Reference System of Belgium

European Datum |[46]

Following World War II, survey data covering the central area of mainland Europe was used in a united adjustment to provide *a common datum for military mapping* in Europe. The completed work was known as the European Datum of 1950 (ED50).

In 1979, the RETrig Commission of the IAG defined and implemented ED79, and after incorporating a number of Transit (Doppler) derived positions, ED87. Since 1989, the successor of RETrig Commission, *EUREF*, has been involved in the definition and implementation of a new 3-D reference system and datum consistent with 3-D geodetic techniques such as GPS for the Eurasian tectonic plate: EUREF89. See [6.3.1], Footnotes About the UTM.

Universal Transverse Mercator - Belgium

Zone Parameters:	Zone 31		
Reference Ellipsoid:	International 1924	False Easting:	**500 000 m**
Semi-Major axis:	**6 378 388**	False Northing:	**0 m**
Recipr. Flattening:	**297**	Parallel of Origin φ_0:	**0° 00' 00" N**
Scale Factor:	**0.9996**	C.M. of zone λ_0:	**3° 00' 00" E**

Conversion of ED50 Geographicals to UTM-Grid

The Direct calculation is to convert Geodetic Co-ordinates into GK-Planar Co-ordinates:

Input: Latitude Longitude		Output: Easting Northing Convergence	Scale Factor
Latitude:	**50° 40' 46".4610 N**	Easting=	698 339.6852
Longitude:	**5° 48' 26".5330 E**	Northing=	5 618 067.9186
Convergence=	+ 2° 10' 21".0919	Scale Factor=	1.000 083 059 912
Latitude:	**50° 47' 25".7956 N**	Easting=	706 941.9111
Longitude:	**5° 56' 09".8964 E**	Northing=	5 630 752.8418
Convergence=	+ 2° 16' 32".8635	Scale Factor=	1.000 125 860 308

Conversion of UTM-Grid to ED50 Geographicals

The Inverse process is to convert GK-Planar Co-ordinates into Geodetic Co-ordinates:

Input: Latitude Longitude		Output: Easting Northing Convergence	Scale Factor
Easting:	**698 339.6852**	Latitude=	50° 40' 46".4610 N
Northing:	**5 618 067.9186**	Longitude=	5° 48' 26".5330 E
Convergence=	+ 2° 10' 21".0919	Scale Factor=	1.000 083 059 753
Easting:	**706 941.9111**	Latitude=	50° 47' 25".7956 N
Northing:	**5 630 752.8418**	Longitude=	5° 56' 09".8964 E
Convergence=	+ 2° 16' 32".8635	Scale Factor=	1.000 125 860 101

[46] For Belgium, see also *Lambert1972,* [7.2.3]

Universal Transverse Mercator - Belgium

Zone Parameters:	Zone 32		
Reference Ellipsoid:	International 1924	False Easting:	**500 000 m**
Semi-Major axis:	**6 378 388**	False Northing:	**0 m**
Recipr. Flattening:	**297**	Parallel of Origin φ_0:	**0° 00' 00" N**
Scale Factor:	**0.9996**	C.M. of zone λ_0:	**9° 00' 00" E**

Conversion of ED50 Geographicals to UTM-Grid

The Direct calculation is to convert Geodetic Co-ordinates into GK-Planar Co-ordinates:

Input: Latitude Longitude		Output: Easting Northing Convergence	Scale Factor
Latitude [47]:	**50° 40' 46".4610 N**	Easting=	274 447.0841
Longitude:	**5° 48' 26".5330 E**	Northing=	5 619 171.2424
Convergence=	- 2° 28' 15".2366	Scale Factor=	1.000 224 723 895
Latitude:	**50° 47' 25".7956 N**	Easting=	284 049.6440
Longitude:	**5° 56' 09".8964 E**	Northing=	5 631 118.6546
Convergence=	- 2° 22' 29".8443	Scale Factor=	1.000 172 643 657

Conversion of UTM-Grid to ED50 Geographicals

The Inverse process is to convert GK-Planar Co-ordinates into Geodetic Co-ordinates:

Input: Latitude Longitude		Output: Easting Northing Convergence	Scale Factor
Easting:	**274 447.0841**	Latitude=	50° 40' 46".4610 N
Northing:	**5 619 171.2424**	Longitude=	5° 48' 26".5330 E
Convergence=	- 2° 28' 15".2366	Scale Factor=	1.000 224 723 551
Easting:	**284 049.6440**	Latitude=	50° 47' 25".7956 N
Northing:	**5 631 118.6546**	Longitude=	5° 56' 09".8964 E
Convergence=	- 2° 22' 29".8443	Scale Factor=	1.000 172 643 390

[47] (Herrewegen, 1989)

6.3.4 UTM Grid Reference System of East Africa

Zone Parameters:	Zone 36		
Reference Ellipsoid:	Clarke 1880G	False Easting:	**500 000 m**
Semi-Major axis:	**6 378 249.145**	False Northing:	**10 000 000 m**
Recipr. Flattening:	**293.465000**	Parallel of Origin φ_o:	**0° 00' 00" S**
Scale Factor:	**0.9996**	C.M. of zone λ_o:	**33° 00' 00" E**

Conversion of Arc50 Geographicals to UTM-Grid

The Direct calculation is to convert Geodetic Co-ordinates into GK-Planar Co-ordinates:

Input:	Latitude	Longitude	Output:	Easting	Northing	Convergence	Scale Factor [48]

Latitude	φ_A:	**15° 52' 29".9092 S**	Easting	E_A=	746 793.6730
Longitude	λ_A:	**35° 18' 16".7558 E**	Northing	N_A=	8 243 690.6272
Convergence	γ_A=	- 0° 37' 50".6420	Scale Factor	k_A=	1.000 353 332 146
Latitude	φ_B:	**15° 39' 22".5964 S**	Easting	E_B=	756 612.7015
Longitude	λ_B:	**35° 23' 37".4538 E**	Northing	N_B=	8 267 790.3423
Convergence	γ_B=	- 0° 38' 46".8347	Scale Factor	k_B=	1.000 414 499 909
Latitude	φ_C:	**15° 44' 56".7567 S**	Easting	E_C=	764 991.4110
Longitude	λ_C:	**35° 28' 22".7070 E**	Northing	N_C=	8 257 418.2729
Convergence	γ_C=	- 0° 40' 17".8375	Scale Factor	k_C=	1.000 468 554 912

Conversion of UTM-Grid to Arc50 Geographicals

The Inverse process is to convert GK-Planar Co-ordinates into Geodetic Co-ordinates:

Input:	Latitude	Longitude	Output:	Easting	Northing	Convergence	Scale Factor

Easting	E_A:	**746 793.6730**	Latitude	φ_A=	15° 52' 29".9092 S
Northing	N_A:	**8 243 690.6272**	Longitude	λ_A=	35° 18' 16".7558 E
Convergence	γ_A=	- 0° 37' 50".6421	Scale Factor	k_A=	1.000 353 332 375
Easting	E_B:	**756 612.7015**	Latitude	φ_B=	15° 39' 22".5964 S
Northing	N_B:	**8 267 790.3423**	Longitude	λ_B=	35° 23' 37".4538 E
Convergence	γ_B=	- 0° 38' 46".8347	Scale Factor	k_B=	1.000 414 500 200
Easting	E_C:	**764 991.4110**	Latitude	φ_C=	15° 44' 56".7567 S
Northing	N_C:	**8 257 418.2729**	Longitude	λ_C=	35° 28' 22".7070 E
Convergence	γ_C=	- 0° 40' 17".8376	Scale Factor	k_C=	1.000 468 555 264

[48] (Allan, 1968)

Universal Transverse Mercator - East Africa

(t - T) Corrections

Input:	Easting and Northing	Output:	Arc-to-chord or (t - T) Correction	
E_A:	**746 793.6730**	(t - T)	δ_{A-B}=	- 15".3713
N_A:	**8 243 690.6272**	(t - T)	δ_{B-A}=	+15".5723
E_B:	**756 612.7015**	(t - T)	δ_{A-C}=	- 8".8534
N_B:	**8 267 790.3423**	(t - T)	δ_{C-A}=	+9".0656
E_C:	**764 991.4110**	(t - T)	δ_{B-C}=	+6".8624
N_C:	**8 257 418.2729**	(t - T)	δ_{B-C}=	- 6".9362

Ellipsoidal and Grid Calculation (manual)

Grid bearing	t_{A-B}	22° 10' 03".4404	Grid bearing	t_{A-C}	52° 58' 14".0102
+ convergence	γ_A	- 0° 37' 50".6420	+ convergence	γ_A	- 0° 37' 50".6420
- (t - T)	δ_{A-B}	- 0° 00' 15".3713	- (t - T)	δ_{A-C}	- 0° 00' 08".8534
True Azimuth	α_{A-B}:	21° 32' 28".1697	True Azimuth	α_{A-C}:	52° 20' 32".2216
Grid bearing	t_{B-A}	202° 10' 03".4404	Grid bearing	t_{B-C}	141° 04' 05".5733
+ convergence	γ_B	- 0° 38' 46".8347	+ convergence	γ_B	- 0° 38' 46".8347
- (t - T)	δ_{B-A}	+0° 00' 15".5723	- (t - T)	δ_{B-C}	+ 0° 00' 06".8624
True Azimuth	α_{B-A}:	201° 31' 01".0334	True Azimuth	α_{B-C}:	140° 25' 11".8762
Grid bearing	t_{C-A}	232° 58' 14".0102	Grid bearing	t_{C-B}	321° 04' 05".5733
+ convergence	γ_C	- 0° 40' 17".8375	+ convergence	γ_C	- 0° 40' 17".8375
- (t - T)	δ_{C-A}	+0° 00' 09".0656	- (t - T)	δ_{C-B}	- 0° 00' 06".9362
True Azimuth	α_{C-A}:	232° 17' 47".1071	True Azimuth	α_{C-B}:	320° 23' 54".6720

6.3.5 UTM Grid Reference System of North Africa

Zone Parameters:	Zone 32 |[49]		
Reference Ellipsoid:	Clarke 1880DoD	False Easting:	**500 000 m**
Semi-Major axis:	**6 378 249.145**	False Northing:	**0 m**
Recipr. Flattening:	**293.465**	Parallel of Origin φ_o:	**0^g.00 00 N**
Scale Factor:	**0.9996**	C.M. of zone λ_o:	**10^g.00 00 E**

Conversion of Clarke 1880DoD Geographicals to UTM-Grid

The Direct calculation is to convert Geodetic Co-ordinates into GK-Planar Co-ordinates: 400^g System

Input: Latitude Longitude		Output: Easting Northing Convergence	Scale Factor
Latitude:	**40^g.98 17 0099 N**	Easting=	378 451.1731
Longitude:	**8^g.48 45 3370 E**	Northing=	4 082 529.0484
Convergence=	- 0^g.90 96 7961	Scale Factor=	0.999 782 007 143
Latitude:	**40^g.75 76 6179 N**	Easting=	413 396.9069
Longitude:	**8^g.92 30 4674 E**	Northing=	4 059 731.7642
Convergence=	- 0^g.64 33 8147	Scale Factor=	0.999 692 399 044

Conversion of UTM-Grid to Clarke 1880DoD Geographicals

The Inverse process is to convert GK-Planar Co-ordinates into Geodetic Co-ordinates: 400^g System

Input: Latitude Longitude		Output: Easting Northing Convergence	Scale Factor
Easting:	**378 451.1731**	Latitude=	40^g.98 17 0099 N
Northing:	**4 082 529.0484**	Longitude=	8^g.48 45 3370 E
Convergence=	- 0^g.90 96 7961	Scale Factor=	0.999 782 007 142
Easting:	**413 396.9069**	Latitude=	40^g.75 76 6179 N
Northing:	**4 059 731.7642**	Longitude=	8^g.92 30 4674 E
Convergence=	- 0^g.64 33 8147	Scale Factor=	0.999 692 399 043

[49] (Levallois, 1970)

6.3.6 UTM Grid Reference System of Norway

UTM zones of Norway	Zone 32, 33, 34, 35 and 36 (Harsson, 1995, 1996).		
Exemplified zones	*UTM Zone 32, 36, see also Norway,* [6.2.8], (Figure 23).		
Reference Ellipsoid:	International 1924	False Easting:	**500 000 m**
Semi-Major axis:	**6 378 388**	False Northing:	**0 m**
Reciprocal Flattening:	**297**	Parallel of Origin φ_o:	**0° 00' 00" N**
Scale Factor:	**0.9996**	CM Zone 32 λ_o:	**9° 00' 00" E**
		CM Zone 36 λ_o:	**33° 00' 00" E**

Conversion of ED50 Geographicals to UTM-Grid

The Direct calculation is to convert Geodetic Co-ordinates into GK-Planar Co-ordinates:

Input: Latitude	Longitude	Output: Easting	Northing Convergence	Scale Factor
Sta. Eigeberg	Zone 32			
Latitude	φ_1: **58° 51' 05".1414 N**	Easting	E_1=	303 121.8047
Longitude	λ_1: **5° 35' 13".5894 E**	Northing	N_1=	6 528 679.2902
Convergence	γ_1= - 2° 55' 18".4102	Scale Factor	k_1=	1.000 075 105 368
Sta. Vigra	Zone 32			
Latitude	φ_2: **62° 33' 14".5385 N**	Easting	E_2=	349 818.8135
Longitude	λ_2: **6° 04' 44".2295 E**	Northing	N_2=	6 939 469.4355
Convergence	γ_2= - 2° 35' 33".9025	Scale Factor	k_2=	0.999 876 241 624
Sta. SK Tårn	Zone 32			
Latitude	φ_3: **60° 08' 38".3586 N**	Easting	E_3=	569 445.8888
Longitude	λ_3: **10° 15' 01".7040 E**	Northing	N_3=	6 668 260.5243
Convergence	γ_3= + 1° 05' 04".3887	Scale Factor	k_3=	0.999 659 094 216
Sta. Vassfjell	Zone 32			
Latitude	φ_4: **63° 15' 44".3780 N**	Easting	E_4=	568 176.4804
Longitude	λ_4: **10° 21' 29".4183 E**	Northing	N_4=	7 015 710.6519
Convergence	γ_4= + 1° 12' 46".7872	Scale Factor	k_4=	0.999 656 918 614
Sta. Bodø	Zone 32			
Latitude	φ_5: **67° 17' 02".0750 N**	Easting	E_5=	732 692.7500
Longitude	λ_5: **14° 24' 18".2909 E**	Northing	N_5=	7 473 356.3269
Convergence	γ_5= + 4° 59' 16".8613	Scale Factor	k_5=	1.000 262 638 121
Sta. Tromsø	Zone 32			
Latitude	φ_6: **69° 39' 44".3645 N**	Easting	E_6=	884 176.8550
Longitude	λ_6: **18° 56' 30".7636 E**	Northing	N_6=	7 759 772.5932
Convergence	γ_6= + 9° 20' 00".2166	Scale Factor	k_6=	1.001 405 841 014
Sta. Domen	Zone 36			
Latitude	φ_6: **70° 20' 00".7516 N**	Easting	E_6=	426 211.9130
Longitude	λ_6: **31° 02' 07".0631 E**	Northing	N_6=	7 804 469.4730
Convergence	γ_6= - 1° 51' 00".6527	Scale Factor	k_6=	0.999 666 593 930

6.4 Round-Trip Errors

6.4.1 Latitude and Longitude Round-Trip Errors

The "round-trip errors" are the differences in degrees between the starting and ending co-ordinates, illustrated in Figure 27 and Figure 28 for a longitudinal distance 0° to 9° E from the central meridian, and are calculated as follows: latitudes and longitudes are converted to the eastings and northings, which are converted back to latitudes and longitudes.

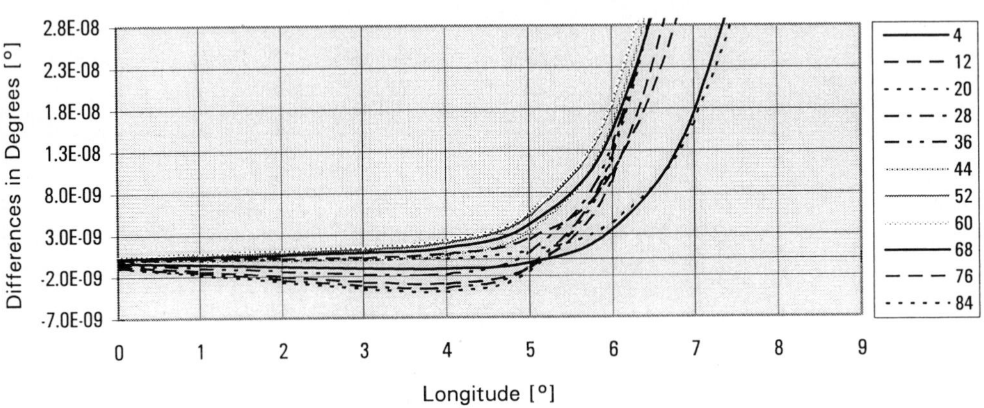

Figure 27: GK round-trip error of latitudes

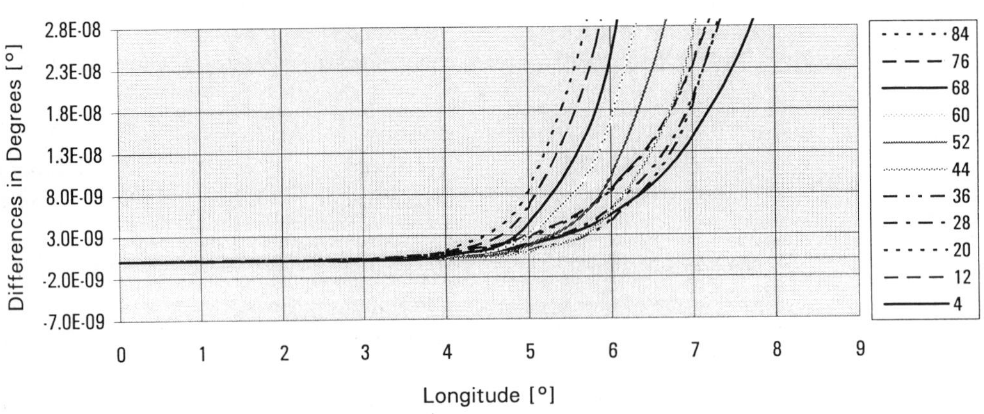

Figure 28: GK round-trip error of longitudes

6.4.2 Easting and Northing Round-Trip Errors

The "round-trip errors" are the differences in metres between the starting and ending co-ordinates, illustrated in Figure 29 and Figure 30 for a longitudinal distance 0° to 9° E from the central meridian, and are calculated as follows: the eastings and northings are converted to latitudes and longitudes, which are converted back to the eastings and northings.

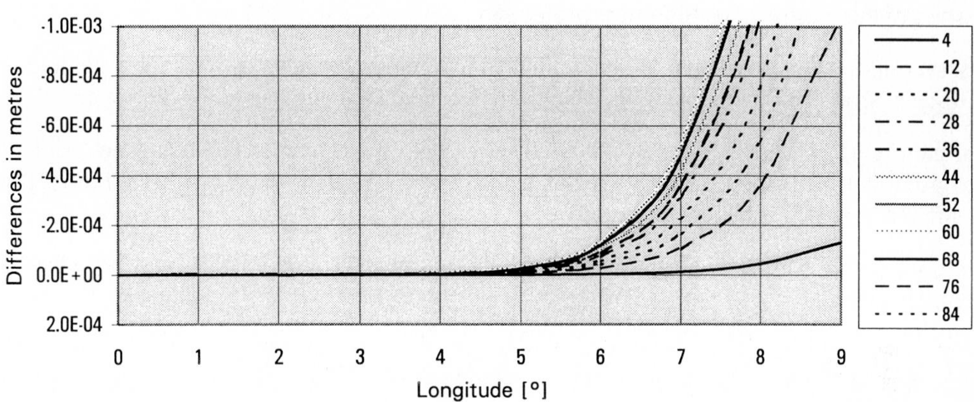

Figure 29: GK round-trip error of eastings

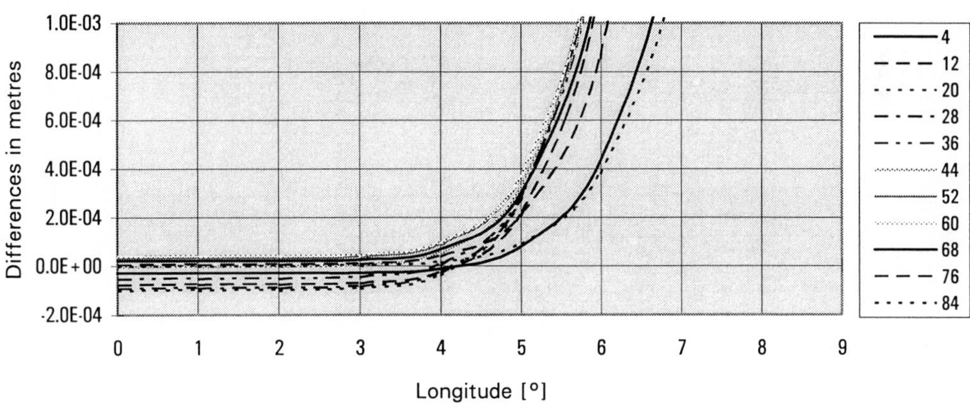

Figure 30: GK round-trip error of northings

Note

Verification of historical data and information is not a function of a data-processing system. Co-ordinates of historical data points must be taken at face value, with the realisation that such co-ordinates could be significantly in error. In practice, exact co-ordinates in any reference system are not obtained. Starting with co-ordinates known to be correct in the reference system is the crux of the problem

(Floyd, 1985).

Boundaries of the Transverse Mercator projection

The Transverse Mercator projection should be limited to a region bounded by a maximum latitude and a longitudinal distance from the central meridian, which will depend on the purpose of the projection.

The algorithm used in the program GK000000.BAS will easily accommodate wide zones with submillimetre accuracy. An excellent example of such a region is the United Kingdom, see Great Britain [6.2.5], with a longitudinal difference of 11° width, approximately.

Such accuracy is unimportant, but can serve as a standard by which approximations may be judged (Bomford, 1977).

7. Lambert's Conformal Conical Projection

For mapping an area of considerable extent in longitude, a geometrical system based on the notion of an east-west centre line can be used. Johann Heinrich Lambert devised and published such a projection in “Beiträge zum Gebrauche der Mathematik und deren Anwendung” in the year 1772. The Lambert projection remained almost unknown until the beginning of the World War I. The system was introduced and has been brought to conspicuous attention by the French Military Survey ("Le Service Géographique de l'Armée") under the name "Quadrillage kilomètrique système Lambert" (Tardi, 1934).

The grid of the military mapping system presented various properties useful for the rapid computation of distances and azimuths in military operations. These properties are (Adams, 1921):

- uniquely defined by the property of conformality
- an unique reference datum
- using the centesimal system instead of the sexagesimal system
- grid system of rectangular co-ordinates.

7.1 Quadrillage Kilomètrique Système Lambert

A quadrillage is a grid system of squares determined by the rectangular co-ordinates of the projection. Referring to one origin, the grid system is projected over the whole area of each zone of the original projection. So every point on that map is fixed both regarding its position in a given quadrilateral as well as to its position in curvilinear co-ordinates. Quick computations of distances are possible between points whose grid co-ordinates are given. Determinations of the azimuth of lines joining any two points within artillery range are of great value to military operations.

The French Institut Géographique National (IGN) divides the circumference of the circle into the centesimal system (400^g grades) instead of the sexagesimal system (360° degrees). During World War I and II the essential tables were produced for the conversion of degrees, minutes, and seconds into grades, the conversion of miles, feet, and inches into the metric system, and vice versa for the Allied Military forces. In the Lambert projection for the map of France employed by the Allied Forces in their military operations, the maximum scale errors are practically negligible, while the angles measured on the map are practically equal to those in the field [7.2; 10.1.1].

Since 1947 the system has been superseded by the Universal Transverse Mercator (UTM) for world-wide military mapping, UTM was developed by the US DoD National Imagery and Mapping Agency (NIMA) Topographic Center (SDHG), Washington DC [6.3].

Lambert Conformal Conical projection

The Lambert Conformal Conical (LCC) projection is often used where the area of concern lies with its predominating, longer dimension in an east-west direction. The LCC projection is not suited for mapping areas of very wide latitudes.

Using the defining parameters of an ellipsoid, the fundamental parameters are the Central Parallel (CP), central or Reference Meridian (RM) and the mapping radius of the Equator. In the LCC projection, all meridians are represented by straight lines that meet a common centre S outside the limits of the map projection.

All parallels are sections of concentric circles centred at the point S of intersection of the meridians. Accordingly, their radii are functions of latitude. All elements retain their authentic forms and meridians and parallels intersect at right angles. With it, the angles formed by any two lines on the earth's surface are correctly represented on this projection.

The algorithm LCC00000.BAS handles three cases for defining a projection:

- secant projection with upper and lower Standard Parallels known
- secant projection with Standard Parallel and its imposed Scale Factor known
- tangent projection with one Standard Parallel and its scale factor of unity known (Figure 31).

The Reference Meridian (RM) orients the cone with respect to the ellipsoid selected. The latitude of the standard parallel and the longitude of the RM would be chosen to run through the centre of the zone to be computed. Consequently, the geometry of a conical projection is described by the value of the constant of the cone which is defined either by a standard parallel with a desired scale factor or two (upper and lower) standard parallels.

Usually, a scale factor of unity can be applied to make the scale true along two selected parallels. For the equal distribution of the scale factors, the standard parallels are placed at distances from its northern and southern limits of the projection, each equal to $^1/_6$ of the total meridional distance to be portrayed. The scale will be less than the nominal value between these parallels.

Moving these parallels close together results in the LCC with one standard parallel. In that case a Reference Meridian and a Standard Parallel are assumed with a cone tangent along the standard parallel.

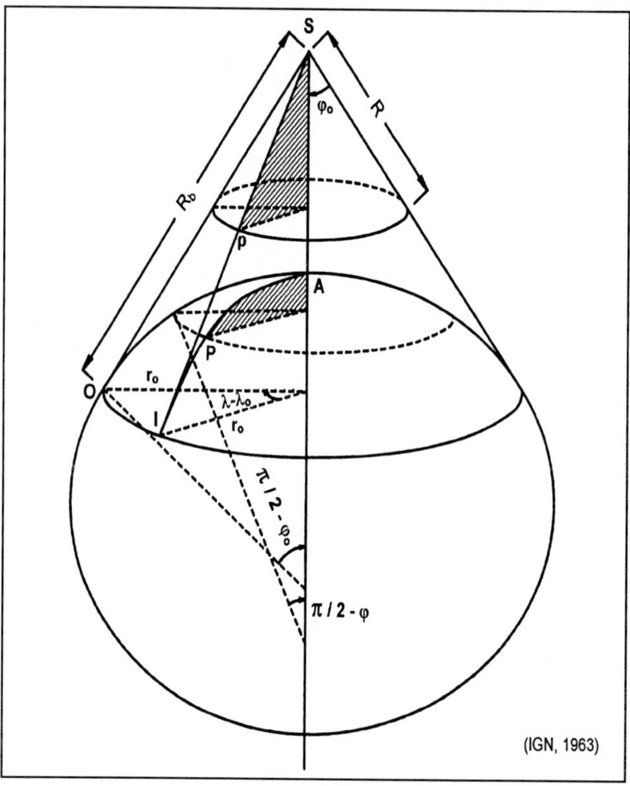

Figure 31: Lambert's conformal conical projection

Secant Conical Projection

The LCC projection employs a cone intersecting the ellipsoid at two parallels known as the upper and lower standard lines or parallels - automecoic - for the section to be represented. Bringing them closer together in order to have greater accuracies in the centre of the map at the expense of the upper and lower border areas may be advisable in some localities, or for special reasons. Instead of two standard parallels - the lines of secancy - some computer programs require the input of a scale factor ($k_0<1$) and one standard parallel, which lies close to the midpoint between these standard parallels (IGN, 1994). For further reading, see (Otero, 1990).

Tangent Conical Projection

This conical projection employs a standard parallel that represents a line of tangency between the cone and the surface of the ellipsoid. The projection parameters required are the reference meridian, a standard parallel, a scale factor of unity and defining parameters of the ellipsoid selected (Pearson II, 1990).

Scale Factor

A reference meridian and a standard parallel are assumed with a cone either tangent along the standard parallel ($k_0=1$) or secants along two standard parallels - with an imposed scale factor, $k_0<1$. Nevertheless, on two selected parallels, arcs of longitude are usually represented to a scale factor $k_0=1$. Between these parallels the scale factor k is <1 and outside them the scale factor k is >1. See [7.1.1].
Remember that in conical projections, the scale errors vary increasingly with the range of latitude north or south of the standard parallels.

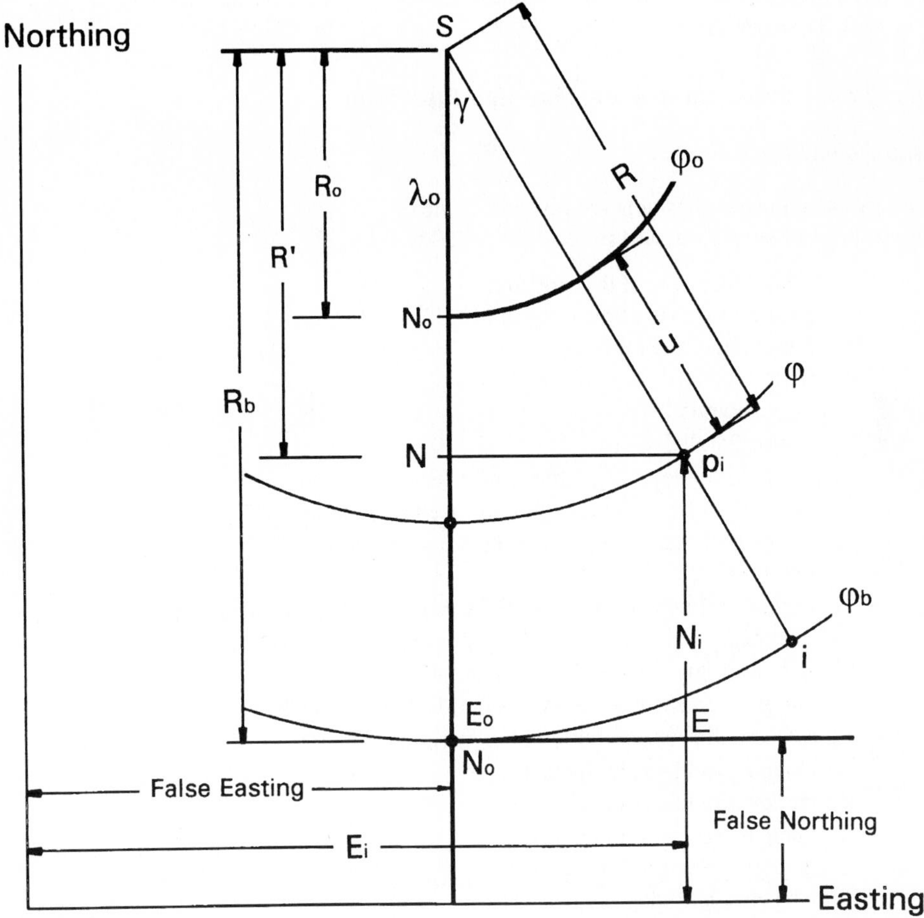

Figure 32: Lambert's grid system

Origins

The true origin is at the intersection of the Reference Meridian (RM) with the apex of the cone.

The false grid origin is specified by the intersection of the RM longitude, λ_o, and the latitude of the Standard Parallel, φ_b. Adding the false northing to the false grid origin furnishes the true grid origin, which lies on the RM. A false easting and a false northing may have been assigned to the false grid origin allowing the zone to be extended farther southwest to maintain positive-valued co-ordinates (Adams, 1921).

(t - T) Correction

The arc-to-chord or (t - T) correction requires the input of two sets of grid co-ordinates (E_1, N_1, E_2, N_2) and the Northing constant, N_o of the true projection origin. Using the latitude of CP (φ_b) and the longitude of RM (λ_o), program LCC00000.BAS furnishes that value N_o.

The Lambert conformal conical projection has unquestionably superior merits for areas of extended longitudes. Consequently, many LCC grids exist, e.g. in Central and North America, Europe, Asia, and North Africa. Each is designed for a particular state or country of the world with its own set of constants. A few Lambert conformal cases follow this introduction (Adams, 1921; Bomford, 1971; Claire, 1968; Geodetic Glossary, 1986; IGN, 1994; Otero, 1990; Tardi, 1934).

7.1.1 Lambert's Conformal Conical Mapping Equations

Symbols and Definitions.

The equations for the ellipsoid constants are given in Table 6.
All angles are expressed in *radians.* See program LCC00000.BAS, [11.7.6].

a	semi-major axis of the ellipsoid
b	semi-minor axis of the ellipsoid
f	flattening of the ellipsoid
e^2	first eccentricity squared
φ_u	Upper Parallel
φ_l	Lower Parallel
φ_o	Central Parallel (CP), latitude of projection origin
φ_b	Latitude of (false) grid origin, in case of 2 parallels
φ_b	Latitude of Standard Parallel, in case of 1 parallel
k_o	Point scale factor at CP
λ_o	Central, Reference Meridian (RM, λ_o), longitude of true and grid origin (Figure 32)
E_0	false easting constant at grid and projection origin, (RM, λ_o)
N_b	false northing constant for φ_b at the RM, λ_o
N_o	Northing constant at intersection of RM, $\lambda_{o,}$ with CP, φ_o
M_o	scaled radius of curvature in the meridian at φ_o
R	mapping radius at latitude, φ
R_b	mapping radius at latitude, φ_b
R_o	mapping radius at latitude, φ_o
K	mapping radius at the equator
Q	isometric latitude
φ	parallel of geodetic latitude, positive north
λ	meridian of geodetic longitude, positive east
E	easting co-ordinate
N	northing co-ordinate
γ	convergence angle
δ_{12}	(t - T), arc-to-chord correction for a line from p_1 to p_2
k	grid scale factor at a general point
k_{12}	line scale factor of a line from p_1 to p_2

Computation of Projection Zone and Ellipsoid Constants

Constants and expressions within Lambert's conical mapping equations are ellipsoid and zone specific.

Q_l $= ½ [(\ln ((1 + \sin \varphi_l) / (1 - \sin \varphi_l)) - e \ln ((1 + e \sin \varphi_l) (1 - e \sin \varphi_l))]$ (7.01)
W_l $= (1 - e^2 \sin^2 \varphi_l)^{½}$ (7.02)

Similarly for Q_u, W_u, Q_b, Q_o, and W_0 upon substitution of the appropriate latitude.

$\sin \varphi_o$ $= \ln [W_u \cos \varphi_l / (W_l \cos \varphi_n)] / (Q_u - Q_l)$ (7.03)
K $= (a \cos \varphi_l \exp (Q_l \sin \varphi_o)) / (W_l \sin \varphi_o)$ (7.04)
$= (a \cos \varphi_u \exp (Q_u \sin \varphi_o)) / (W_u \sin \varphi_o)$ (7.05)

R_b $= K / \exp (Q_b \sin \varphi_o)$ (7.06)
R_0 $= K / \exp (Q_0 \sin \varphi_0)$ (7.07)
k_0 $= (W_0 \tan \varphi_0 R_0) / a$ (7.08)
N_0 $= R_b + N_b - R_0$ (7.09)

Note: $\exp (x) = \varepsilon^x$, in which ε= 2.71828 18284 59045 23536 02875 (base of natural logarithms)

Recognise in (7.01, 7.11): $\ln (\tan ((\pi / 4)+(\varphi / 2)))=½ \ln ((1+\sin (\varphi)) / (1 - \sin (\varphi)))$ (7.10)
(Gretschel, 1873; Le Pape, 1994), and [8.1.1].

Direct Conversion Computation

Input: geodetic co-ordinates of point P (φ, λ) on meridian API, projected on Spi.
Output: grid co-ordinates of point p (E, N), convergence angle (γ), grid scale factor (k).

Q $= ½ [\ln ((1 + \sin \varphi) / (1 - \sin \varphi)) - e \ln ((1 + e \sin \varphi) / (1- e \sin \varphi))]$ (7.11)
R $= K / \exp (Q \sin \varphi_0)$ (7.12)
E $= E_0 + R \sin \gamma$ (7.13)
N $= R_b + N_b - R \cos \gamma$ (7.14)
γ $= (\lambda_0 - \lambda) \sin \varphi_0$ (7.15)
k $= (1 - e^2 \sin^2 \varphi)^{½} (R \sin \varphi_0) / (a \cos \varphi)$ (7.16)

Inverse Conversion Computation

Input: grid co-ordinates of point p (E, N) on a line Spi (Figure 31, Figure 32).
Output: geodetic co-ordinates of point P (φ, λ), convergence (γ), grid scale factor (k).

R' $= R_b - N + N_b$ (7.17)
E' $= E - E_0$ (7.18)
γ $= \tan^{-1} (E' / R')$ (7.19)
λ $= \lambda_0 - \gamma / \sin \varphi_0$ (7.20)
R $= (R'^2 + E'^2)^{½}$ (7.21)
Q $= [\ln (K / R)] / \sin \varphi_0$ (7.22)

Use an approximation for φ as follows (an iterative solution is not used in program LCC00000.BAS):

$\sin \varphi$ $= (\exp (2 Q) - 1) / (\exp (2 Q) + 1)$ and iterate $\sin \varphi$ as follows: (7.23a)
f_1 $= ½ [\ln ((1 + \sin \varphi) / (1 - \sin \varphi)) - e \ln ((1 + e \sin \varphi) / (1 - e \sin \varphi))] - Q$ (7.24)
f_2 $= 1 / (1 - \sin^2 \varphi) - (e^2 / (1 - e^2 \sin^2 \varphi))$ (7.25)
$\sin \varphi$ $= \sin \varphi+(- f_1 / f_2)$ and iterate to obtain φ with sufficient accuracy (7.23b)

In program LCC00000.BAS, *latitude* is obtained *without iteration* using the five constants F_0, F_2, F_4, F_6, and F_8 as computed in [8.1.1], Oblique Mercator section (8.02):

k $= (1 - e^2 \sin^2 \varphi)^{½} (R \sin \varphi_0) / (a \cos \varphi)$ (7.26)

Arc-to-Chord Correction $\delta = (t - T)$

Grid azimuth *(t)*, geodetic azimuth *(α)*, convergence angle *(γ)*, and arc-to-chord correction *(δ)* at any given point are related as follows: $t = \alpha - \gamma + \delta$ (5.01).

Input: p_1 (E_1, N_1), and p_2 (E_2, N_2)
Output: δ_{12}

$$p_1 = N_1 - N_0 \quad (7.27)$$
$$p_2 = N_2 - N_0 \quad (7.28)$$
$$q_1 = E_1 - E_0 \quad (7.29)$$
$$q_2 = E_2 - E_0 \quad (7.30)$$
$$R'_1 = R_0 - p_1 \quad (7.31)$$
$$R'_2 = R_0 - p_2 \quad (7.32)$$
$$\Delta N = N_2 - N_1 \quad (7.33)$$
$$M_0 = k_0\, a\, (1 - e^2) / (1 - e^2 \sin^2 \varphi)^{3/2} \quad (7.34)$$

$$u_1 = p_1 - q_1^2 / 2\, R'_1 \quad (7.35)$$
$$\varphi_3 = \varphi_0 + (u_1 + \Delta N / 3) / M_0 \quad (7.36)$$
$$\delta_{12} = (\sin \varphi_3 / \sin \varphi_0 - 1)\,(q_2 / R'_2 - q_1 / R'_1) / 2 \quad (7.37)$$

The size of δ varies linearly with the length of the ΔE (Δλ) component of the line and with the distance of the point from the central parallel.

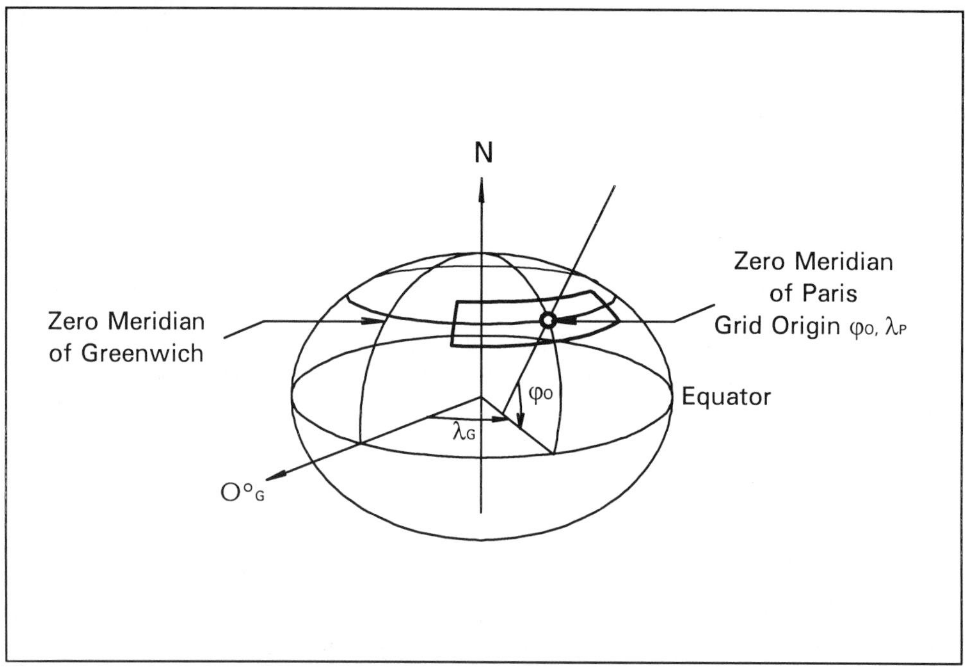

Figure 33: Lambert's conical IGN zone

Note

The length of the lines are 20 km, the orientation of the lines in azimuths of 90°, 135°, and 180°. Dividing the line into several traverse legs results in a proportional decrease in the required correction to a direction. However, it does nothing to diminish the closure error in azimuth because errors due to omission of δ are cumulative

(Vincenty, 1986a).

The examples below (Table 19; Table 20), referring to [7.2.4.3], are computed for a hypothetical zone with a central parallel of approximately φ_o=32° on the GRS80 ellipsoid.

Values of (t - T) and computational errors in δ determination for φ_o= 32° (in seconds of arc)					
$\varphi_1 - \varphi_0$ $\lambda_1 - \lambda_0$		1° 5°	2° 5°	1° 10°	2° 10°
True δ, azimuth= Error	90°	5.67 0.00	11.44 0.01	5.67 0.01	11.44 0.06

Table 19: Values of computational errors in δ for φ_o=32° from the LCC equations

The examples are computed for a hypothetical zone with a central parallel of approximately 42° (standard parallels 41° and 43°), on the GRS80 ellipsoid (Stem, 1989a).

Values of (t - T) and computational errors in δ determination for φ_o= 42° (in seconds of arc)					
$\varphi_1 - \varphi_0$ $\lambda_1 - \lambda_0$		1° 5°	2° 5°	1° 10°	2° 10°
True δ, azimuth Error	= 90°	5.67 0.00	11.44 0.02	5.67 0.03	11.44 0.11
True δ, azimuth= Error	135°	3.83 0.00	7.91 0.01	3.83 0.02	7.91 0.08
True δ, azimuth= Error	180°	0.00 0.00	0.00 0.00	0.00 0.00	0.00 0.00

Table 20: Values of computational errors in δ for φ_o=42° from the LCC equations

The equations in this section are found in (Berry, 1970; Burkholder, 1985; Boucher, 1979; IGN, 1963, 1994) |[50] and (Floyd, 1985; Stem, 1989a), based on (Vincenty, 1985a, 1985b, 1986a, 1986b).

[50] Explanations in full, see (Boucher, 1979)

7.2 Lambert's Conical Projection Applications

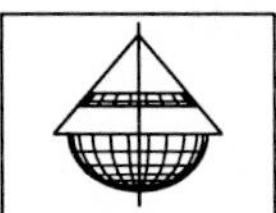

7.2.1 Reference and LCC-projection Systems of France

Lambert's conical projection with two standard parallels. Modern countries like France with a predominantly east-west extent employ the Lambert Conformal Conical projection.

In France, the Lambert's conformal IGN-grid with two standard parallels is a Lambert grid with *one standard parallel* to which a grid scale constant has been applied (IGN, 1994).

The longitude of the *Meridian of Paris,* $2^g.59\ 68\ 98$ E of Greenwich, is used as the datum for National surveys and maps. Here, geodetic co-ordinates - expressed in *centesimal units* - are used for the civilian surveys in France since 1920. Several Lambert projection systems are also in use in Northern Africa, see [7.2.2].

Lambert - One or Two Standard Parallels - France

Ellipsoid	Lambert System	Origin	C.M.	Boundary	False Easting False Northing	Scale Factor	Upper Parallel Lower Parallel
Clarke 1880	Lambert I [51] Zone Nord	55^g	0^g E Paris	$56^g.5$ $53^g.5$	x= 600 000 y= 200 000	0.99987 734	56^g approx. 54^g approx.
Clarke 1880	Lambert II Zone Central	52^g	0^g E Paris	$53^g.5$ $50^g.5$	x= 600 000 y= 200 000	0.99987 742	53^g approx. 51^g approx.
Clarke 1880	Lambert II Zone Etendu	52^g	0^g E Paris	$53^g.5$ $50^g.5$	x= 600 000 y= 2 200 000	0.99987 742	53^g approx. 51^g approx.
Clarke 1880	Lambert III Zone Sud	49^g	0^g E Paris	$50^g.5$ $47^g.5$	x= 600 000 y= 200 000	0.99987 750	50^g approx. 48^g approx.
Clarke 1880	Lambert IV Corse	$46^g.85$	0^g E Paris	$47^g.8$ $45^g.9$	x= 234 358 y= 185 861.369	0.99994 471	$47^g.52$ approx. $46^g.18$ approx.
International	Eurolambert	52^g	3^g E Gr.	- -	x= 500 000 y= 2 200 000	0.99987 752	
Plessis	Nord de Guerre [52]	55^g	6^g E Paris	$57^g.5$ $52^g.5$	x= 500 000 y= 300 000	0.99950 908	57^g approx. 53^g approx.
Clarke 1880	Grand Champ	47°	0°E Paris	- -	x= 600 000 y= 600 000	1 (exactly)	49° 47°

Examples of *the French Survey System* follow in [7.2.1.1 - 7.2.1.5; 10.1.1] (IGN, 1994).

IGN defined the Clarke 1880IGN ellipsoid *uniquely* as (IGN, 1994):

- semi-major axis a = 6 378 249.2000 m
- semi-minor axis b = 6 356 515.0000 m.

The adapted and validated algorithms are given in program listing LCC00000.BAS.

[51] Systems also in use in France : Lambert I Carto, Lambert II Carto, Lambert III Carto, Lambert IV Carto, Eurolambert, Lambert Nord de Guerre, Lambert Grand Champ.

[52] This system was the first Lambert conformal conical projection system applied to practical geodesy. It has been in use from the outbreak of World War I at the battle-field in "Nord-Est France" in 1915 and it remained in operational use until the end of World War II in 1945. Certain maps are still in use.

7.2.1.1 Lambert I - Zone Nord - France

Secant projection with Central Parallel and its Scale Factor

Zone Parameters:	*Lambert I,* (IGN, 1994).	L2 = .760405965600031
Reference Ellipsoid:	Clarke 1880IGN	L3 = 5457616.674047252
Semi-Major axis a:	**6 378 249.2**	L4 = 5657616.674047252
Recipr. Flattening f^{-1}:	**293.46602 12936 294** |[53]	L5 = 5457616.674047252
Lat. Grid Origin φ_o:	**55^g N**	L6 = 11603796.97665211
Lon. Grid Origin λ_p:	**0^g E** of Paris	L7 = .99987734
Scale Factor k_o:	**0.99987 734**	F0 = 6.795783595977233D-03
False Easting E_o:	**600 000**	F2 = 5.371938835736265D-05
False Northing N_b:	**200 000**	F4 = 5.819180402086291D-07
		F6 = 7.164660897964055D-09
		F8 = 9.657615214881198D-11

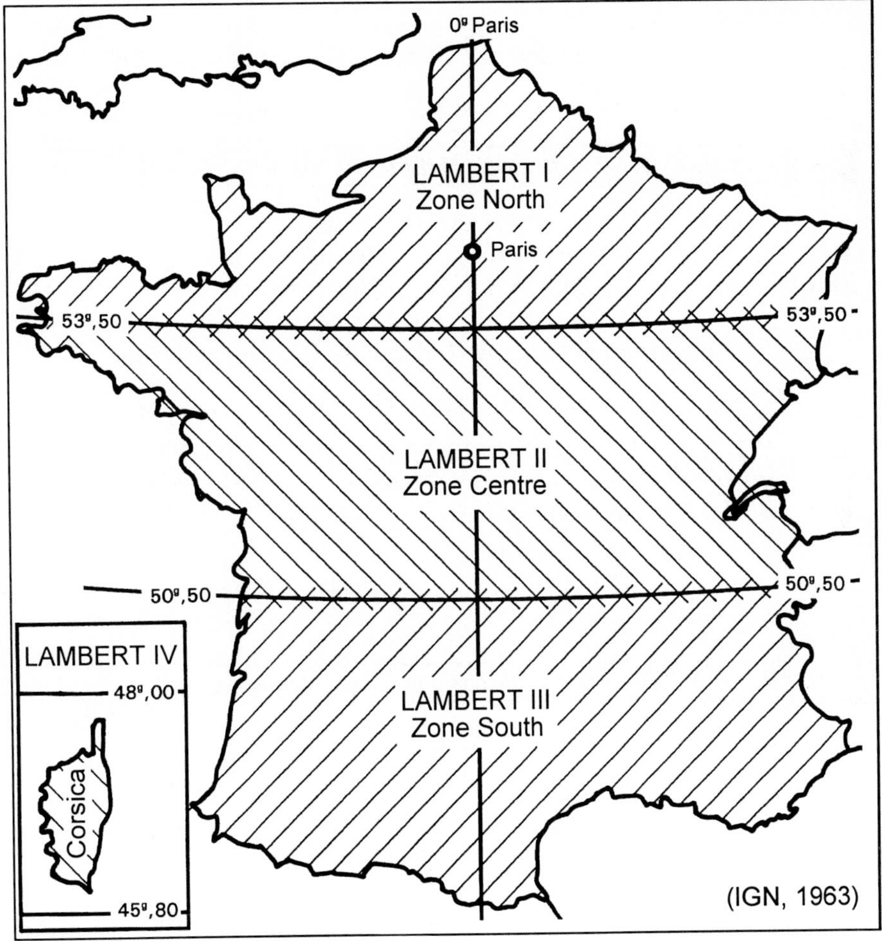

Figure 34: Lambert's conformal conical projection of France

---

[53] See Table 21: Derived ellipsoidal parameters for Clarke 1880IGN

7.2.1.2 Lambert I - Zone Nord - France

Two Standard Parallels

Zone Parameters:		*Lambert I,* (IGN, 1994).	L2 = .7604059656000921
Reference Ellipsoid:		Clarke 1880IGN	L3 = 5457616.674046539
Semi-Major axis	a:	**6 378 249.2**	L4 = 5657616.674046539
Recipr. Flattening	f^{-1}:	**293.46602 12936 294**	L5 = 5457616.674045939
Lower Parallel	φ_l:	**53^g.99 83 58719 082 N**	L6 = 11603796.9766513
Upper Parallel	φ_u:	**55^g.99 54 57368 531 N**	L7 = .9998773399999499
Lat. Grid Origin	φ_b:	**55^g N**	F0 = 6.795783595977233D-03
Central Parallel	φ_o=	55^g.00 00 00000 006 N	F2 = 5.371938835736265D-05
Lon. Grid Origin	λ_p:	**0^g E** of Paris	F4 = 5.819180402086291D-07
Scale Factor	k_o=	0.999 877 340 000	F6 = 7.164660897964055D-09
False Easting	E_o:	**600 000**	F8 = 9.657615214881198D-11
False Northing	N_b:	**200 000**	

Conversion of IGN1922 Geographicals to Lambert Grid

The Direct calculation is to convert Geodetic Co-ordinates into LCC-Planar Co-ordinates:

Input:	Latitude	Longitude	Output:	Easting	Northing	Convergence	Scale Factor

Latitude:	**54^g.77 46 5390 N**	Easting=	401 986.9164
Longitude:	**3^g.02 57 2180 W**	Northing=	181 025.0244
Convergence=	- 2^g.30 07 7691	Scale Factor=	0.999 883 577 354

Conversion of Lambert Grid to IGN1922 Geographicals

The Inverse process is to convert LCC-Planar Co-ordinates into Geodetic Co-ordinates:

Input:	Easting	Northing	Output:	Latitude	Longitude	Convergence	Scale Factor

Easting:	**401 986.9164**	Latitude=	54^g.77 46 5390 N
Northing:	**181 025.0244**	Longitude=	3^g.02 57 2180 W
Convergence=	- 2^g.30 07 7691	Scale Factor=	0.999 883 577 354

Clarke 1880IGN - Institute Géographique National		
a	=	**6 378 249.2000** m (unique)
b	=	**6 356 515.0000** m (unique)
a / b	=	1.00341 92006 1|5431 56902 79972
1 / f	=	293.46602 12936 2|9395 14681 92986
f	=	0.00340 75495 20015 61|807 90176
e^2	=	0.00680 34876 46299 87|748 88800
e'^2	=	0.00685 00921 637|11 70567 63915
c	=	6 400 057.71359 001|59 12807 56829 80375

Table 21: Derived ellipsoidal parameters for Clarke 1880IGN

7.2.1.3 Lambert II - Zone Centre - France

Two Standard Parallels

Zone Parameters:		*Lambert II,* (IGN, 1994).	L2 = .7289686274214495
Reference Ellipsoid:		Clarke 1880IGN	L3 = 5999695.768001665
Semi-Major axis	a:	**6 378 249.2**	L4 = 6199695.768001665
Recipr. Flattening	f^{-1}:	**293.46602 12936 294**	L5 = 5999695.768001312
Lower Parallel	φ_l:	**50^g.99 87 98849 354 N**	L6 = 11745793.39343506
Upper Parallel	φ_u:	**52^g.99 55 71668 931 N**	L7 = .9998774200000195
Lat. Grid Origin	φ_b:	**52^g N**	F0 = 6.795783595977233D-03
Central Parallel	φ_o=	52^g.00 00 00000 00353 N	F2 = 5.371938835736265D-05
Lon. Grid Origin	λ_p:	**0^g E of Paris**	F4 = 5.819180402086291D-07
Scale Factor	k_o=	0.999 877 420 000	F6 = 7.164660897964055D-09
False Easting	E_o:	**600 000**	F8 = 9.657615214881198D-11
False Northing	N_b:	**200 000**	

Conversion of IGN1922 Geographicals to Lambert Grid

The Direct calculation is to convert Geodetic Co-ordinates into LCC-Planar Co-ordinates:

Input:	Latitude	Longitude	Output:	Easting	Northing	Convergence	Scale Factor

Latitude:	**51^g.80 72 3130 N**	Easting=	632 542.0576
Longitude:	**0^g.47 21 6690 E**	Northing=	180 804.1446
Convergence=	+ 0^g.34 41 9486	Scale Factor=	0.999 881 984 229

Conversion of Lambert Grid to IGN1922 Geographicals

The Inverse process is to convert LCC-Planar Co-ordinates into Geodetic Co-ordinates:

Input:	Easting	Northing	Output:	Latitude	Longitude	Convergence	Scale Factor

Easting:	**632 542.0576**	Latitude=	51^g.80 72 3130 N
Northing:	**180 804.1446**	Longitude=	0^g.47 21 6690 E
Convergence=	+0^g.34 41 9486	Scale Factor=	0.999 881 984 230

Note

The N_b (False Northing, or Y_0 values) of the Lambert projection in France receive the prefixed numbers 1, 2, 3, and 4 for the Lambert Zones 1, 2, 3, and 4, respectively.

Thus N_b = 4 185 861.369 for Lambert IV, Corsica (IGN, 1986).

The "round-trip errors" are calculated in [7.3.1; 7.3.2] for the area between latitude 49^g N to 55^g N and longitude 9^g W to 9^g E of Paris.

7.2.1.4 Lambert III - Zone Sud - France

Two Standard Parallels

Zone Parameters:		*Lambert III,* (IGN, 1994).	L2 = .6959127965923538
Reference Ellipsoid:		Clarke 1880IGN	L3 = 6591905.084692603
Semi-Major axis	a:	**6 378 249.2**	L4 = 6791905.084692603
Recipr. Flattening	f^{-1}:	**293.46602 12936 294**	L5 = 6591905.084692252
Lower Parallel	φ_l:	**47^g.99 92 12528382 N**	L6 = 11947992.52465187
Upper Parallel	φ_u:	**49^g.99 56 59793901 N**	L7 = .9998774999999359
Lat. Grid Origin	φ_b:	**49^g N**	F0 = 6.795783595977233D-03
Central Parallel	φ_o=	49^g.00 00 000000035 N	F2 = 5.371938835736265D-05
Lon. Grid Origin	λ_p:	**0^g E** of Paris	F4 = 5.819180402086291D-07
Scale Factor	k_o=	0.999 877 500 000	F6 = 7.164660897964055D-09
False Easting	E_o:	**600 000**	F8 = 9.657615214881198D-11
False Northing	N_b:	**200 000**	

Conversion of IGN1922 Geographicals to Lambert Grid

The Direct calculation is to convert Geodetic Co-ordinates into LCC-Planar Co-ordinates:

Input:	Latitude	Longitude	Output:	Easting	Northing	Convergence	Scale Factor

Latitude:	**48^g.56 17 2170 N**	Easting=	959 716.8169
Longitude:	**4^g.96 14 7110 E**	Northing=	165 935.2167
Convergence=	+ 3^g.45 27 5123	Scale Factor=	0.999 901 059 055

Conversion of Lambert Grid to IGN1922 Geographicals

The Inverse process is to convert LCC-Planar Co-ordinates into Geodetic Co-ordinates:

Input:	Easting	Northing	Output:	Latitude	Longitude	Convergence	Scale Factor

Easting:	**959 716.8169**	Latitude=	48^g.56 17 2170 N
Northing:	**165 935.2167**	Longitude=	4^g.96 14 7110 E
Convergence=	+3^g.45 27 5123	Scale Factor=	0.999 901 059 055

7.2.1.5 Lambert IV - Zone Corse - France

Two Standard Parallels

Zone Parameters:		*Lambert IV,* (IGN, 1994).	L2 = .6712679322465734
Reference Ellipsoid:		Clarke 1880IGN	L3 = 7053300.173162977
Semi-Major axis	a:	**6 378 249.2**	L4 = 7239161.542162977
Recipr. Flattening	f^{-1}:	**293.46602 12936 294**	L5 = 7053300.17316296
Lower Parallel	φ_l:	**46^g.17 82 08712164 N**	L6 = 12136281.98624047
Upper Parallel	φ_u:	**47^g.51 96 26077389 N**	L7 = .9999447099999506
Lat. Grid Origin	φ_b:	**46^g.85 N**	F0 = 6.795783595977233D-03
Central Parallel	φ_o=	46^g.85 00 0000000017 N	F2 = 5.371938835736265D-05
Lon. Grid Origin	λ_p:	**0^g E of Paris**	F4 = 5.819180402086291D-07
Scale Factor	k_o=	0.999 944 710 000	F6 = 7.164660897964055D-09
False Easting	E_o:	**234.358**	F8 = 9.657615214881198D-11
False Northing	N_b:	**185861.369**	

Conversion of IGN1922 Geographicals to Lambert Grid

The Direct calculation is to convert Geodetic Co-ordinates into LCC-Planar Co-ordinates:

Input:	Latitude	Longitude	Output:	Easting	Northing	Convergence	Scale Factor

Latitude:	**47^g.07 43 9140 N**	Easting=	523 142.3259
Longitude:	**7^g.05 99 5040 E**	Northing=	227 764.4389
Convergence=	+4^g.73 91 1831	Scale Factor=	0.999 950 904 966

Conversion of Lambert Grid to IGN1922 Geographicals

The Inverse process is to convert LCC-Planar Co-ordinates into Geodetic Co-ordinates:

Input:	Easting	Northing	Output:	Latitude	Longitude	Convergence	Scale Factor

Easting:	**523 142.3259**	Latitude=	47^g.07 43 9140 N
Northing:	**227 764.4389**	Longitude=	7^g.05 99 5040 E
Convergence=	+4^g.73 91 1831	Scale Factor=	0.999 950 904 966

7.2.2 Reference and LCC-projection Systems of North Africa

(IGN, 1963, 1994)

Country	Lambert System	Origin [54]	C.M.	Boundary	False Easting False Northing	Scale Factor	Upper Parallel Lower Parallel
Algeria[55]	North- Algeria	40^{g}	3^{g} E Gr.	$41^{g}\,5$ $38^{g}\,5$	x= 500 000 y= 300 000	0.99962 5544	$41^{g}\,75$ approx. $38^{g}\,25$ approx.
Algeria	South-Algeria	37^{g}	3^{g} E Gr.	$38^{g}\,5$ $34^{g}\,5$	x= 500 000 y= 300 000	0.99962 5769	$38^{g}\,75$ approx. $35^{g}\,25$ approx.
Tunisia[56]	North -Tunisia	40^{g}	11^{g} E Gr.	$41^{g}\,5$ $38^{g}\,0$	x= 500 000 y= 300 000	0.99962 5544	$41^{g}\,75$ approx. $38^{g}\,25$ approx.
Tunisia	South-Tunisia	37^{g}	11^{g} E Gr.	$38^{g}\,0$ $34^{g}\,5$	x= 500 000 y= 300 000	0.99962 5769	$38^{g}\,75$ approx. $35^{g}\,25$ approx.
Morocco	North-Morocco	37^{g}	6^{g} W Gr.	$39^{g}\,0$ $35^{g}\,5$	x= 500 000 y= 300 000	0.99962 5769	$38^{g}\,75$ approx. $35^{g}\,25$ approx.
Morocco	South-Morocco	33^{g}	6^{g} W Gr.	$35^{g}\,0$ $30^{g}\,5$	x= 500 000 y= 300 000	0.99961 5596	$34^{g}\,77$ approx. $31^{g}\,23$ approx.

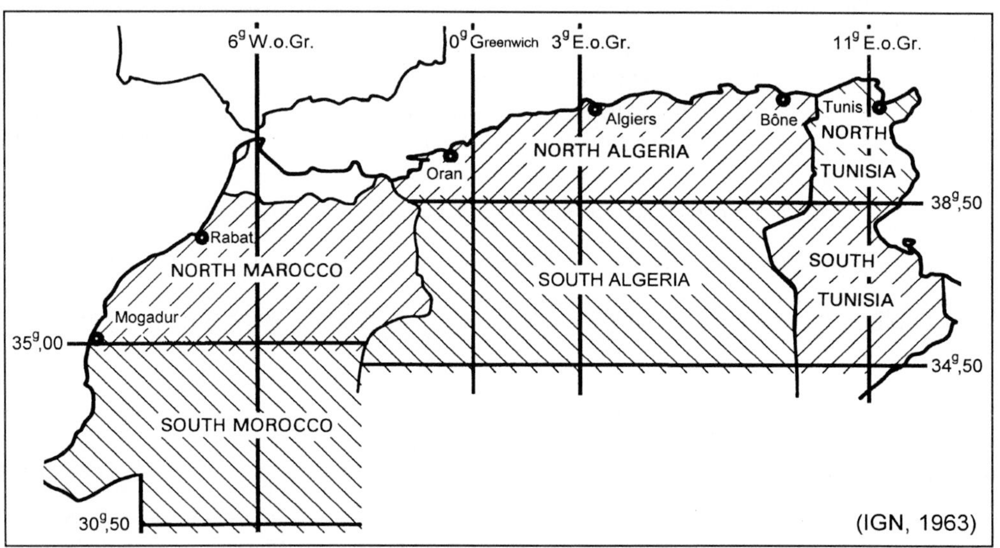

Figure 35: Lambert's projection of North Africa

[54] Using ellipsoid Clarke 1880IGN

[55] An important scale factor has been calculated based on recent computations, which is not yet in use

[56] Origin Carthage, instead of Voirol 1960!

7.2.2.1 Lambert's Conical Projection of Morocco

Secant projection with Standard Parallel and its Scale Factor - South Morocco

Sta.: Bou Ourioul

Zone Parameters:		Zone South-Morocco	L2 = .4954586684324076
Reference Ellipsoid:		Clarke 1880IGN	L3 = 11187308.68177676
Semi-Major axis	a:	**6 378 249.2**	L4 = 11487308.68177676
Recipr. Flattening	f^{-1}:	**293.46602 12936 294**	L5 = 11187308.68177676
Standard Parallel	φ_o:	**33^g N**	L6 = 14618332.47695414
Lon. Grid Origin	λ_p:	**6^g W** of Greenwich	L7 = .999615596
Scale Factor	k_o:	**0.99961 5596**	F0 = 6.795783595977233D-03
False Easting	E_o:	**500 000**	F2 = 5.371938835736265D-05
False Northing	N_b:	**300 000**	F4 = 5.819180402086291D-07
			F6 = 7.164660897964055D-09
			F8 = 9.657615214881198D-11

Conversion of Merchich Geographicals to Lambert Grid

The Direct calculation is to convert Geodetic Co-ordinates into LCC-Planar Co-ordinates:

Input: Latitude Longitude		Output: Easting Northing Convergence	Scale Factor
Latitude:	**34^g.77 71 4770 N**	Easting=	305 571.1730
Longitude:	**8^g.26 91 6830 W**	Northing=	478 977.4806
Convergence=	- 1^g.12 42 7910	Scale Factor=	1.000 005 329 198

Conversion of Lambert Grid to Merchich Geographicals

The Inverse process is to convert LCC-Planar Co-ordinates into Geodetic Co-ordinates:

Input: Easting Northing		Output: Latitude Longitude Convergence	Scale Factor
Easting:	**305 571.1730**	Latitude=	34^g.77 71 4770 N
Northing:	**478 977.4806**	Longitude=	8^g.26 91 6830 W
Convergence=	- 1^g.12 42 7910	Scale Factor=	1.000 005 329 198

7.2.3 Reference and LCC-projection Systems of Belgium

Caution:

> *Lambert's Projection with two standard parallels of Belgium requires a transformation as explained in "Referentiesystemen en Transformatieformules". Therefore, it is not possible to use the formulae without alteration* (Herrewegen, 1989).

A suggestion for the corrections is given in program listing LCC00000.BAS [11.7.6; 11.7.10]. The values of co-ordinates computed are defined as precisely as possible in an absolute sense.

The formulae in the publication mentioned above will insure that the co-ordinates computed will agree as closely as possible: thus, in a relative sense, the plane co-ordinates are well defined.

Lambert1972 - Two Standard Parallels - Belgium |[57]

Zone Parameters		*Lambert1972*	L2 = .7716421928
Reference Ellipsoid		International 1924	L3 = 5234715.481521557
Semi-Major axis	a:	**6 378 388**	L4 = 5400088.437801557
Recipr. Flattening	f^{-1}:	**297**	L5 = 5267839.669502068
Lower Parallel	φ_l:	**49° 50' 00" N**	L6 = 11565915.812935
Upper Parallel	φ_u:	**51° 10' 00" N**	L7 = .9999324916433006
Lat. Grid Origin	φ_b:	**50° 47' 57".704 N**	F0 = 6.715147793101421D-03
Central Parallel	φ_o=	50° 30' 05".71040 30454 N	F2 = 5.245398813629418D-05
Lon. Grid Origin	λ_o:	**4° 21' 24".983 E** of Greenwich	F4 = 5.61493409847134D-07
Scale Factor	k_o=	0.999 932 491 643 300 6	F6 = 6.831509004398433D-09
False Easting	E_o:	**150 000.01256**	F8 = 9.097474960471413D-11
False Northing	N_b:	**165 372.95628**	

Conversion of ED50 Geographicals to Lambert1972 Grid

The Direct calculation is to convert Geodetic Co-ordinates into LCC-Planar Co-ordinates:

Input: Latitude Longitude. Output: Easting Northing Convergence Scale Factor

Latitude	φ_1:	**50° 40' 46".4610 N**	Easting	E_1=	251 763.2030
Longitude	λ_1:	**5° 48' 26".5330 E**	Northing	N_1=	153 034.1740
Convergence	γ_1=	+ 1° 07' 09".1683	Scale Factor	K_1=	0.999 937 309 295
Latitude	φ_2:	**50° 47' 25".7956 N**	Easting	E_2=	260 597.9204
Longitude	λ_2:	**5° 56' 09".8964 E**	Northing	N_2=	165 555.2202
Convergence	γ_2=	+ 1° 13' 06".7190	Scale Factor	k_2=	0.999 945 195 738
Latitude	φ_m:	**50° 44' 06".1282 N**	Easting	E_m=	256 185.9548
Longitude	λ_m:	**5° 52' 18".2150 E**	Northing	N_m=	159 292.7922
			Scale Factor	k_m=	0.999 940 782 928

[57] See UTM Belgium, [6.3.3]

Conversion of Lambert1972 Grid to ED50 Geographicals - Belgium

The Inverse process is to convert LCC-Planar Co-ordinates into Geodetic Co-ordinates |[58] :

Input:	Easting	Northing.	Output:	Latitude	Longitude	Convergence	Scale Factor

Easting	E_1:	**251 763.2030**	Latitude	φ_1=	50° 40' 46".4610 N
Northing	N_1:	**153 034.1740**	Longitude	λ_1=	5° 48' 26".5330 E
Convergence	γ_1=	+ 1° 07' 09".1683	Scale Factor	k_1=	0.999 937 309 295
Easting	E_2:	**260 597.9204**	Latitude	φ_2=	50° 47' 25".7956 N
Northing	N_2:	**165 555.2202**	Longitude	λ_2=	5° 56' 09".8964 E
Convergence	γ_2=	+ 1° 13' 06".7190	Scale Factor	k_2=	0.999 945 195 738
Easting	E_m:	**256 185.9548**	Latitude	φ_m=	50° 44' 06".1282 N
Northing	N_m:	**159 292.7922**	Longitude	λ_m=	5° 52' 18".2150 E
			Scale Factor	k_m=	0.999 940 782 928

Input:	Easting	Northing	Output: Arc-to-chord or (t - T) Correction

N_0:	**132 248.8214** (calc.)					
E_1:	**251 763.2030**	E_2:	**260 597.9204**	(t - T)	$\delta_{1\text{-}2}$=	+0".5532
N_1:	**153 034.1740**	N_2:	**165 555.2202**	(t - T)	$\delta_{2\text{-}1}$=	- 0".6451

Ellipsoidal and Grid Calculation (manual)

Using GBD00000.BAS gives:

True Distance	$S_{1\text{-}2}$:	*15 325.0308*
True Azimuth	$\alpha_{1\text{-}2}$:	*36° 19' 02".3152*
True Azimuth	$\alpha_{2\text{-}1}$:	*216° 25' 01".0641*
Grid Distance	$D_{1\text{-}2}$:	15 324.1257
True Distance	$S_{calc\ 1\text{-}2}$:	15 325.0308
See (5.03), using		
Scale Factor	$k_{1\text{-}2}$:	0.999 940 939 458

Grid bearing	$t_{1\text{-}2}$	35° 12' 22".9978
+ convergence	γ_1	+1° 07' 09".1683
- *ϕ (Belgium only)*		*0° 00' 29".2985*
- (t - T)	$\delta_{1\text{-}2}$	+0° 00' 00".5532
True Azimuth	$\alpha_{1\text{-}2}$:	36° 19' 02".3144
Grid bearing	$t_{2\text{-}1}$	215° 12' 22".9978
+ convergence	γ_2	+1° 13' 06".7190
- *ϕ (Belgium only)*		*0° 00' 29".2985*
- (t - T)	$\delta_{2\text{-}1}$	- 0° 00' 00".6451
True Azimuth	$\alpha_{2\text{-}1}$:	216° 25' 01".0634

[58] (Herrewegen, 1989)

7.2.4 Reference and SPC Systems of the USA

In the United States of America the State Plane Co-ordinate System (SPCS) was introduced in 1933. The individual projections forming the State Co-ordinate Systems were based upon the Clarke 1866 ellipsoid: North American Datum of 1927 (NAD27).

Each State has its own system of co-ordinates. Several States have more than one Transverse Mercator or Lambert zone, and a few States have a combination of both Lambert and Transverse Mercator, see [6.2.9].
Alaska is a State in which three systems are employed; a small part of Alaska makes use of the *Oblique Mercator projection*, see [8.2.2].

Referring to the redefinition of the North American Datum from NAD27 into NAD83, which includes the change from the Clarke 1866 ellipsoid to the GRS80 ellipsoid gave the National Geodetic Survey (NGS) the possibility of considering changing the projection system to provide the best projection within the boundaries of a particular State (Claire, 1968; Osterhold, 1993; Vincenty, 1976a).

Units of Measurement.

The following States use *international feet and metres:*

- Arizona, Michigan, Montana, Oregon, South Carolina and Utah.

The following States use US survey feet and metres:

- California, Colorado, Connecticut, Indiana, Maryland, Nebraska, North Carolina, Texas and Wyoming
- All other States use *only metres.*

US Lambert State Plane Co-ordinate Systems

The US Lambert State Plane Co-ordinate System has the following specifications:

• Projection:	Lambert conformal conic		
• Reference Ellipsoid	Clarke 1866	______________	NAD27
	GRS80	______________	NAD83
• Lower Parallel	φ_l,	see *specs of a State:* __	for 2 parallels only
• Upper Parallel	φ_u,	see *specs of a State:* __	for 2 parallels only
• Standard Parallel	φ_o,	see *specs of a State:* __	for 1 parallel given
• Central Parallel	φ_b,	computed grid origin	
• Reference Meridian	λ_o,	see *specs of a State:* __	for 1 or 2 parallels, W of Greenwich
• Scale Factor at the RM	see *specs of a State:* ________		for 1 parallel given
• Unit	see *specifications of a State:*		
	♦	US survey feet and / or metre (SI)	
	♦	international feet and / or metre (SI)	
	♦	metre (SI)	
• False Northing	see *specifications of a State*		
• False Easting	See *specifications of a State*		
• Latitude Limits of SPCS	State boundaries.		

The specifications for the *SPCS27* are given in e.g. (Mitchell, 1945).

The specifications for the *SPCS83* are given in e.g. (Stem, 1989a).

7.2.4.1 SPCS27 - Lambert - State Alaska - Zone 10 - USA

Two Standard Parallels

Zone Parameters		*SPCS Alaska, Zone 10*	L2 = .7969223939103863
Reference Ellipsoid		Clarke 1866 (NAD27)	L3 = 16564628.76943945
Semi-Major axis	a:	**20 925 832.16 ft**	L4 = 16564628.76943945
Recipr. Flattening	f^{-1}:	**294.97869 82**	L5 = 15893950.36808007
Lower Parallel	φ_l:	**51° 50' 00" N**	L6 = 37726989.54287059
Upper Parallel	φ_u:	**53° 50' 00" N**	L7 = .999848064156409
Lat. Grid Origin	φ_b:	**51° 00' 00" N**	F0 = 6.761032571864926D-03
Central Parallel	φ_o=	52° 50' 13".95423 98662 N	F2 = 5.317220426784075D-05
Lon. Grid Origin	λ_o:	**176° 00' 00 W** of Greenwich	F4 = 5.73056251095528D-07
Scale Factor	k_o=	0.999848064156409	F6 = 7.019628022792133D-09
False Easting	E_o:	**3 000 000 ft**	F8 = 9.412928282121681D-11
False Northing	N_b:	**0 ft**	

Conversion of NAD27 Geographicals to Lambert Grid

The Direct calculation is to convert Geodetic Co-ordinates into LCC-Planar Co-ordinates:

Input:	Latitude	Longitude.	Output:	Easting	Northing	Convergence	Scale Factor

Latitude:	**54° 27' 30" N**	Easting |[59] =	5 533 424.3919 ft
Longitude:	**164° 02' 30" W**	Northing=	1 473 805.1224 ft
Convergence=	+9° 31' 47".5091	Scale Factor=	1.000 252 581 841

Conversion of Lambert Grid to NAD27 Geographicals

The Inverse process is to convert LCC-Planar Co-ordinates into Geodetic Co-ordinates:

Input:	Easting	Northing.	Output:	Latitude	Longitude	Convergence	Scale Factor

Easting:	**5 533 424.3919 ft**	Latitude=	54° 27' 30".0000 N
Northing:	**1 473 805.1224 ft**	Longitude=	164° 02' 30".0000 W
Convergence=	+9° 31' 47".5091	Scale Factor=	1.000 252 581 841

[59] US survey feet

7.2.4.2 SPCS27 - Lambert - American Samoa - USA

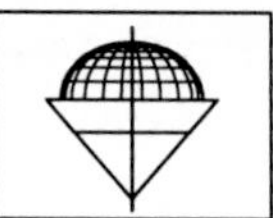

One Standard Parallel

Lambert's Conformal Conical projection with *One Standard Parallel* exists in the SPCS of American Samoa, *southern hemisphere.*

The construction of a conical projection is based on making a Central Parallel, Latitude φ, true to scale. There is no distortion of scale along the circle of tangency.

For the southern hemisphere everything remains the same except the directions of North and East which are inverted. For clarification, and the examples referring to American Samoa (Claire, 1968).

Tangent Projection with Central Parallel

Zone Parameters:		*American Samoa*	L2 =-.2464352204538777
Reference Ellipsoid		Clarke 1866	L3 =- 82312234.62429496
Semi-Major axis	a:	**20 925 832.16 ft**	L4 =- 82000000.02429496
Recipr. Flattening	f^{-1}:	**294.9786982**	L5 =- 82312234.62429496
Standard Parallel	φ_o:	**14° 16' 00" S**	L6 =- 87541693.88196063
Lon. Grid Origin	λ_o:	**170° 00' 00" W** of Greenwich	L7 = 1
Scale Factor	k_o:	**1.0000**, exactly	F0 = 6.761032571864926D-03
False Easting	E_o:	**500 000.0 ft**	F2 = 5.317220426784075D-05
False Northing	N_b:	**312 234.6 ft**	F4 = 5.73056251095528D-07
			F6 = 7.019628022792133D-09
			F8 = 9.412928282121681D-11

Conversion of NAD27 Geographicals to Lambert Grid

The Direct calculation is to convert Geodetic Co-ordinates into LCC-Planar Co-ordinates:

Input: Latitude Longitude. Output: Easting Northing Convergence Scale Factor

Latitude:	**14° 17' 00" S**	Easting [60] =	429 198.4837 ft
Longitude:	**170° 12' 00" W**	Northing=	306 154.5369 ft
Convergence=	+ 0° 02' 57".4334	Scale Factor=	1.000 000 042 040

Conversion of Lambert Grid to NAD27 Geographicals

The Inverse process is to convert LCC-Planar Co-ordinates into Geodetic Co-ordinates:

Input: Easting Northing. Output: Latitude Longitude Convergence Scale Factor

Easting:	**429 198.4837** ft	Latitude=	14° 17' 00".0000 S
Northing:	**306 154.5369** ft	Longitude=	170° 12' 00".0000 W
Convergence=	+0° 02' 57".4334	Scale Factor=	1.000 000 042 040

[60] US survey feet

7.2.4.3 SPCS83 - Lambert - State Texas Central - USA

Two Standard Parallels

Zone Parameters		*SPCS Texas, Zone Central*	L2 = .5150588822350693
Reference Ellipsoid		GRS80	L3 = 10770561.10342463
Semi-Major axis	a:	**6 378 137**	L4 = 13770561.10342463
Recipr. Flattening	f^{-1}:	**298.25722 21008 827** |[61]	L5 = 10622600.32497674
Lower Parallel	φ_l:	**30° 07' 00" N**	L6 = 14219009.88127395
Upper Parallel	φ_u:	**31° 53' 00" N**	L7 = .9998817436292895
Lat. Grid Origin	φ_b:	**29° 40' 00" N**	F0 = 6.686920927320709D-03
Central Parallel	φ_o=	31° 00' 05".00701 57360 N	F2 = 5.201458331082286D-05
Lon. Grid Origin	λ_o:	**100° 20' 00" W** of Greenwich	F4 = 5.544580077481781D-07
Scale Factor	k_o=	0.999 881 743 629 290	F6 = 6.717675115728047D-09
False Easting	E_o:	**700 000**	F8 = 8.907661557192931D-11
False Northing	N_b:	**3 000 000**	

Conversion of NAD83 Geographicals to Lambert Grid

The Direct calculation is to convert Geodetic Co-ordinates into LCC-Planar Co-ordinates:

Input: Latitude Longitude.			Output: Easting Northing	Convergence	Scale Factor
Latitude	φ_1:	**32° 00' 00" N**	Easting	E_1=	117 571.2278
Longitude	λ_1:	**106° 30' 00" W**	Northing	N_1=	3 274 824.8169
Convergence	γ_1=	- 3° 10' 34".3072	Scale Factor	K_1=	1.000 033 424 208
Latitude	φ_2:	**31° 54' 15" N**	Easting	E_2=	127 581.7269
Longitude	λ_2:	**106° 23' 16" W**	Northing	N_2=	3 263 631.5509
Convergence	γ_2=	- 3° 07' 06".2234	Scale Factor	k_2=	1.000 005 662 807

Conversion of Lambert Grid to NAD83 Geographicals

The Inverse process is to convert LCC-Planar Co-ordinates into Geodetic Co-ordinates:

Input: Easting Northing.			Output: Latitude Longitude	Convergence	Scale Factor
Easting	E_1:	**117 571.2278**	Latitude	φ_1=	32° 00' 00" N
Northing	N_1:	**3 274 824.8169**	Longitude	λ_1=	106° 30' 00" W
Convergence	γ_1=	- 3° 10' 34".3072	Scale Factor	k_1=	1.000 033 424 208
Easting	E_2:	**127 581.7269**	Latitude	φ_2=	31° 54' 15" N
Northing	N_2:	**3 263 631.5509**	Longitude	λ_2=	106° 23' 16" W
Convergence	γ_2=	- 3° 07' 06".2234	Scale Factor	k_2=	1.000 005 662 807

[61] Value accepted by NOAA-NGS (Burkholder, 1984).

SPCS83 - Lambert - State Texas Central - USA

Two Standard Parallels

Input:	Easting	Northing	Output: Arc-to-chord or (t - T) correction

N_0: **3 147 960.7783** (calc.)					
E_1: **117 571.2278**	E_2:	**127 581.7269**	(t - T)	δ_{1-2}=	+2".9100
N_1: **3 274 824.8169**	N_2:	**3 263 631.5509**	(t - T)	δ_{2-1}=	- 2".8243

Ellipsoidal and Grid Calculation (manual)

Using GBD00000.BAS gives:

True Distance	S_{1-2}:	*15 016.3439*
True Azimuth	α_{1-2}:	*135° 00' 56".5292*
True Azimuth	α_{2-1}:	*315° 04' 30".3301*
Grid Distance	D_{1-2}:	*15 016.6340*
True Distance	$S_{calc\ 1-2}$:	15 016.3437
See (5.03), using		
Scale Factor	k_{1-2}:	1.000 019 329 958

Grid bearing	t_{1-2}	138° 11' 33".7429
+ convergence	γ_1	- 3° 10' 34".3072
- (t - T)	δ_{1-2}	+0° 00' 02".9100
True Azimuth	α_{1-2}:	135° 00' 56".5257
Grid bearing	t_{2-1}	318° 11' 33".7429
+ convergence	γ_2	- 3° 07' 06".2234
- (t - T)	δ_{2-1}	- 0° 00' 02".8243
True Azimuth	α_{2-1}:	315° 04' 30".3438

Note

Documentation may be found on the Lambert Conformal Conical projection for the scale factor, the use of polynomial coefficients, and the arc-to-chord correction using this example (Vincenty, 1985b, 1986a, 1986b).

Use of Polynomials

Use of *polynomial coefficients* in conversions of co-ordinates on the Lambert conformal conical projection.

Conversions of co-ordinates with *hand calculators and small computers* from geodetic co-ordinates to Lambert grid co-ordinates and vice versa can be greatly simplified by using computed constants for a specific area. The method, and the constants for the *Contiguous USA* can be found in (Stem, 1989a). Formulae are given in (Pearson II, 1990).
Nevertheless, these formulae are unsuitable without modification for application to Lambert1972 (Belgium).

With *10 significant figures* and some care in handling large numbers of the input, the method gives results correctly to the millimetre. The method, a polynomial approach, is simple in application, efficient in calculation, and sufficiently precise in zones of wide north-south extent.

7.3 Accuracy

The term accuracy refers to the closeness between calculated values and their correct or true values, see [2.4.1], Accuracy and Precision.

7.3.1 Latitude and Longitude Round-Trip Errors

The "round-trip errors" are the differences in grades between the starting and ending co-ordinates, illustrated in Figure 36 and Figure 37 for a latitudinal distance 49^g N to 55^g N, a longitudinal distance 9^g W to 9^g E from the reference meridian (longitude grid origin), and are calculated as follows:

Latitudes and longitudes are converted to the eastings and northings, which are converted back to latitudes and longitudes.

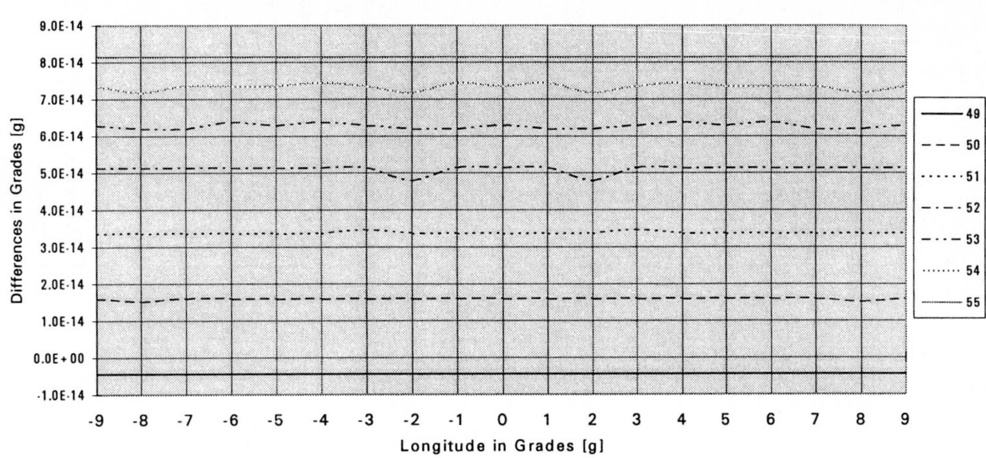

Figure 36: LCC round-trip error of latitudes

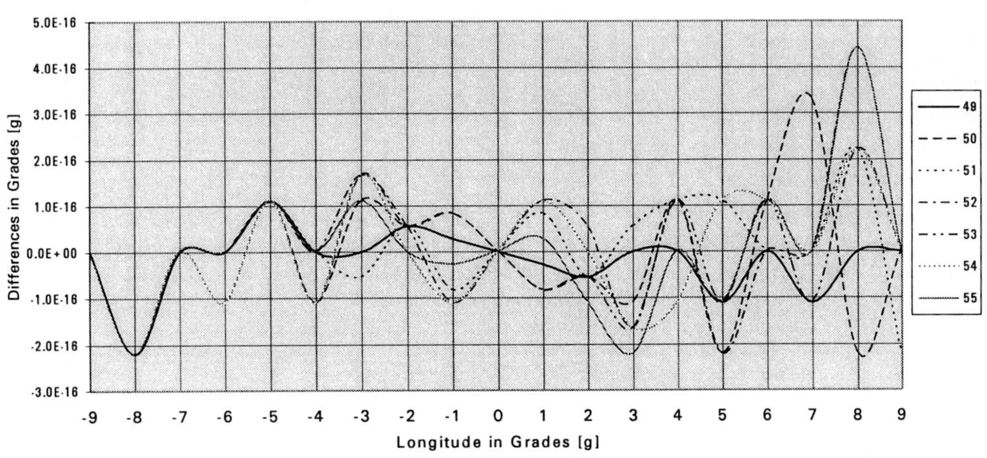

Figure 37: LCC round-trip error of longitudes

7.3.2 Easting and Northing Round-Trip Errors

The "round-trip errors" are the differences in metres between the starting and ending co-ordinates, illustrated in Figure 38 and Figure 39 for a latitudinal distance 49^g N to 55^g N, longitudinal distance 9^g W to 9^g E from the reference meridian (longitude grid origin), and are calculated as follows:

The eastings and northings are converted to latitudes and longitudes, which are converted back to the eastings and northings.

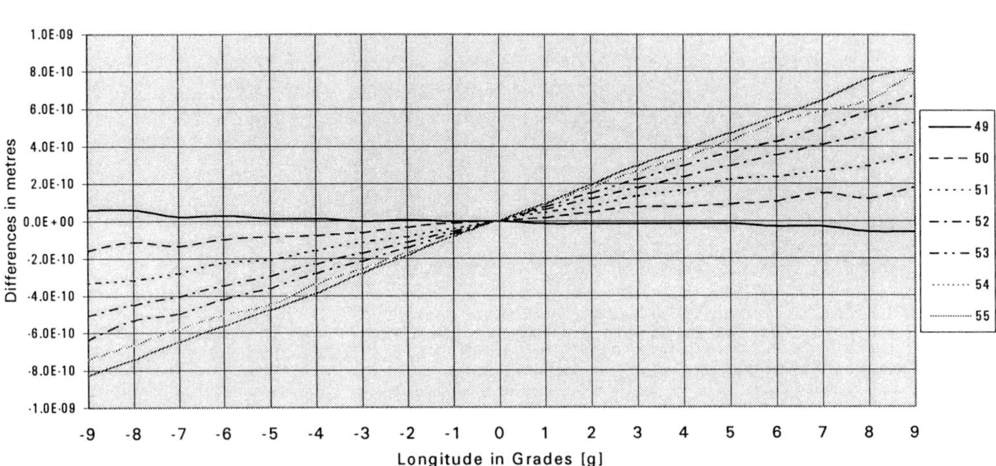

Figure 38: LCC round-trip error of eastings

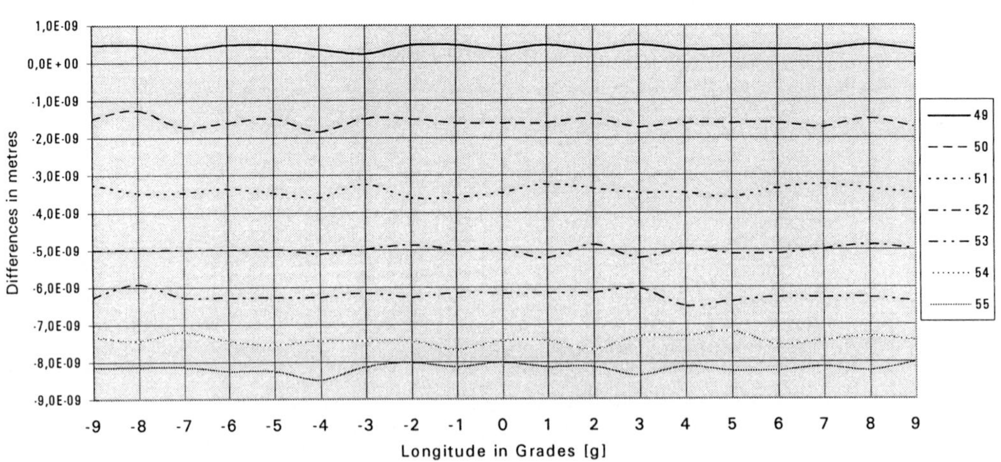

Figure 39: LCC round-trip error of northings

Note

Co-ordinates of historical data points must be taken at face value, with the realisation that such co-ordinates could be significantly in error. In practice, exact co-ordinates in any reference system are not obtained (Floyd, 1985).

8. Oblique Mercator Projection

The oblique Mercator (OM) projection for which the basic formulae were originally developed by Brigadier M. Hotine is used where the area of interest is oblong, and the longer axis through the territory is skewed with respect to the meridians (Hotine, 1946-1947).

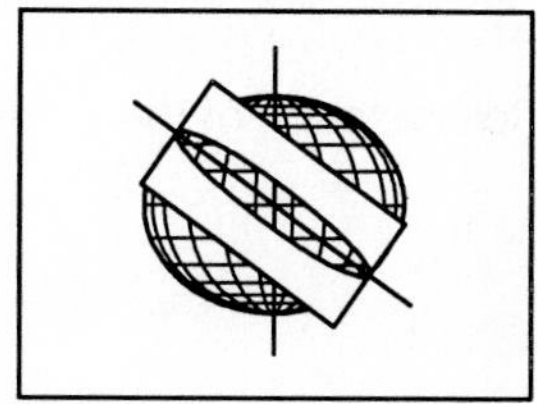

Figure 40: Oblique Mercator

8.1 RSO and HOM Projections

The fact that the *Rectified Skew Orthomorphic (RSO) projection* is skewed introduces complications, but in certain areas it has advantages over the Gauss-Krüger (GK) and the Lambert (LCC) projection. Its errors are very similar to those of the Gauss-Krüger projection. The projection is actually made from the ellipsoid onto an auxiliary surface of an *aposphere* - the general surface of constant Gaussian curvature and an intermediate surface between the ellipsoid and the plane - and finally to a plane.

The process is complicated, but orthomorphism is preserved throughout. Grids constructed on the skew orthomorphic projection exist for e.g. Borneo (Brunei Darussalam, East-Malaysia), and West-Malaysia - *the Rectified Skew Orthomorphic projection* - and the State Alaska, United States of America - *the Hotine Oblique Mercator (HOM) projection* (Bomford, 1977).

8.1.1 Oblique Mercator Mapping Equations

Symbols and Definitions

See Figure 41, and [8.1.2], pp 159.
All angles are expressed in *radians.* See program OM000000.BAS [11.7.7].

a	equatorial radius of the ellipsoid
b	semi-minor axis of the ellipsoid
f	flattening of the ellipsoid
k_c	grid scale factor at the local origin
φ_c	latitude of local origin
λ_c	longitude of local origin
α_c	azimuth of positive skew axis (u-axis) at local origin, Initial Line
α_0	azimuth of positive skew axis at equator, Line of Projection
E_0	false easting
N_0	false northing
M_0	conversion factor
Q	isometric latitude
X	conformal latitude
λ_0 (ω_0)	basic longitude of the true origin |[62]
φ	parallel of geodetic latitude, positive north
λ	meridian of geodetic longitude, positive east
E (X)	easting rectified co-ordinate |[62]
N (Y)	northing rectified co-ordinate
γ	convergence
k	point grid scale factor
k_{12}	line scale factor between P_1 and P_2
δ_{12}	(t - T)", arc-to-chord correction from P_1 to P_2

[62] Borneo formulae use the constants: x, y, ω_o, E and N; for Alaska instead: u, v, λ_o, X and Y

e^2 first eccentricity squared
e'^2 second eccentricity squared

Computation of the Constants

$$c_2 = e^2/2 + 5e^4/24 + e^6/12 + 13e^8/360 + 3e^{10}/160$$
$$c_4 = 7e^4/48 + 29e^6/240 + 811e^8/11520 + 81e^{10}/2240$$
$$c_6 = 7e^6/120 + 81e^8/1120 + 3029e^{10}/53760 \qquad (8.01)$$
$$c_8 = 4279e^8/161280 + 883e^{10}/20160$$
$$c_{10} = 2087e^{10}/161280$$

$$F_0 = 2(c_2 - 2c_4 + 3c_6 - 4c_8 + 5c_{10})$$
$$F_2 = 8(c_4 - 4c_6 + 10c_8 - 20c_{10})$$
$$F_4 = 32(c_6 - 6c_8 + 21c_{10}) \qquad (8.02)$$
$$F_6 = 128(c_8 + 8c_{10})$$
$$F_8 = 512c_{10}$$

Computation of Projection Zone Constants

Constants and expressions within the oblique Mercator mapping equations are ellipsoid and zone specific.

$$B = (1 + e'^2 \cos^4 \varphi_c)^{1/2} \qquad (8.03)$$
$$W_c = (1 - e^2 \sin^2 \varphi_c)^{1/2} \qquad (8.04)$$
$$A = a B (1 - e^2)^{1/2} / W_c^2 \qquad (8.05)$$
$$Q_c = \tfrac{1}{2} [\ln\{(1 + \sin\varphi_c)/(1 - \sin\varphi_c)\} - e \ln\{(1 + e\sin\varphi_c)/(1 - \sin\varphi_c)\}] \qquad (8.06)$$
$$C = \cosh^{-1}[B(1 - e^2)^{1/2}/(W_c \cos\varphi_c)] - B Q_c \qquad (8.07)$$
$$D = k_c A / B \qquad (8.08)$$
$$\sin\alpha_0 = (a \sin\alpha_c \cos\varphi_c)/(A W_c) \qquad (8.09)$$
$$\lambda_0\ (\omega_0) = \lambda_c + (\sin^{-1}[\sin\alpha_0 \sinh(B Q_c + C)/\cos\alpha_0])/B \qquad (8.10)$$
$$F = \sin\alpha_0 \qquad (8.11)$$
$$G = \cos\alpha_0 \qquad (8.12)$$
$$I = k_c A / a \qquad (8.13)$$

Direct Conversion Computation

Input: geodetic co-ordinates of a point P (φ, λ).
Output: grid co-ordinates of a point P (E, N), convergence angle (γ), the grid scale factor (k).

$$L = (\lambda - \lambda_0) B \qquad (8.14)$$
$$Q = \tfrac{1}{2}[\ln\{(1 + \sin\varphi)/(1 - \sin\varphi)\} - e\ln\{(1 + e\sin\varphi)/(1 - e\sin\varphi)\}] \qquad (8.15)$$
$$J = \sinh(B Q + C) \qquad (8.16)$$
$$K = \cosh(B Q + C) \qquad (8.17)$$
$$u\ (x) = D \tan^{-1}[(J G - F \sin L)/\cos L] \quad \text{(skew co-ordinates along axis of projection)} \qquad (8.18)$$
$$v\ (y) = \tfrac{1}{2} D \ln[(K - F J - G \sin L)/(K + F J + G \sin L)]$$
(skew co-ordinates perpendicular to axis of projection) (8.19)

$$E\ (X) = u \sin\alpha_c + v \cos\alpha_c + E_0 \qquad (8.20)$$
$$N\ (Y) = u \cos\alpha_c - v \sin\alpha_c + N_0 \qquad (8.21)$$
$$\gamma = \tan^{-1}((F - J G \sin L)/(K G \cos L)) - \alpha_c \qquad (8.22)$$
$$k = I(1 - e^2 \sin^2\varphi)^{1/2} \cos(u/D)/(\cos\varphi \cos L) \qquad (8.23)$$

Recognise in (8.06, 8.15): $\ln(\tan((\pi/4)+(\varphi/2))) = \tfrac{1}{2}\ln((1+\sin(\varphi))/(1-\sin(\varphi)))$ (Gretschel, 1873; Le Pape, 1994), and [7.1.1].

Inverse Conversion Computation

Input:	grid co-ordinates of a point P (E, N).
Output:	geodetic co-ordinates of a point P (φ, λ).

To compute the convergence angle (γ), the grid scale factor (k), see *the equations of the direct conversion* computation.

u (x)	$= (E - E_0) \sin \alpha_c + (N - N_0) \cos \alpha_c$	(skew co-ordinates along axis of projection)	(8.24)
v (y)	$= (E - E_0) \cos \alpha_c - (N - N_0) \sin \alpha_c$	(skew co-ordinates perpendicular to axis of projection)	(8.25)
R	$= \sinh (v / D)$		(8.26)
S	$= \cosh (v / D)$		(8.27)
T	$= \sin (u / D)$		(8.28)
Q	$= [½ \ln \{ (S - R F + G T) / (S + R F - G T) \} - C] / B$		(8.29)
X	$= 2 \tan^{-1} [(\exp (Q) - 1) / (\exp (Q) + 1)]$		(8.30)
φ	$= X + (\sin X \cos X) (F_0 + F_2 \cos^2 X + F_4 \cos^4 X + F_6 \cos^6 X + F_8 \cos^8 X)$		(8.31)
λ	$= \lambda_0 - (1 / B) \tan^{-1} [(R G + T F) / (\cos (u / D))]$		(8.32)

Note

cosh x	$= (\varepsilon^x + \varepsilon^{-x}) / 2$
$\cosh^{-1}$ x	$= \ln [x + (x^2 - 1)^{½}]$
sinh x	$= (\varepsilon^x - \varepsilon^{-x}) / 2$
exp (Q)	$= \varepsilon^Q$
cosh x	$= (\varepsilon^x + \varepsilon^{-x}) / 2$, in which ε = 2.71828 18284 59045 23536 02875 (base of natural logarithms)

Arc-to-Chord Correction δ = (t - T) and Line Grid Scale Factor

Grid azimuth (t), geodetic azimuth (α), convergence angle (γ), and arc-to-chord correction (δ) at any given point are related as follows: $t = \alpha - \gamma + \delta$ (5.01).

Input:	$P_1 (u_1, v_1)$, and $P_2 (u_2, v_2)$.
Output:	(t - T) correction (δ_{12}) and line scale facto.r (k_{12}) for the line from P_1 to P_2.

φ_m	$= (\varphi_1 + \varphi_2) / 2$	(8.33)
Q	$= ½ [\ln \{ (1 + \sin \varphi_m / (1 - \sin \varphi_m) \} - e \ln \{ (1 + e \sin \varphi_m) / (1 - e \sin \varphi_m) \}]$	(8.34)
δ_{12}	$= (u_1 - u_2) (2 v_1 + v_2) / (6 D^2)$	(8.35)
k_{12}	$= [k_c \{ 1 + (v_1^2 + v_1 v_2 + v_2^2) / (6 D^2) \} (1 - e^2 \sin^2 \varphi_m)^{½}] / [\cos \varphi_m \cosh (B Q + C)]$	(8.36)

8.1.2 Notes on Oblique Mercator

The equations for the ellipsoid constants are given in Table 6: Defining parameters and associated constants.

The equations in this section are found in (Brazier, 1950; Admiralty, 1965; NOAA-C&GS, 1961) and (Floyd, 1985; Stem, 1989a), based on (Vincenty, 1984b). The equations for the isometric latitude constants are found in (Berry, 1970; Burkholder, 1985). The equations were further manipulated by the author.

The algorithms [11.7.7] for conversion of co-ordinates, local scale factor and the arc-to-chord correction for any point on the grid are based on Hotine's *"Rectified Skew Orthomorphic"* (RSO) projection. Some grids may include a false easting and/or a false northing constant in the formulae.

The Oblique Mercator projection and its conversion equations should not be used under certain circumstances:

- If the centre point of the area of interest lies near either pole: the Stereographic projection may be used
- If a point of the Central Line defining the projection lies at either pole, use equations for the Transverse Mercator projection instead
- If the two points of the Central Line both lie on or close to the Equator, the normal Mercator projection may be used
- In general, if the two points of the Central Line lie on the same parallel of latitude other than the Equator, use equations for the Lambert Conformal Conical projection instead.

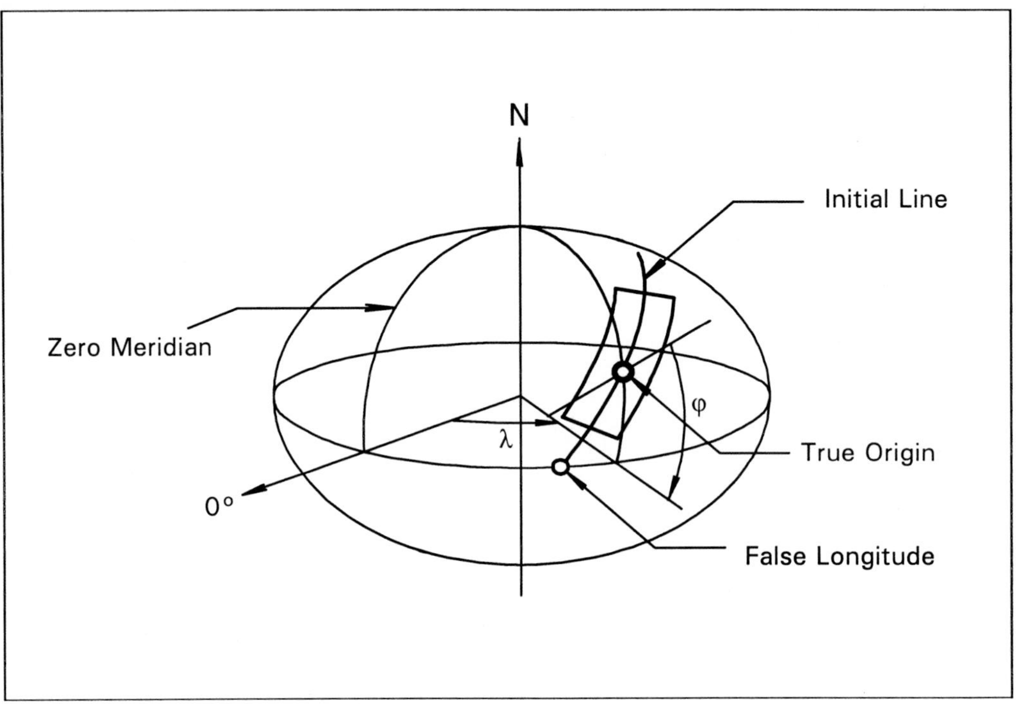

Figure 41: Initial Line of the Borneo RSO grid

8.1.3 Description of the Borneo Rectified Skew Orthomorphic Grid

The Central Line - called *the Initial Line* - of the projection runs in the direction of the longer dimension of the area of interest. The projection parameters for the Oblique Mercator projection are provided by *latitude and longitude* of a centre point, the *azimuth* of the Initial Line, and the *scale factor* at the centre point of the skewed Initial Line, which is in this projection a *geodesic,* positive between 0° and 180°.

The program is using the following constants [63]:

Everest's Figure of the Earth for Borneo			
Semi-Major axis	a	=	6 974 310.600 indian yards or
	a	=	6 377 276.3458 m
Reciprocal Flattening	f	=	300.8017
scale factor at Origin	k_c	=	0.99984
Defined True Origin O:	φ_c	=	4° 00' N
	λ_c	=	115° 00' E
Co-ordinates at origin:	E_0	=	29 352.4763 chains west from true origin
	N_0	=	22 014.3572 chains south from true origin
False Origin:	Lat	=	0° 00' 00".61 S (calculated value)
	Lon	=	109° 41' 07".87 E (calculated value)

Figure 42 shows a part of the Borneo Rectified Skew Orthomorphic grid. The Initial Line O'OQ makes an angle θ'=53° 07'48".3685 with the true meridian S'O'N' at the false point of origin O'. At O, the true origin, the Initial Line of projection makes an angle θ= 53° 18' 56".9537 E of true North with the true meridian SON. The difference between the angles θ and θ' is the convergence of these meridians.

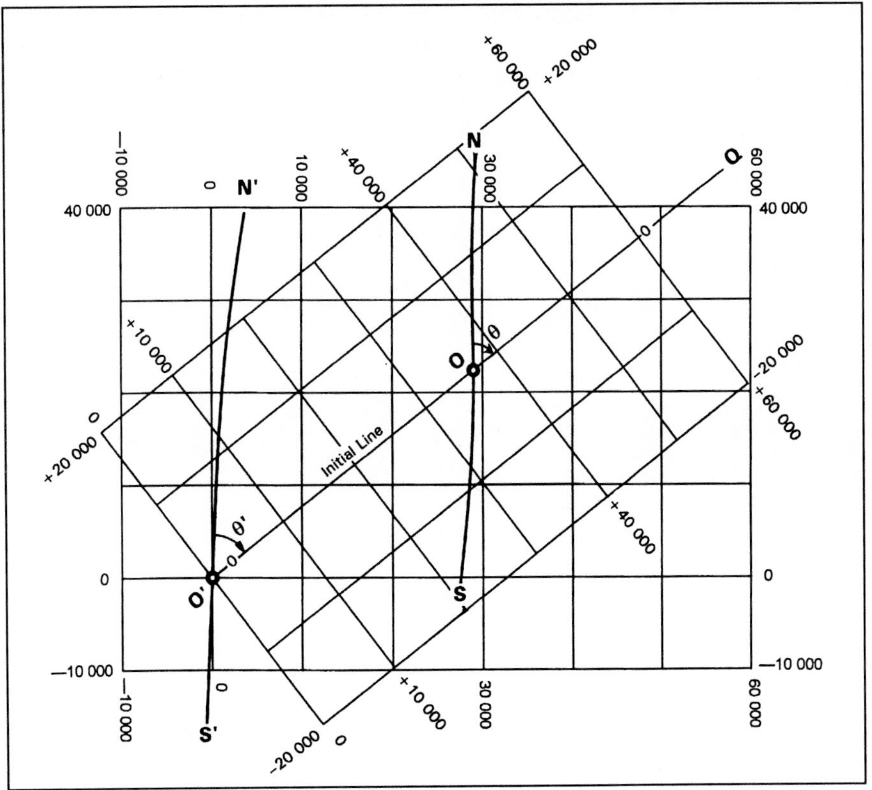

Figure 42: Borneo Rectified Skew Orthomorphic grid

[63] Borneo formulae use the constants: x, y, ω_o, E and N; for Alaska instead: u, v, λ_o, X and Y

The skew co-ordinates x y, parallel to the Initial Line, and at right angles to the Initial Line, respectively, and the map co-ordinates E, N are interrelated by the formulae:

$$E = 0.8x + 0.6y, \text{ and}$$
$$N = 0.6x - 0.8y, \text{ whence it follows that:}$$

$$x = 0.8E + 0.6N, \text{ and}$$
$$y = 0.6E - 0.8N$$

The x, y terms are referred to as skew co-ordinates and the E, N terms as rectified, planar mapping co-ordinates.

The semi-major axis, a, is entered in metres, thus all calculations are made in metres. The resulting plane co-ordinates are then converted into the unit required, for Borneo into *indian chains* (Brazier, 1950).

Footnotes

It will be observed that the Initial Line - a geodesic - of the projection has a scale factor that is nearly constant throughout its length.

The interested reader may consult a report about development of algorithms to calculate the oblique Mercator projection of the ellipsoid of revolution by Grafarend:

> *"An investigation of the conformal projection is the basis of a paper about the ellipsoidal oblique Mercator projection as devised by Hotine 50 years ago. Research in the Oblique Mercator was raised by Hotine's cryptic procedure to derive the mapping equations which should be based on similar concepts known for the normal Mercator and the transverse Mercator projection, thus extending the exercises of Hotine. Enhanced modern mapping equations of the conformal oblique Mercator projection are given in the publication"*
>
> (Grafarend / Engels, 1995e).

8.2 Oblique Mercator Projection Applications

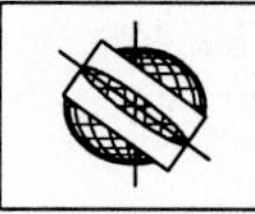

8.2.1 Reference and OM-projection Systems of Borneo

Zone Parameters:	*Borneo RSO* [64]		
Reference Ellipsoid:	Everest 1830	a=	6 377 276.843658787
Semi-Major axis:	**6 377 276.3458**	b=	1.003303209179641
Reciprocal Flattening:	**300.8017**	c=	2.991328941527593D-06
Scale Factor:	**0.99984**	d=	6 355 263.713924926
Latitude Centre φ_c:	**4° 00' 00" N**	f=	0.7999999999286045
Longitude Centre λ_c:	**115° 00' 00" E**	g=	0.600000000095194
Skew Angle $_{\text{True Origin}}$:	**+53° 18' 56".95370 00000 4016**	h=	1.333333333002799
Skew Angle $_{\text{False Origin}}$:	**+53° 07' 48".36847 49615 2333**	i=	0.9998400780551293
Atan $_{\text{True Origin}}$=	1.342376 31592 6919	ω_o (λ_o)=	109°.6855202029758 E =basic longitude
Sin $_{\text{False Origin}}$=	0.8		
Cos $_{\text{False Origin}}$=	0.6		
False Easting (x):	**0**		
False Northing (y):	**0**		
Input unit:	*m*	Output unit= indian chains	
Conversion Factor:	**0.04970 99546 8** [65]		

Conversion of Timbalei1948 Geographicals to Oblique Mercator Grid

The Direct calculation is to convert Geodetic Co-ordinates into OM-Planar Co-ordinates:

Input:	Latitude	Longitude	Output:	Easting	Northing	Convergence	Scale Factor

Sta.: Ulu Membakut

Latitude	φ_1: **5° 23' 14".1129 N**	Easting	E_1=	33 765.15703 i/c
Longitude	λ_1: **115° 48' 19".8196 E**	Northing	N_1=	29 655.00567
Convergence	γ_1= + 0° 14' 36".8379	Scale Factor	k_1=	0.999 900 131 343

Sta.:Kinandukan

Latitude	φ_2: **5° 43' 08".3444 N**	Easting	E_2=	34 253.65351
Longitude	λ_2: **115° 53' 44".2988 E**	Northing	N_2=	31 480.41516
Convergence	γ_2= + 0° 14' 55".8046	Scale Factor	k_2=	0.999 947 468 405

Conversion of Oblique Mercator Grid to Timbalei1948 Geographicals

The Inverse process is to convert OM-Planar Co-ordinates into Geodetic Co-ordinates:

Input:	Easting	Northing	Output:	Latitude	Longitude
Easting	E_1:	**33 765.15703** i/c	Latitude	φ_1=	5° 23' 14".1129 N
Northing	N_1:	**29 655.00567**	Longitude	λ_1=	115° 48' 19".8196 E
Easting	E_2:	**34 253.65351**	Latitude	φ_2=	5° 43' 08".3444 N
Northing	N_2:	**31 480.41516**	Longitude	λ_2=	115° 53' 44".2988 E

[64] See (Brazier, 1950)

[65] Conversion to indian chains

Borneo Rectified Skew Orthomorphic Grid - Borneo (cont'd)

(t - T) corrections

Input:	Easting	Northing		Output:	Arc-to-chord or (t - T) correction	
E_1:	**33 765.15703**	E_2:	**34 253.65351**	(t - T)	$\delta_{1\text{-}2}$=	+5".9181
N_1:	**29 655.00567**	N_2:	**31 480.41516**	(t - T)	$\delta_{2\text{-}1}$=	- 6".5156

Borneo RSO - Ellipsoidal and Grid Calculation

Ellipsoidal and Grid Calculation (manual)

Using GBD00000.BAS gives[66] *:*

True Distance	$S_{1\text{-}2}$:	*38016.3011 m*
True Distance	$S_{1\text{-}2}$:	*1 889.78860*
True Azimuth	$\alpha_{1\text{-}2}$:	*15° 13' 25".4956*
True Azimuth	$\alpha_{2\text{-}1}$:	*195° 13' 56".8951*
Grid Distance	$D_{1\text{-}2}$:	1 889.64246
Distance	$S_{calc\ 1\text{-}2}$:	1 889.78855
See (5.03), using		
Scale Factor	$k_{1\text{-}2}$:	0.999 922 694 259

Grid bearing	$t_{1\text{-}2}$	14° 58' 54".5749
+ convergence	γ_A	+0° 14' 36".8379
- (t - T)	$\delta_{1\text{-}2}$	+0° 00' 05".9181
True Azimuth	$\alpha_{1\text{-}2}$:	15° 13' 25".4947
Grid bearing	$t_{2\text{-}1}$	194° 58' 54".5749
+ convergence	γ_2	+0° 14' 55".8046
- (t - T)	$\delta_{2\text{-}1}$	- 0° 00' 06".5156
True Azimuth	$\alpha_{2\text{-}1}$:	195° 13' 56".8951

Note

The "round-trip errors" are calculated in [8.3.1; 8.3.2] for the limits of the Borneo tables of the area between: latitude 0° 30' N to 8° 00' N, and longitude 109° 30' E to 119° 30' E of Greenwich.

[66] For Borneo: all values are expressed in indian chains

8.2.2 Reference and OM-projection Systems of Alaska - USA

The defining elements of the HOM projection are:

- Parameters of the Clarke ellipsoid of 1866 for SPCS27 (old datum) and parameters of the Geodetic Reference System of 1980 for SPCS83 (new datum)
- The geodesic passing through the point φ_c=57° 00' N, λ_c=133° 40' W, designated the centre of the projection, at an azimuth α_c=arctan (-¾, by definition)
- A grid scale factor k_c = 0.9999 (exactly)
- A rotation of the skew plane co-ordinate system u, v |[63] to minimise the map convergence of meridians (γ_c=0) and a shift of origin, as defined by the equations:
    - X (Easting) = - 0.6u+0.8v +5 000 000
    - Y (Northing) = +0.8u+0.6v - 5 000 000
- Because all computations were made in meters, the resulting X (Easting), Y (Northing) co-ordinates are then converted as a final step, into *US survey feet* by means of the relation 1.00 m= 39.37 inches (on NAD27, old datum only)
- For Alaska Zone 1 the true origin is in the vicinity of the 0° 15' N latitude, 101° 31' W Longitude. The grid origin lies at a specified point near the area of interest, for Alaska Zone 1, that point is 5 000 000 metres to the north and 5 000 000 metres to the west of the true origin.

The calculation for SPCS27, State Alaska Zone 1, United States of America, are exemplified below, SPCS83 in [8.2.2.2], and in (C&GS SP No. 62-4, 1968; C&GS SP No. 65-1 Part 49, 1961).

8.2.2.1 SPCS27 - Oblique Mercator - State Alaska Zone 1 - USA

Zone Parameters:	Alaska, Zone 1 *(old datum)*		
Reference Ellipsoid:	Clarke 1866	a=	6 388 906.015131144
Semi-Major axis:	**6378206.4**	b=	1.00029977273354
Reciprocal Flattening:	**294.9786982**	c=	4.475991314604633D-03
Scale Factor:	**0.9999**	d=	6 386 352.670132357
Latitude Centre φ_c:	**57° 00' 00" N**	f=	- 0.3270155171758155
Longitude Centre λ_c:	**133° 40' 00" W**	g=	0.9450189688711195
Skew Angle $_{\text{True Origin}}$:	**- 36° 52' 11".63152 50384 7826** |[67]	h=	- 0.3460412202799005
Skew Angle $_{\text{False Origin}}$:	**- 36° 52' 11".63152 50384 7826**	i=	1.00157735951123
Atan $_{\text{True Origin}}$=	- 0.75	λ_o=	(-) 101°.514031441744 W
Sin $_{\text{False Origin}}$=	- 0.6		
Cos $_{\text{False Origin}}$=	0.8		
False Easting:	**5 000 000**		
False Northing:	**- 5 000 000**		
Input unit:	*m*	Output unit=US survey feet	
Conversion Factor:	**3.28083 33333 33333**		

[67] Skew angle is 360° - 36° 52' 11".63152503847826

SPCS27 - Oblique Mercator - State Alaska Zone 1 - USA

Conversion of NAD27 Geographicals to Oblique Mercator Grid

The Direct calculation is to convert Geodetic Co-ordinates into OM-Planar Co-ordinates:

Input:	Latitude	Longitude	Output:	Easting	Northing	Convergence	Scale Factor

Latitude	φ_1:	**55° 00' 00".0000 N**	Easting	E_1=	2 615 716.5328 ft [68]
Longitude	λ_1:	**134° 00' 00".0000 W**	Northing	N_1=	1 156 768.9367 ft
Convergence	γ_1=	- 0° 15' 32".0289	Scale Factor	k_1=	1.000 178 235 363
Latitude	φ_2:	**54° 39' 02".6543 N**	Easting	E_2=	3 124 247.9618 ft
Longitude	λ_2:	**131° 35' 45".4321 W**	Northing	N_2=	1 035 731.6759 ft
Convergence	γ_2=	+1° 43' 24".4074	Scale Factor	k_2=	0.999 929 293 798

Conversion of Oblique Mercator Grid to NAD27 Geographicals

The Inverse process is to convert OM-Planar Co-ordinates into Geodetic Co-ordinates:

Input:	Easting	Northing	Output:	Latitude	Longitude

Easting	E_1:	**2 615 716.5328** ft	Latitude	φ_1=	55° 00' 00".0000 N
Northing	N_1:	**1 156 768.9367** ft	Longitude	λ_1=	134° 00' 00".0000 W
Easting	E_2:	**3 124 247.9618** ft	Latitude	φ_2=	54° 39' 02".6543 N
Northing	N_2:	**1 035 731.6759** ft	Longitude	λ_2=	131° 35' 45".4321 W

Input:	Easting	Northing	Output:	Arc-to-chord or (t - T) correction

E_1:	**2 615 716.5328** ft	E_2:	**3 124 247.9618** ft	(t - T)	$\delta_{1\text{-}2}$=	- 36".2674
N_1:	**1 156 768.9367** ft	N_2:	**1 035 731.6759** ft	(t - T)	$\delta_{2\text{-}1}$=	+25".7130

Ellipsoidal and Grid Calculation (manual)

Using GBD00000.BAS gives:

True Distance	$S_{1\text{-}2}$:	*159 325.4451 m* or
True Distance	$S_{1\text{-}2}$:	*522 720.2311 ft*
True Azimuth	$\alpha_{1\text{-}2}$:	*103° 08' 21".1405*
True Azimuth	$\alpha_{2\text{-}1}$:	*285° 06' 15".7628*
Grid Distance	$D_{1\text{-}2}$:	522 737.2502 ft
True Distance	$S_{calc\ 1\text{-}2}$:	522 720.1422 ft
See (5.03), using		
Scale Factor	$k_{1\text{-}2}$:	1.000 032 728 860

Grid bearing	$t_{1\text{-}2}$	103° 23' 17".0040
+ convergence	γ_1	- 0° 15' 32".0289
- (t - T)	$\delta_{1\text{-}2}$	- 0° 00' 36".2674
True Azimuth	$\alpha_{1\text{-}2}$:	103° 08' 21".2425
Grid bearing	$t_{2\text{-}1}$	283° 23' 17".0040
+ convergence	γ_2	+1° 43' 24".4074
- (t - T)	$\delta_{2\text{-}1}$	+0° 00' 25".7130
True Azimuth	$\alpha_{2\text{-}1}$:	285° 06' 15".6984

[68] For ConUS: all values are expressed in US survey feet

8.2.2.2 SPCS83 - Oblique Mercator - State Alaska Zone 1 - USA

Zone Parameters:	SPCS83 Alaska, Zone 1 *(new datum)*		
Reference Ellipsoid:	GRS80	a=	6 388 718.862305024
Semi-Major axis:	**6 378 137**	b=	1.000296461404355
Reciprocal Flattening:	**298.25722 21008 827** |[69]	c=	4.42683392640783D-03
Scale Factor:	**0.9999**	d=	6 386 186.732531594
Latitude Centre φ_c:	**57° 00' 00" N**	f=	- 0.3270129554499785
Longitude Centre λ_c:	**133° 40' 00" W**	g=	0.9450198553299663
Skew Angle $_{\text{True Origin}}$:	**- 36° 52' 11".63152 50384 7826** |[70]	h=	- 0.3460381849181334
Skew Angle $_{\text{False Origin}}$:	**- 36° 52' 11".63152 50384 7826**	i=	1.001558917661818
Atan $_{\text{True Origin}}$=	- 0.75	λ_o=	(-) 101°.513839559969 W
Sin $_{\text{False Origin}}$=	- 0.6		
Cos $_{\text{False Origin}}$=	0.8		
False Easting:	**5 000 000**		
False Northing:	**- 5 000 000**		
Input unit:	metre	Output unit=	metre
Conversion Factor:	**1**		

Conversion of NAD83 Geographicals to Oblique Mercator Grid

The Direct calculation is to convert Geodetic Co-ordinates into OM-Planar Co-ordinates:

Input:	Latitude	Longitude	Output:	Easting	Northing	Convergence	Scale Factor

Latitude	φ:	**58° 00' 12".0000 N**	Easting	E=	806 060.8446
Longitude	λ:	**133° 52' 48".0000 W**	Northing	N=	686 847.5239
Convergence	γ=	- 0° 10' 33".7951	Scale Factor	k=	0.999 939 757 868

Conversion of Oblique Mercator Grid to NAD83 Geographicals

The Inverse process is to convert OM-Planar Co-ordinates into Geodetic Co-ordinates:

Input:	Easting	Northing	Output:	Latitude	Longitude

Easting	E:	**806 060.8446**	Latitude	φ=	58° 00' 12".0000 N
Northing	N:	**686 847.5239**	Longitude	λ=	133° 52' 48".0000 W

[69] (Burkholder, 1985)

[70] Skew angle is 360° - 36° 52' 11".63152503847826

8.3 Accuracy

The term accuracy refers to the closeness between calculated values and their correct or true values, see [2.4.1], Accuracy and Precision.

8.3.1 Latitude and Longitude Round-Trip Errors

The "round-trip errors" are the differences in degrees between the starting and ending co-ordinates, illustrated in Figure 43 and Figure 44 for a latitudinal distance 0° 30' N to 8° 00' N, longitudinal distance 109° 30' E to 119° 30' E, and are calculated as follows:

Latitudes and longitudes are converted to the eastings and northings, which are converted back to latitudes and longitudes.

Oblique Mercator Differences of Latitude

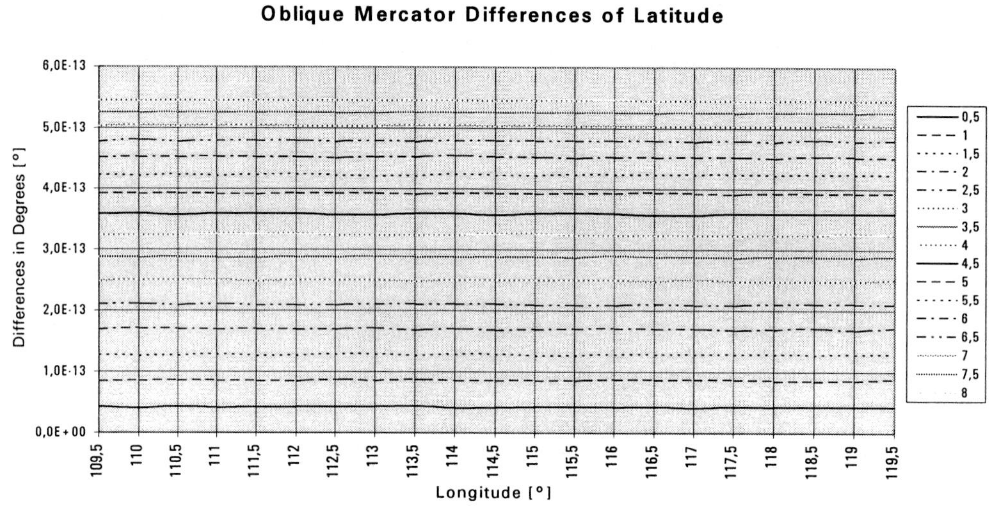

Figure 43: OM round-trip error of latitudes

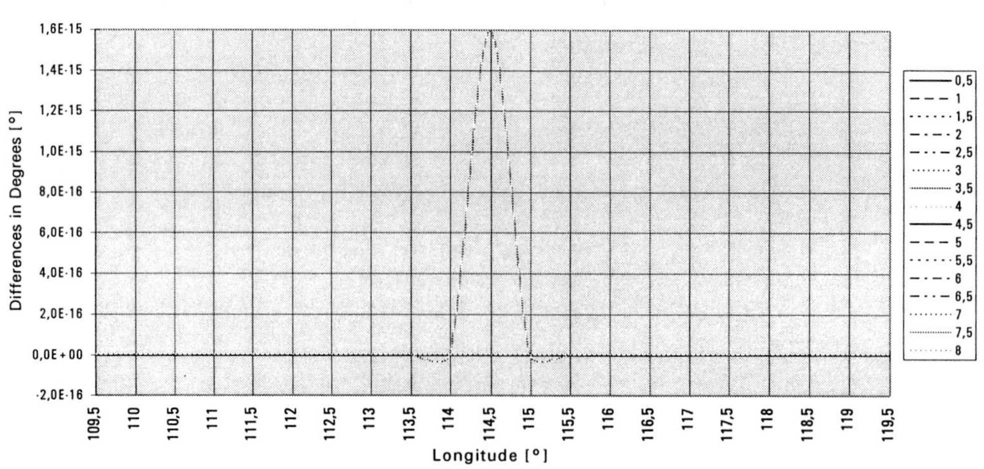

Figure 44: OM round-trip error of longitudes

8.3.2 Easting and Northing Round-Trip Errors

The "round-trip errors" are the differences in metres between the starting and ending co-ordinates, illustrated in Figure 45 and Figure 46 for a latitudinal distance 0°.5 N to 8° N, longitudinal distance 109°.5 E to 119°.5 E, and are calculated as follows:

The eastings and northings are converted to latitudes and longitudes, which are converted back to the eastings and northings.

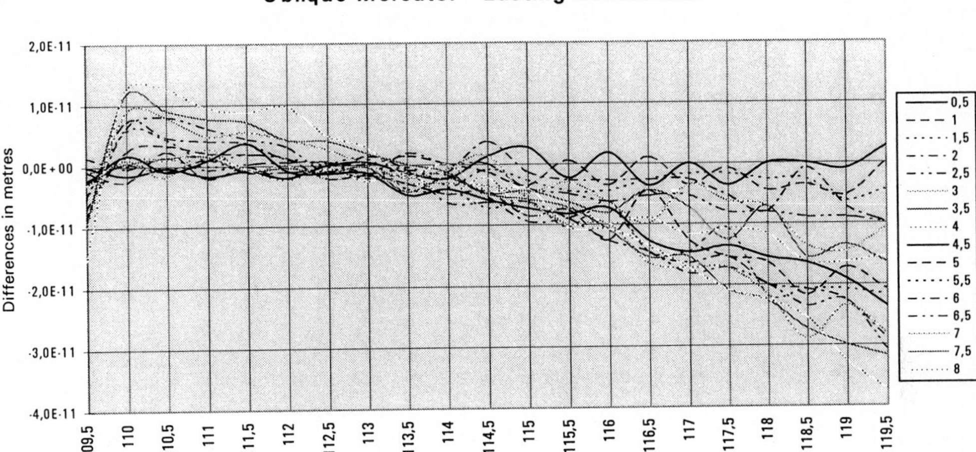

Figure 45: OM round-trip error of eastings

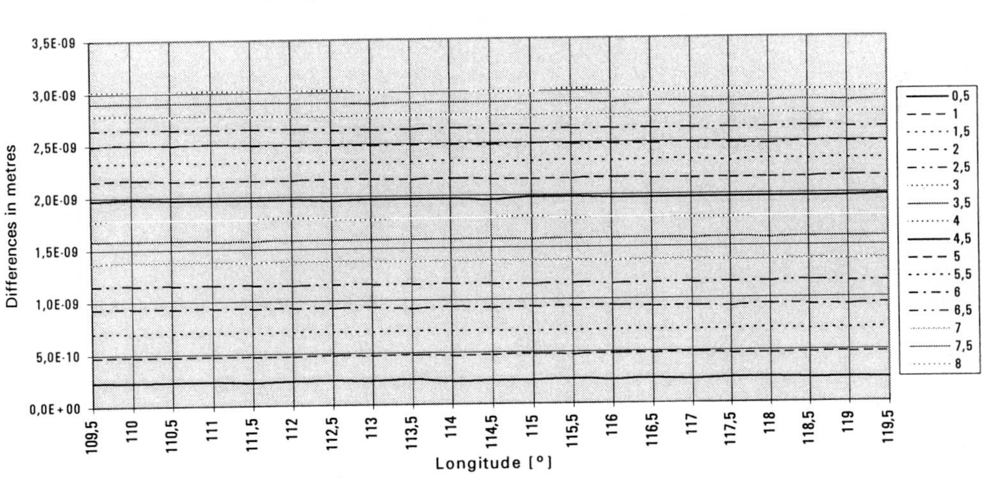

Figure 46: OM round-trip error of northings

Note

Co-ordinates of historical data points must be taken at face value, with the realisation that such co-ordinates could be significantly in error. In practice, exact co-ordinates in any reference system are not obtained

(Floyd, 1985).

9. Spatial Co-ordinate Calculations

Changing from one datum to another implies a mathematical operation with datum transformation formulae that fit the geoid as closely as possible. Thus, the geoidal heights will be small.

Employing analytical and exact techniques for the transformation of data from one datum to another are widely accepted in geodetic and surveying organisations.
A transformation can be accomplished in curvilinear and in rectangular space. The procedure is only possible if the datum shifts, e.g. $d\varphi$, $d\lambda$ and dh, are measured or calculated. NIMA has prepared methods for the transformation of data from various local geodetic datums into the *WGS84, the only global and geocentric datum right now* (Ayres, 1995; NIMA, 1991).

Several types of datum transformation formulae have been developed, such as:

- the standard Molodensky formulae or the abridged Molodensky formulae
- Multiple Regression Equation (MRE) technique
- two-dimensional, high-order polynomials fitting datum shifts
- affine transformation that relies upon actual and observed co-ordinates in both datums
- the 3-D transformation parameters for local regions.

Eliminating the conversion from geodetic to rectangular co-ordinates, the 3-parameter rectangular case is embedded mathematically in the Molodensky formulae. Depending on the availability of accurate transformation parameters, the 3-D similarity transformation is the most commonly used technique in the industry.

Representing the same point on the earth's surface, data with two different, distinct co-ordinate values is all that is required to convert data between two datums. The defining parameters provide the basis for computation of all other positions in the geodetic network. TRM00000.BAS [11.7.8] is such a program designed to transform or convert data. Readers requiring more details are requested to consult the references (Brouwer, 1989; Floyd, 1985; Ihde, 1995; Paggi, 1994a, 1994b; Rapp, 1981).

The Earth's Average Spin Axis

Datum differences attributed to differences in the ellipsoid can be more than 200 m (e.g. in SE Asia) in amplitude due to differences in adjustment and survey methodologies implicit within classical datums.
Redefining the deflection of the vertical at the origin and the orientation of the reference ellipsoid requires determination of the geodetic and astronomic positions of many monumented points throughout the geodetic network, making the geodetic datum a close fit to the Earth.

A perfect geodetic datum is not necessarily geocentric. If a datum is created with care, the equatorial plane of the ellipsoid is nearly coincident with the equatorial plane of the Earth, and the minor axis of the ellipsoid is nearly parallel with the average spin axis of the Earth *at a designated epoch.* See [3.7.1], Reference Systems.

Space Co-ordinates

In real time positioning, the user is fixed in inertial space. Co-ordinates are expressed *in the Earth-Centred-Inertial* (ECI) co-ordinate system and True-Time. Assuming an exact user's clock, the satellite and user co-ordinates are both expressed in an ECI co-ordinate system - which is *nonrotating* with the X-axis aligned with a vector from the Sun's centre to the Earth's position at the Vernal Equinox, with the *Origin* at the Earth's centre (Parkinson, 1996).

An XYZ-space co-ordinate system is simply a three-dimensional system of cartesian co-ordinates in an Euclidean space and time. *Earth-Centred Earth-Fixed* (ECEF) co-ordinates rotate with the Earth, and the XZ-plane contains the meridian of zero longitude - the Greenwich meridian. When used for satellite surveying, the origin of the co-ordinate system is at the Earth's centre of mass, providing a reference ellipsoid upon which a geodetic datum is based with truly geocentric co-ordinates. The XY-plane of the system is coincident with the equatorial plane of the ellipsoid, the Z-axis is initially coincident with the minor axis of the ellipsoid (Figure 47, [3.5], Geodetic Reference Systems).

9.1 Curvilinear Geodetic Datum Transformation

Methods for transforming cartesian co-ordinates into three-dimensional data are widely accepted in geodetic and surveying organisations. One technique used here is to carry out a similarity or S-transformation:

- geodetic co-ordinates in the *old datum* (φ_o, λ_o, h_o) are converted to old XYZ-co-ordinates: X_o ,Y_o ,Z_o ($\mathbf{X}_o$)
- old XYZ-co-ordinates ($\mathbf{X}_o$) are transformed to new XYZ-co-ordinates, X_n ,Y_n ,Z_n ($\mathbf{X}_n$), by translation (**T**) of the origin O, rotation of the axes Z, X, Y, (**R**), and a change of the scale (**1+k**)
- new XYZ-co-ordinates ($\mathbf{X}_n$) are converted to *new* geodetic co-ordinates in the new datum: φ_n, λ_n, h_n of the origin O'.

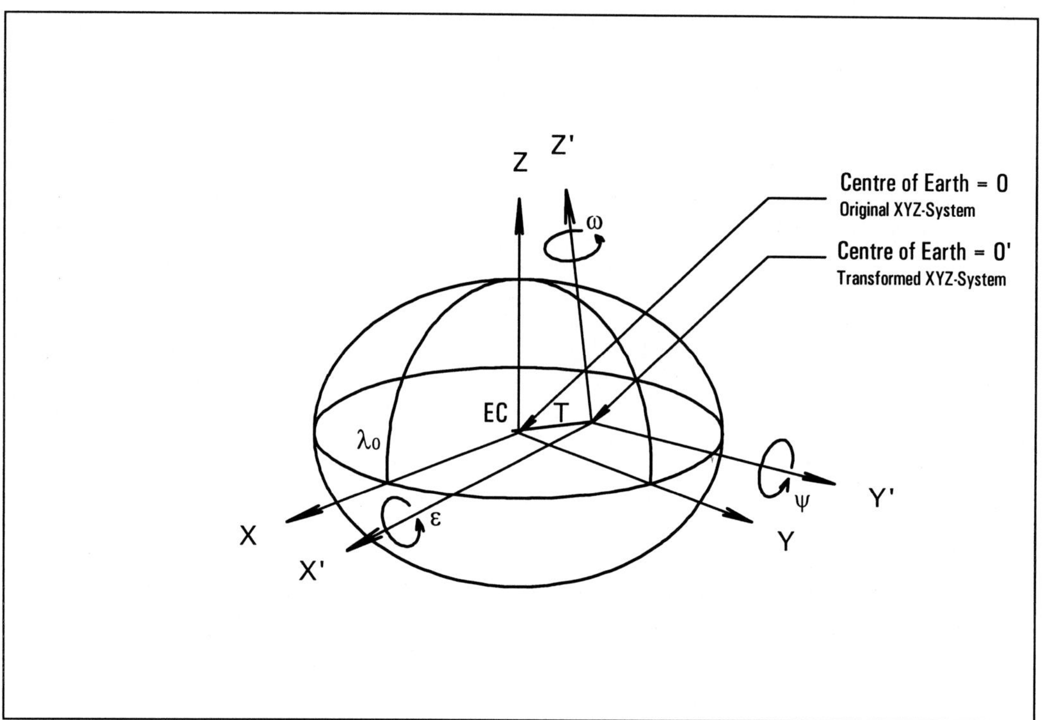

Figure 47: Translated and rotated 3-D co-ordinate system

Conversion of geodetic co-ordinates of a point P (φ_p, λ_p, h_p) to XYZ-co-ordinates X_p ,Y_p ,Z_p initially requires knowledge of the height h_p of that point P above the ellipsoid, which is the sum of orthometric height (H=elevation above MSL) and geoidal separation (N_s).

9.1.1 Transformation Equations

For computation of the ellipsoid, constants and expressions (Table 6) may contain subscripts "$_o$" and "$_n$", which denote "OLD" and "NEW", respectively.

All angles are expressed in *radians*. See Figure 48, Figure 49.

Program: TRM00000.BAS, [11.7.8]

Symbols and Definitions of S-transformation Equations

a	semi-major axis of the ellipsoid	
b	semi-minor axis of the ellipsoid	
f	flattening	
e^2	first eccentricity of ellipsoid squared	
e'^2	second eccentricity of ellipsoid squared	
H	geoidal orthometric height	
h	elevation of the datum	
NS	geoidal separation of the datum (N)	
RN	radius of curvature in the meridian	
φ	geodetic latitude	
λ	geodetic longitude	
$\mathbf{X}_o$	old three-dimensional cartesian co-ordinate system	
$\mathbf{X}_n$	new three-dimensional cartesian co-ordinate system	
O	origin of the old three-dimensional system	
O'	origin of the new three-dimensional system	
ΔX	component of origin translation X______	direction new minus old datum
ΔY	components of translation Y ________	direction new minus old datum or $(Y_n - Y_o)$
ΔZ	components of translation Z ________	direction new minus old datum or $(Z_n - Z_o)$
T	O-O' translation vector (ΔZ, ΔX, ΔY)	
R	rotation matrix of XYZ system ________	about ω, ε, ψ-angles (rad)
ω (γ)	omega ________________________	rotation of XYZ system about Z-axis
ε (α)	epsilon ______________________	rotation of XYZ system about X-axis
ψ (β)	psi __________________________	rotation of XYZ system about Y-axis
$\Delta \mathbf{k}$	change in scale of the datum ________	from old to new datum in ppm (10^{-6})
$\Delta \mathbf{k}'$	change in scale Δk_Z, Δk_X, Δk_Y________	for 9-parameter transformation

The similarity transformation in three dimensions provides an accurate methodology for the transformation of original XYZ-co-ordinates X_o, Y_o, Z_o to new XYZ-co-ordinates X_n, Y_n, Z_n.

Here, the ω, ε, ψ Eulerian angles for reference frame orientation, are rotated successively about the Z, X, Y axes respectively from the old datum to the new datum. The specific sequence of individual rotations is here as follows:

1. ω-rotation of XYZ system about Z-axis |[71] rotation, in counter-clockwise direction
2. ε-rotation of XYZ system about X-axis after application of ω, rotation in same sense as ω
3. ψ-rotation of XYZ system about Y-axis after application of ω and ε.

Note that, in principle, the formulae are different for any other sequence of Eulerian angles.

Assuming a transformation for nearly aligned co-ordinate systems, the Eulerian angles ω, ε, and ψ are very small. Hence the sines of the angles are approximately equal to the angles themselves. Consequently the sines are approximately equal to zero, and the cosines are approximately equal to one. So, it is allowable to simplify the matrix of rotation (**R**).

6-Parameters Transformation Algorithm

$$\mathbf{X}_n = \mathbf{X}_\Delta + \mathbf{R}\mathbf{x}_o \tag{9.01}$$

or:

[71] Rotation is counter-clockwise direction as seen from the positive axes toward the origin

$$\begin{vmatrix} X \\ Y \\ Z \end{vmatrix}_{new} = \begin{vmatrix} \Delta X \\ \Delta Y \\ \Delta Z \end{vmatrix} + \begin{vmatrix} 1 & \omega & -\psi \\ -\omega & 1 & \varepsilon \\ \psi & -\varepsilon & 1 \end{vmatrix} \begin{vmatrix} X \\ Y \\ Z \end{vmatrix}_{old} \tag{9.02}$$

7-Parameters Transformation Algorithm

$$\mathbf{X_n} = \mathbf{X_\Delta} + \mathbf{RX_o}\ \mathbf{(1+k)} = \mathbf{X_\Delta} + \mathbf{R_kX_o} \tag{9.03}$$

or:

$$\begin{vmatrix} X \\ Y \\ Z \end{vmatrix}_{new} = \begin{vmatrix} \Delta X \\ \Delta Y \\ \Delta Z \end{vmatrix} + \begin{vmatrix} (1+\Delta k) & \omega & -\psi \\ -\omega & (1+\Delta k) & \varepsilon \\ \psi & -\varepsilon & (1+\Delta k) \end{vmatrix} \begin{vmatrix} X \\ Y \\ Z \end{vmatrix}_{old} \tag{9.04}$$

See also: General 7-parameters transformation algorithm, (9.07).

9-Parameters Transformation Algorithm

$$\mathbf{X_n} = \mathbf{X_\Delta} + \mathbf{RX_o(1+k')} = \mathbf{X_\Delta} + \mathbf{R'_kX_o} \tag{9.05}$$

or:

$$\begin{vmatrix} X \\ Y \\ Z \end{vmatrix}_{new} = \begin{vmatrix} \Delta X \\ \Delta Y \\ \Delta Z \end{vmatrix} + \begin{vmatrix} (1+\Delta k_x) & \omega & -\psi \\ -\omega & (1+\Delta k_y) & \varepsilon \\ \psi & -\varepsilon & (1+\Delta k_z) \end{vmatrix} \begin{vmatrix} X \\ Y \\ Z \end{vmatrix}_{old} \tag{9.06}$$

General 7-Parameters Transformation Algorithm

See TRM00000.BAS: Menu: R/G (="regular" or "general").

The cartesian co-ordinates $(X_i, Y_i, Z_i)_o$ of a point P on the old datum are to be transformed into its co-ordinates $(X_i, Y_i, Z_i)_n$ on the new datum. X_0, Y_0, Z_0 are cartesian co-ordinates of a defining, Initial Point on the old datum, about which the scale, shift and rotation changes are applied.

In the *general 7-parameter transformation algorithm,* the cartesian co-ordinates X_0, Y_0 and Z_0 of the Initial Point on old datum may be set in such a way that the Initial Point is at the *centre of earth* - regular case - or at the *Origin* (φ_o, λ_o) of the old Datum - general case. Exemplified in [9.1.5.2].

$$\begin{vmatrix} X_i \\ Y_i \\ Z_i \end{vmatrix}_{new} = \begin{vmatrix} X_i \\ Y_i \\ Z_i \end{vmatrix}_{old} + \begin{vmatrix} \Delta X \\ \Delta Y \\ \Delta Z \end{vmatrix} + \begin{vmatrix} \Delta k & \omega & -\psi \\ -\omega & \Delta k & \varepsilon \\ \psi & -\varepsilon & \Delta k \end{vmatrix} \begin{vmatrix} X_i - X_0 \\ Y_i - Y_0 \\ Z_i - Z_0 \end{vmatrix}_{old} \tag{9.07}$$

in which the Cartesian co-ordinates of P_i on the old datum $(X_i, Y_i, Z_i)_o$ are to be transformed into its co-ordinates $(X_i, Y_i, Z_i)_n$ on the new datum and the parameters are, using program TRM00000.BAS, [11.7.8]:

Input and Output Data

Input: geodetic co-ordinates of a point P (φ_o, λ_o, H, N_o).
Output: geodetic co-ordinates of a point P (φ_n, λ_n, h_n, N_n).

Compute Cartesian Co-ordinates in the Old System

$$h_o = H + Ns_o \tag{9.08}$$
$$RN = a / (1 - e^2 \sin^2 \varphi)^{½} \tag{9.09}$$
$$X_0 = (RN_0 + h_0) \cos \varphi_0 \cos \lambda_0 \tag{9.10}$$
$$Y_0 = (RN_0 + h_0) \cos \varphi_0 \sin \lambda_0 \tag{9.11}$$
$$Z_0 = (RN_0 (1 - e_0^2) + h_0) \sin \varphi_0 \tag{9.12}$$

Compute Cartesian Co-ordinates in the New System

Using the following equations, the S-transformation can be performed in spatial co-ordinates:

$$\begin{aligned} X_n &= \Delta X + (X_o \quad +\omega Y_o \quad - \psi Z_o) \quad (1+\Delta k) \\ Y_n &= \Delta Y + (Y_o \quad - \omega X_o \quad +\varepsilon Z_o) \quad (1+\Delta k) \\ Z_n &= \Delta Z + (Z_o \quad +\psi X_o \quad - \varepsilon Y_o) \quad (1+\Delta k), \end{aligned} \tag{9.13}$$

in which subscripts "$_o$" and "$_n$" denote "old" and "new", respectively.

ΔX, ΔY, ΔZ, ω, ε, ψ, and Δk are the seven transformation parameters needed to compute X_n's, Y_n's, Z_n's co-ordinates. In a well-designed geodetic datum, Eulerian angles ω, and especially ε and ψ are (almost) equal to zero, the reference ellipsoid is properly oriented to physical Earth. Then the datum transformation can be used for geodetic- and precise positioning applications.

Figure 48 and Figure 49 illustrate the relation between h, Ns and H. In Figure 49 is N the radius in the Prime Vertical, and perpendicular to the meridional plane as shown by the shaded area.

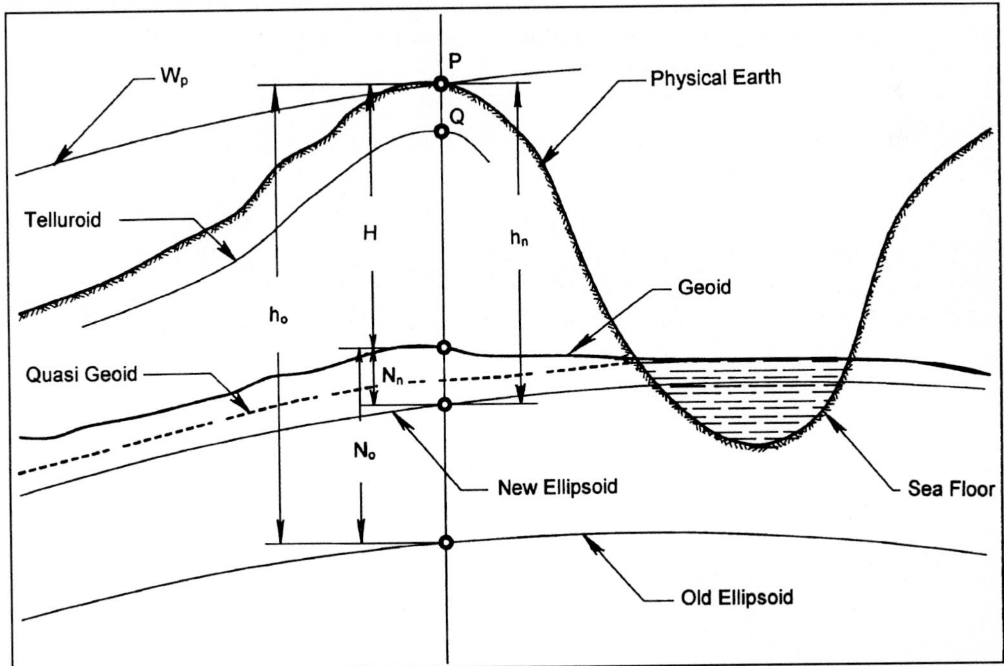

Figure 48: Geoid separation of the old and the new datum

Computation of Geodetic Co-ordinates:

Computation of latitude is obtained without iteration (Bowring, 1976):

$$p = (X_n^2 + Y_n^2)^{½} \tag{9.14}$$
$$\theta = \tan^{-1} [a_n Z_n / (b_n p) \tag{9.15}$$
$$\varphi_n = \tan^{-1} [(Z_n + e_n'^2 b_n \sin^3\theta) / (p - e_n^2 a_n \cos^3\theta)] \tag{9.16}$$

- Use a corrected value of θ and iterate if $[(\varphi_n' - \varphi_n) > 1.10^{-11}]$ before obtaining φ_n (Schuhr, 1996):

$$\varphi_n' = \varphi_n$$
$$\theta = \tan^{-1} [b_n / a_n . \tan \varphi_n') \tag{9.17}$$
$$\varphi_n = \tan^{-1} [(Z_n + e_n'^2 b_n \sin^3\theta) / (p - e_n^2 a_n \cos^3\theta)] \tag{9.18}$$

$$\lambda_n = \tan^{-1} (Y_n / X_n) \tag{9.19}$$
$$Rn_n = a_n / (1 - e^2_n \sin^2\varphi_n)^{½} \tag{9.20}$$
$$h_n = p / \cos \varphi_n - RN_n \tag{9.21}$$
$$Ns_n = h_n - H \tag{9.22}$$

Footnote

For more features on datum transformation in an Euclidean three-dimensional space, the reader may consult the report of Grafarend:

Conformal group $C_7^{(3)}$ - Curvilinear Geodetic Datum transformations

"Positioning by global satellite positioning systems results in the problem of curvilinear geodetic datum transformations between stations in a local (2+1)-dimensional geodetic reference system and a global 3-dimensional geodetic reference system.

In the process of (2+1)-D towards 3-D geodesy the datum parameters of the seven parameter global conformal group $C_7^{(3)}$ with translation, rotation, and scale observational equations are solved by a least squares adjustment"

(Grafarend, 1995b).

9.1.2 Differential GPS

Using Global Positioning System observations in the single point method, together with the satellite precise ephemeris in the WGS84 co-ordinate system will result in a horizontal accuracy of one metre and a vertical accuracy of two metres. However, several techniques are available for accomplishing a datum transformation, such as Differential GPS (DGPS) and Wide Area Differential GPS (WADGPS).

DGPS

DGPS offers several significant advantages over standalone GPS single point positioning. A DGPS can perform geodetic positioning for precise determination of new reference station co-ordinates. Typically a DGPS reference system will consist of a reference system at a selected *reference location* and a *mobile system*.

Conventional DGPS systems operate on the principle that the main error sources (i.e. the relativistic time delay, satellite ephemeris and the ionospheric and tropospheric distortions / atmospheric delays) are very closely correlated. Pseudo range measurements made at precisely located reference stations are compared with corresponding ranges computed from the known co-ordinates, and the errors derived are transmitted as differential corrections for application by DGPS users within a specified range.
The estimated position is substracted from its known location to learn the position error. This deviation is used as a sign of the quality of the corrections being received. If this error exceeds a given threshold, various schemes are used to correct the reference, or to warn DGPS users in the area.

The main drawback of DGPS systems is the limited range over which the differential corrections are valid, due to the rapid decorrelation of the error sources between the reference-station-to-user distance and the positioning accuracies achieved. These include higher accuracies, the provision of a measure of integrity, and an effective countering of most of the *selected availability* effects by an actual increase in the number of reference stations. DGPS systems have been largely limited to areas of commercial interests, such as hydrographic activities and busy sea ways (Johnston, 1993). An example is given in [9.1.5].

WADGPS

Any proposed WADGPS must maintain the advantages offered by a conventional DGPS system. Eliminating the correlation between achievable accuracy and reference-station-to-user distance is achieved by separating the combined differential corrections into their component parts, dealing with those which are dependent on range separately from those which are not.

Nevertheless, a full treatment of WADGPS is beyond the scope of this book. For further reading, see (Ashkenazi, 1992).

9.1.3 Similarity Transformations of Italy

GPS technique is an important element in the study about geophysical models used to describe the dynamics of the earth. The theory of plate tectonics, crustal deformation of the Earth's surface, and identification of earthquake hazard zones have been in focus at a regional scale, and here DGPS techniques can give positions an accuracy of a few centimetres.

In 1990 the Italian Geodetic Agencies decided to set up the *TYRGEONET Project* |[72] with the purpose of performing first order geodetic GPS surveys in Italy and surrounding countries. For this project many primary stations within the Italian Geodetic Network were adopted as sites for the "TYRGEONET" stations.

For the first time, the World Geodetic System of 1984 and a regional geodetic datum were compared by using up to 25 receivers installed in 33 stations recording for six hours per day over a period of six days in 1990.

---

[72] TYRGEONET=Tyrrhenian geodetic network

In 1991, another campaign was carried out. The observation schedule consisted of up to 32 dual-frequency receivers installed in 33 stations recording for 24 hours per day over a period of ten consecutive days.

[9.1.3.2] shows six points of TYRGEONET, in WGS84 co-ordinates, which also belong to the Italian first order geodetic network, IGM1940 Datum (Hayford ellipsoid oriented in Rome, M. Mario, Astronomical definition). Using all EDM determinations of the past twenty years with orientation by Laplace azimuths resulted in a single block adjustment in 1983.

Using seven parameters to transform the co-ordinates, [9.1.3.3] shows improved co-ordinates of the same six points in the Italian Geodetic Network, IGM1983 Datum, Hayford ellipsoid. These data are the result of a one-block adjustment in 1983, with orientation by computing Laplace azimuths and by using all the EDM determinations of the past twenty years.

The comparison of the shifts between a set of GPS co-ordinates and IGM1940 / IGM1983 transformed co-ordinates and IGM1940 / IGM1983 co-ordinates in the IGM-catalogues for certified users show minor differences. The differences originate in the reference systems, but are also due to differences in adjustment and survey methodologies (Achilli, 1994; Anzidei, 1995).

See also [9.3], Using Bi-linear Interpolation.

9.1.3.1 Transformation Parameters of Italy

The following transformation parameters have been computed in Italy. Relationships between the *national datums:* IGM1940 - IGM1983 - GENOA, and WGS84 were developed and transformation parameters between International 1924 - Bessel and WGS84 were derived (Surace, 1995):

7-Parameters Transformation, WGS84 → IGM1940, referring to [9.1.3.2]

ΔX	=	+144.2 m
ΔY	=	+72.8 m
ΔZ	=	- 11.4 m
ε	=	+1.8 sec
ψ	=	- 1.2 sec
ω	=	+2.3 sec
Δk	=	+8.3 ppm

7-Parameters Transformation, WGS84 → IGM1983, referring to [9.1.3.3]

ΔX	=	+169.5 m
ΔY	=	+79.0 m
ΔZ	=	+12.9 m
ε	=	+0.6 sec
ψ	=	- 1.5 sec
ω	=	+1.2 sec
Δk	=	+2.8 ppm

6-Parameters Transformation WGS84 → IGM1940 (Figure 50)

ΔX	=	+265.358 m
ΔY	=	+177.209 m
ΔZ	=	- 86.405 m
ε	=	- 1.7199D - 05 rad
ψ	=	+1.2662D - 05 rad
ω	=	+8.4928D - 06 rad

7-Parameters Transformation, WGS84 → IGM1940 (Figure 51)

ΔX	=	- 139.052 m
ΔY	=	+90.470 m
ΔZ	=	- 471.337 m
ε	=	- 1.7199D - 05 rad
ψ	=	+1.2662D - 05 rad
ω	=	+8.4928D - 06 rad
Δk	=	+8.8720D - 05

9-Parameters Transformation, WGS84 → IGM1940 (Figure 52)

ΔX	=	+655.941 m
ΔY	=	+98.669 m
ΔZ	=	- 1232.988 m
ε	=	- 2.8062D-05 rad
ψ	=	+9.7721D-05 rad
ω	=	- 1.7656D-06 rad
Δk_x	=	- 2.5246D-06
Δk_y	=	+8.0713D-05
Δk_z	=	+1.7246D-04

7-Parameters Transformation, WGS84 → Bessel-Genoa, referring to [9.1.3.5]

$\Delta X=$	- 899.956 m
$\Delta Y=$	- 72.052 m
$\Delta Z=$	- 984.521 m
$\varepsilon=$	- 2.3884D-06 rad
$\psi=$	+1.3451D-05 rad
$\omega=$	+3.0252D-05 rad
$\Delta k=$	+8.8961D-05

7-Parameters transformation algorithms in which the Cartesian co-ordinates of P_i on the old datum (X, Y, Z) are to be transformed into its co-ordinates (X, Y, Z) on the new datum are used in [9.1.3.2 - 9.1.3.5].

Using the 6-,7- or 9-parameter data mentioned above in the S-transformation algorithm result in calculated differences as shown in Figure 50, Figure 51, and Figure 52.

Exact co-ordinate values of the *six IGM1940 primary control stations* in Italy are given in [6.2.7], Reference and GK-projection Systems of Italy.

Footnotes

For additional facts on datum transformation in an Euclidean three-dimensional space, the reader may consult the report of Grafarend:

> Ten parameter conformal group $C_{10}^{(3)}$
>
> *"Three-dimensional geodetic datum transformations with two data sets of three-dimensional Cartesian co-ordinates which leave angles and distance ratios equivariant (covariant, form invariant) are generated by the ten parameter conformal transformation group $C_{10}^{(3)}$ in an Euclidean three-dimensional space.*
>
> *A geodetic datum transformation whose ten parameters are determined by effective adjustment techniques will play a central role in contemporary Euclidean point positioning"*
>
> (Grafarend, 1996b).

or consult (Baarda, 1981; Bowring, 1976; Burša, 1966; Leick, 1975; Molenaar, 1981; Paggi, 1994; Rapp, 1981; Schuhr, 1996) for more information about mathematical developments and derivations.

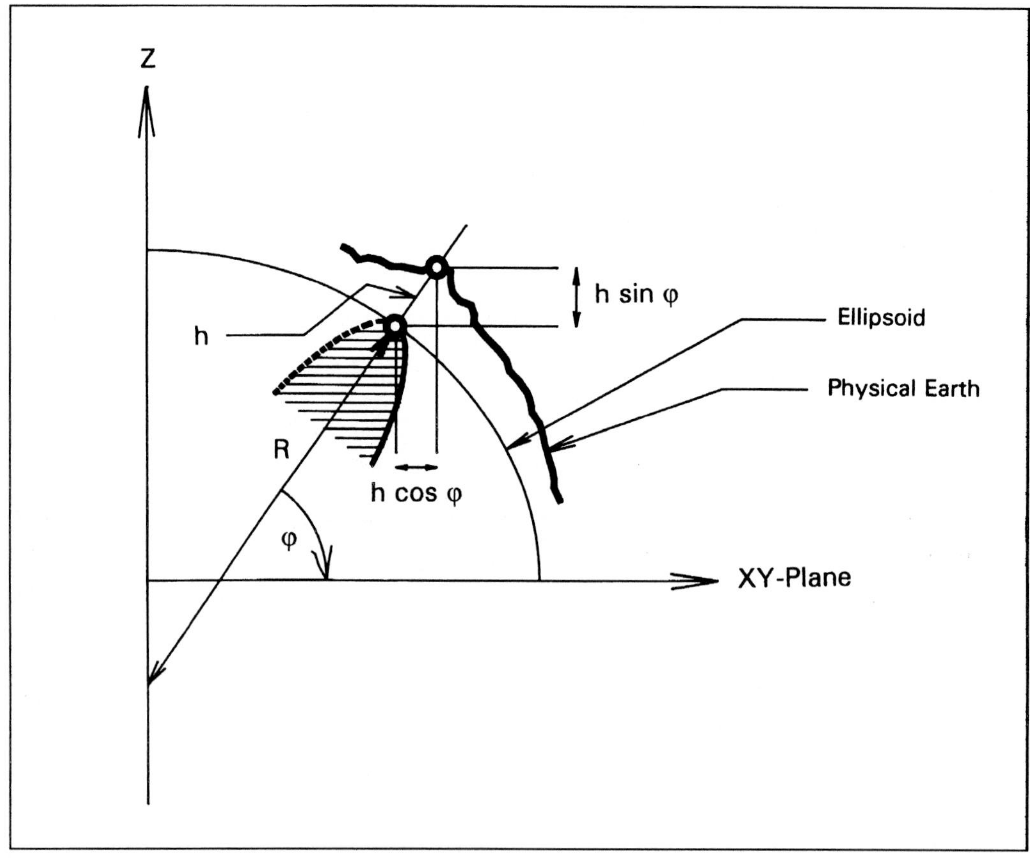

Figure 49: Ellipsoidal and physical earth

9.1.3.2 WGS84 to IGM1940 - Transformation - Italy

S-transformation from WGS84 to IGM1940 (TYRGEONET) (Achilli, 1994)
Datum: Roma, M. Mario, Sistema Geodetico Nationale

Ellipsoid Old:	WGS84	Ellipsoid New:	International 1924
Semi-Major axis a_0:	**6 378 137**	Semi-Major axis a_n:	**6 378 388**
Recipr. flattening f^{-1}_0:	**298.25722 3563**	Recipr. flattening f^{-1}_n:	**297**
Delta X_m:	**144.2**	Omega ["]:	**2.3** (γ)
Delta Y_m:	**72.8**	Epsilon ["]:	**1.8** (α)
Delta Z_m:	**- 11.4**	Psi ["]:	**- 1.2** (β)
		Delta k:	**0.00000 83**

Datum Transformation

S-transformation of WGS84 Geodetic Co-ordinates into IGM1940 Geodetic Co-ordinates:

Input: Latitude Longitude H N_0		Output: Latitude Longitude h_n	
Sta.Firenze			
Latitude-old:	**43° 46' 39".4948 N**	Latitude-new=	43° 46' 37".1619 N
Longitude-old:	**11° 15' 37".0789 E**	Longitude-new=	11° 15' 38".1697 E
H:	**114.20**		
Nsep:	**0**	h-new=	64.045
Sta. Tremiti			
Latitude-old:	**42° 07' 19".7685 N**	Latitude-new=	42° 07' 17".3518 N
Longitude-old:	**15° 30' 23".7096 E**	Longitude-new=	15° 30' 24".0550 E
H:	**124.48**		
Nsep:	**0**	h-new=	77.050
Sta. S.Angelo			
Latitude-old:	**41° 42' 27".0395 N**	Latitude-new=	41° 42' 24".6277 N
Longitude-old:	**15° 57' 07".7978 E**	Longitude-new=	15° 57' 08".0511 E
H:	**900.40**		
Nsep:	**0**	h-new=	853.327
Sta. Lucera			
Latitude-old:	**41° 31' 00".4650 N**	Latitude-new=	41° 30' 58".0800 N
Longitude-old:	**15° 28' 21".9479 E**	Longitude-new=	15° 28' 22".2585 E
H:	**114.9**		
Nsep:	**0**	h-new=	67.705
Sta. Taburno			
Latitude-old:	**41° 05' 33".3983 N**	Latitude-new=	41° 05' 31".0665 N
Longitude-old:	**14° 36' 12".8313 E**	Longitude-new=	14° 36' 13".2385 E
H:	**1442.79**		
Nsep:	**0**	h-new=	1395.401
Sta. Solaro			
Latitude-old:	**40° 32' 42".3430 N**	Latitude-new=	40° 32' 40".0515 N
Longitude-old:	**14° 13' 24".6588 E**	Longitude-new=	14° 13' 25".0839 E
H:	**638.73**		
Nsep:	**0**	h-new=	591.368

9.1.3.3 WGS84 to IGM1983 - Transformation - Italy

S-transformation from WGS84 to IGM1983 (TYRGEONET) (Achilli, 1994)
Datum: Roma, M. Mario, Sistema Geodetico Nationale

Ellipsoid Old:	WGS84	Ellipsoid New:	International 1924
Semi-Major axis a_0:	**6 378 137**	Semi-Major axis a_n:	**6 378 388**
Recipr. flattening $f^{-1}{}_0$:	**298.25722 3563**	Recipr. flattening $f^{-1}{}_n$:	**297**
Delta X_m:	**169.5**	Omega ["]:	**1.2** (γ)
Delta Y_m:	**79.0**	Epsilon ["]:	**0.6** (α)
Delta Z_m:	**12.9**	Psi ["]:	**- 1.5** (β)
		Delta k:	**0.00000 28**

Datum Transformation

S-transformation of WGS84 Geodetic Co-ordinates into IGM1983 Geodetic Co-ordinates:

Input: Latitude Longitude H N_0		Output: Latitude Longitude h_n	
Sta. Firenze			
Latitude-old:	**43° 46' 39".4948 N**	Latitude-new=	43° 46' 37".0909 N
Longitude-old:	**11° 15' 37".0789 E**	Longitude-new=	11° 15' 38".1448 E
H:	**114.20**		
Nsep:	**0**	h-new=	64.617
Sta. Tremiti			
Latitude-old:	**42° 07' 19".7685 N**	Latitude-new=	42° 07' 17".4056 N
Longitude-old:	**15° 30' 23".7096 E**	Longitude-new=	15° 30' 24".0099 E
H:	**124.48**		
Nsep:	**0**	h-new=	77.636
Sta. S.Angelo			
Latitude-old:	**41° 42' 27".0395 N**	Latitude-new=	41° 42' 24".6995N
Longitude-old:	**15° 57' 07".7978 E**	Longitude-new=	15° 57' 08".0139 E
H:	**900.40**		
Nsep:	**0**	h-new=	853.899
Sta. Lucera			
Latitude-old:	**41° 31' 00".4650 N**	Latitude-new=	41° 30' 58".1451 N
Longitude-old:	**15° 28' 21".9479 E**	Longitude-new=	15° 28' 22".2378 E
H:	**114.90**		
Nsep:	**0**	h-new=	68.282
Sta. Taburno			
Latitude-old:	**41° 05' 33".3983 N**	Latitude-new=	41° 05' 31".1212 N
Longitude-old:	**14° 36' 12".8313 E**	Longitude-new=	14° 36' 13".2508 E
H:	**1442.79**		
Nsep:	**0**	h-new=	1395.967
Sta. Solaro			
Latitude-old:	**40° 32' 42".3430 N**	Latitude-new=	40° 32' 40".1091 N
Longitude-old:	**14° 13' 24".6588 E**	Longitude-new=	14° 13' 25".1241 E
H:	**638.73**		
Nsep:	**0**	h-new=	591.925

9.1.3.4 WGS84 to IGM1940 - Transformation - Italy

S-transformation from WGS84 to IGM1940 (Paggi, 1994)
Datum: Roma, M. Mario, Sistema Geodetico Nationale

Ellipsoid Old:	WGS84	Ellipsoid New:	International 1924
Semi-Major axis a_0:	**6 378 137**	Semi-Major axis a_n:	**6 378 388**
Recipr. flattening $f^{-1}{}_0$:	**298.25722 3563**	Recipr. flattening $f^{-1}{}_n$:	**297**
Delta X_m:	**- 139.052**	Omega $_{rad}$:	**0.00000 84928**
Delta Y_m:	**90.47**	Epsilon $_{rad}$:	**- 0.00001 7199**
Delta Z_m:	**- 471.337**	Psi $_{rad}$:	**0.00001 2662**
		Delta k:	**0.00008 872**

Datum Transformation

S-transformation of WGS84 Geodetic Co-ordinates into IGM1940 Geodetic Co-ordinates:

Input: Latitude Longitude H N_0		Output: Latitude Longitude h_n	
Sta. 122901			
Latitude-old:	**43° 07' 39".6134 N**	Latitude-new=	43° 07' 37".2498 N
Longitude-old:	**12° 03' 20".3498 E**	Longitude-new=	12° 03' 21".0760 E
H:	**381.53**		
Nsep:	**0**	h-new=	330.838
Sta. 122906			
Latitude-old:	**43° 05' 02".9613 N**	Latitude-new=	43° 05' 00".5876 N
Longitude-old:	**12° 06' 09".7856 E**	Longitude-new=	12° 06' 10".5191 E
H:	**308.10**		
Nsep:	**0**	h-new=	257.605
Sta. 122904			
Latitude-old:	**43° 08' 36".3259 N**	Latitude-new=	43° 08' 33".9666 N
Longitude-old:	**12° 09' 33".6158 E**	Longitude-new=	12° 09' 34".3569 E
H:	**307.61**		
Nsep:	**0**	h-new=	257.019
Sta. 122905			
Latitude-old:	**43° 05' 24".9261 N**	Latitude-new=	43° 05' 22".5541 N
Longitude-old:	**12° 09' 18".1767 E**	Longitude-new=	12° 09' 18".9177 E
H:	**308.20**		
Nsep:	**0**	h-new=	257.765
Sta. 122903			
Latitude-old:	**43° 11' 00".5721 N**	Latitude-new=	43° 10' 58".2222 N
Longitude-old:	**12° 08' 07".5398 E**	Longitude-new=	12° 08' 08".2770 E
H:	**307.30**		
Nsep:	**0**	h-new=	256.551
Sta. 122902			
Latitude-old:	**43° 10' 58".7419 N**	Latitude-new=	43° 10' 56".3912 N
Longitude-old:	**12° 01' 30".5887 E**	Longitude-new=	12° 01' 31".3100 E
H:	**307.99**		
Nsep:	**0**	h-new=	257.077

9.1.3.5 WGS84 to Bessel-Genoa - Transformation - Italy

S-transformation from WGS84 to Bessel (Paggi, 1994)
Datum: Genoa, M. del Telegrafico 1902, Sistema Geodetico Catastali

Ellipsoid Old:	WGS84	Ellipsoid New :	Bessel 1841
Semi-Major axis a_0:	**6 378 137**	Semi-Major axis a_n:	**6 377 397.155**
Recipr. flattening $f^{-1}{}_0$:	**298.25722 3563**	Recipr. flattening $f^{-1}{}_n$:	**299.15281 285**
Delta X_m:	**- 899.956**	Omega $_{rad}$:	**0.00003 0252**
Delta Y_m:	**- 72.052**	Epsilon $_{rad}$:	**- 0.00000 23884**
Delta Z_m:	**- 984.521**	Psi $_{rad}$:	**0.00001 3451**
		Delta k:	**0.00008 8961**

Datum Transformation

S-transformation of WGS84 Geodetic Co-ordinates into Bessel-Genoa Geodetic Co-ordinates:

Input: Latitude Longitude H N_0		Output: Latitude Longitude h_n	
Sta. 122901			
Latitude-old:	**43° 07' 39".6134 N**	Latitude-new=	43° 07' 36".7662 N
Longitude-old:	**12° 03' 20".3498 E**	Longitude-new=	12° 03' 19".3983 E
H:	**381.53**		
Nsep:	**0**	h-new=	330.839
Sta. 122902			
Latitude-old:	**43° 10' 58".7419 N**	Latitude-new=	43° 10' 55".9373 N
Longitude-old:	**12° 01' 30".5887 E**	Longitude-new=	12° 01' 29".6197 E
H:	**307.99**		
Nsep:	**0**	h-new=	257.078
Sta. 122903			
Latitude-old:	**43° 11' 00".5721 N**	Latitude-new=	43° 10' 57".7626 N
Longitude-old:	**12° 08' 07".5398 E**	Longitude-new=	12° 08' 06".6521 E
H:	**307.30**		
Nsep:	**0**	h-new=	256.552
Sta. 122904			
Latitude-old:	**43° 08' 36".3259 N**	Latitude-new=	43° 08' 33".4854 N
Longitude-old:	**12° 09' 33".6158 E**	Longitude-new=	12° 09' 32".7422 E
H:	**307.61**		
Nsep:	**0**	h-new=	257.020
Sta. 122905			
Latitude-old:	**43° 05' 24".9261 N**	Latitude-new=	43° 05' 22".0462 N
Longitude-old:	**12° 09' 18".1767 E**	Longitude-new=	12° 09' 17".2952 E
H:	**308.20**		
Nsep:	**0**	h-new=	257.766
Sta. 122906			
Latitude-old:	**43° 05' 02".9613 N**	Latitude-new=	43° 05' 00".0794 N
Longitude-old:	**12° 06' 09".7856 E**	Longitude-new=	12° 06' 08".8650 E
H:	**308.10**		
Nsep:	**0**	h-new=	257.606

Transformation - Primary Stations - Italy

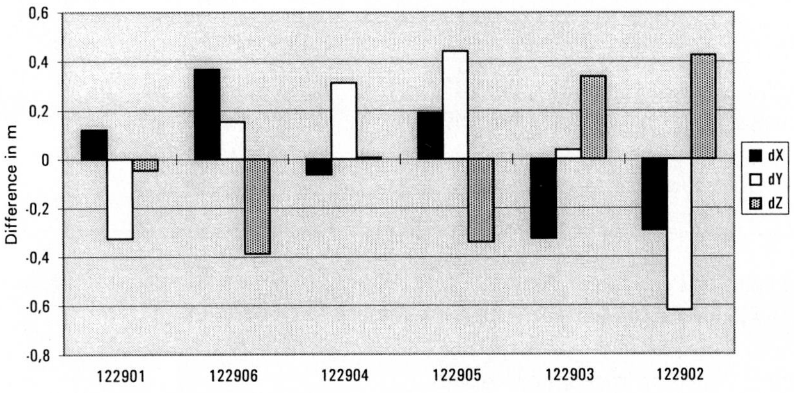

Figure 50: Calculated differences for the 6-parameter transformation solution

Transformation - Primary Stations - Italy

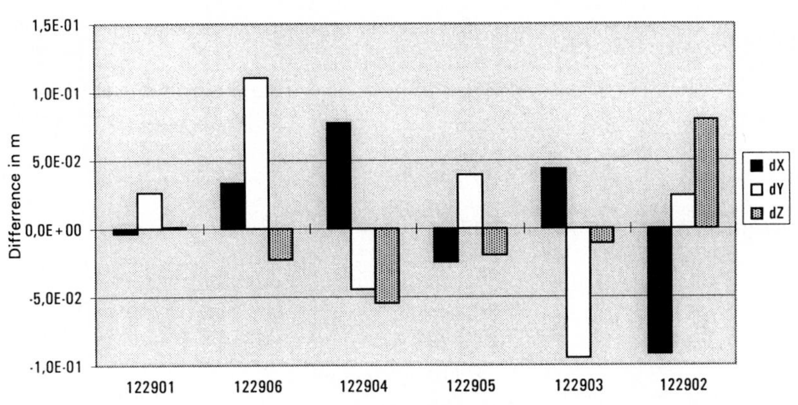

Figure 51: Calculated differences for the 7-parameter transformation solution

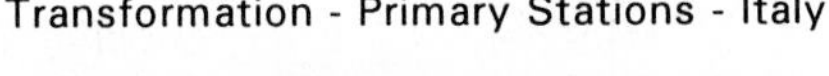

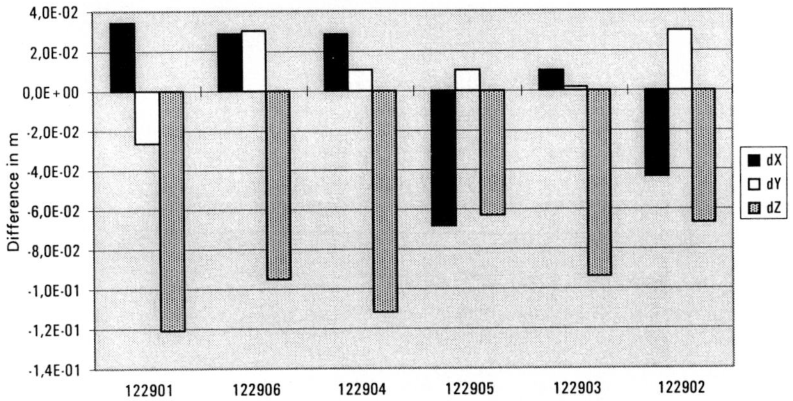

Figure 52: Calculated differences for the 9-parameter transformation solution

9.1.4 Similarity Transformations of Ireland

S-transformation from Ireland(1975) to WGS84 (Codd, 1995)

Ellipsoid Old:	Airy Modified New	Ellipsoid:	WGS84
Semi-Major axis a_0:	**6 377 340.189**	Semi-Major axis a_n:	**6 378 137**
Recipr. flattening f^{-1}_0:	**299.32496 459**	Recipr. flattening f^{-1}_n:	**298.25722 3563**
See specifications below [9.1.4.1]			

Cartesian co-ordinates of P_i on the old datum (X_i, Y_i, Z_i) are to be transformed into its co-ordinates (X_i, Y_i, Z_i) on the new datum and the parameters are, using program TRM00000.BAS, [11.7.8].

9.1.4.1 Transformation Parameters of Ireland

The following transformation parameters have been computed by the British Military Survey in 1992 using the Doppler derived values of 47 primary stations in Great Britain and 7 primary stations in Ireland. An excellent relationship between OSGB(SN)80 and WGS84 was developed and transformation parameters between Ireland(1975) and WGS84 Datum were derived.

3-Parameters Transformation, from Ireland(1975) to WGS84

ΔX	=	+506 m	Internal root mean square (RMS) vector of
ΔY	=	- 122 m	residuals within ± 3 m (Figure 53)
ΔZ	=	+611 m	

5-Parameters Transformation, from Ireland(1975) to WGS84

ΔX	=	+487.1 m	Internal RMS vector of residuals
ΔY	=	- 86.8 m	is 0.6 m (Figure 54)
ΔZ	=	+581.8 m	
ω	=	- 1.719 arc second	
Δk	=	+5.837 ppm	

7-Parameters Transformation, from Ireland(1975) to WGS84

ΔX	=	+498.5 m	Internal RMS of residuals
ΔY	=	- 106.1 m	is 0.4 m (Figure 55)
ΔZ	=	+571.3 m	
ε	=	- 0.654 arc second	
ψ	=	- 0.481 arc second	
ω	=	- 1.549 arc second	
Δk	=	+5.837 ppm	

9.1.4.2 Primary Control Stations in Ireland

The transformation parameters were computed by the British Military Survey using the Doppler derived values of seven primary stations in Ireland.

The geodetic co-ordinates of these stations are given below in Ireland(1975) and in WGS84 Datum.

Ireland1975 Station	Latitude	Longitude	Height MSL
Truskmore	54° 22' 26".7149 N	8° 22' 15".3667 W	646.5
Loughan Leigh	53° 54' 29".3826 N	6° 54' 05".3171 W	340.2
Redmount Hill	53° 13' 22".5173 N	8° 08' 35".1688 W	127.7
Corronher	52° 25' 40".7038 N	8° 52' 03".4609 W	272.8
Doolieve	51° 47' 21".5270 N	8° 27' 28".1354 W	183.8
Forth	52° 18' 56".7890 N	6° 33' 41".3839 W	236.5
Howth	53° 22' 23".1566 N	6° 04' 06".0065 W	170.7

WGS84 Station	Latitude	Longitude	Ellipsoidal Height
Truskmore	54° 22' 27".095 N	8° 22' 17".904 W	702.24
Loughan Leigh	53° 54' 29".972 N	6° 54' 08".563 W	395.12
Redmount Hill	53° 13' 23".415 N	8° 08' 37".807 W	182.05
Corronher	52° 25' 41".960 N	8° 52' 05".743 W	328.81
Doolieve	51° 47' 23".083 N	8° 27' 30".613 W	238.98
Forth	52° 18' 58".115 N	6° 33' 44".719 W	289.98
Howth	53° 22' 24".001 N	6° 04' 09".626 W	224.70

Note

These geodetical "true" co-ordinates are given for comparison with calculated values. Using the datum transformation program will result in slightly different figures.

9.1.4.3 Services of Ordnance Survey

Ireland(1975) and the Irish Grid have been adopted by both Ordnance Surveys in Ireland, Belfast and Dublin, to enable OS to produce a contiguous series of maps covering the whole island.

IRENET95 |[73]

A new geodetic network - labelled IRENET95 - was established for the whole of Ireland using dual-frequency GPS receivers in 1995. The geodetic network was observed, simultaneously with four stations in Great Britain and eight IGS *Fiducial Stations* |[74] in Europe, to incorporate the results into the European Terrestrial Reference Frame of 1989 (ETRF89).

Station co-ordinates in ETRF89 will be available in 1996.

Co-ordinate values of trigonometrical stations, benchmark heights and other technical information may be obtained by writing to the OSNI, Belfast or OSI, Dublin.

[73] (OSNI, 1994)

[74] Fiducial stations = GPS data collected at VLBI and SLR stations

Transformation - Primary Control Stations - Ireland

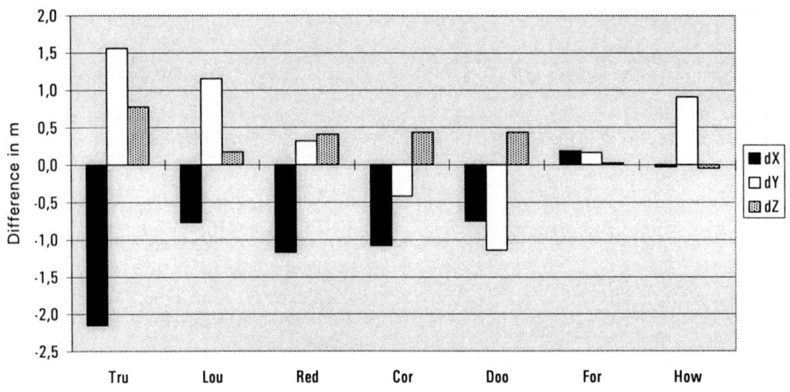

Figure 53: Calculated differences for the 3-parameter transformation solution

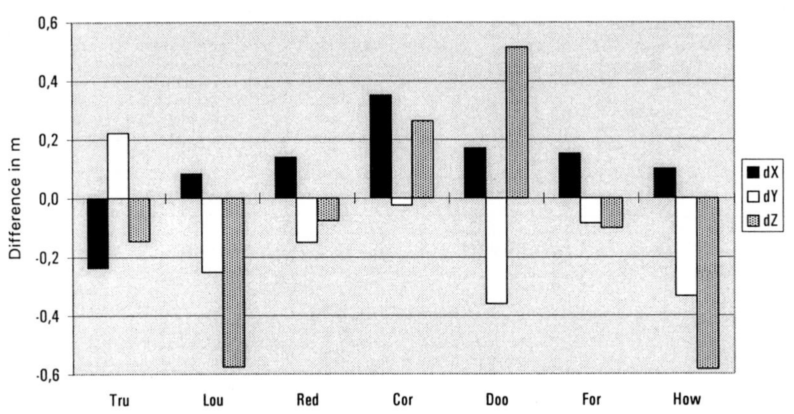

Figure 54: Calculated differences for the 5-parameter transformation solution

Transformation - Primary Control Stations - Ireland

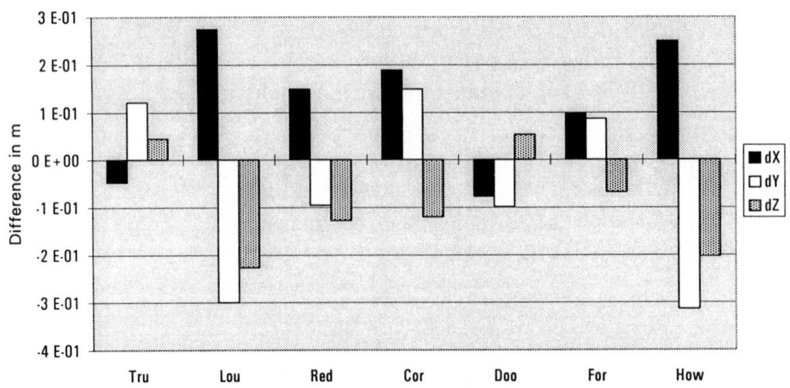

Figure 55: Calculated differences for the 7-parameter transformation solution

9.1.5 Similarity Transformations of the Netherlands

Converting geodetic co-ordinates of a national or continental terrestrial datum into the WGS84 Datum and vice versa requires specific transformation formulae; (NIMA, 1991) gives such formulae for most datums, but an *S-transformation* produces only an average result due to differences in adjustment and distortions. Using a least squares adjustment method, a relationship between WGS84 and Bessel-Amersfoort (RD1918) was developed and significant local distortions were removed. Transformation parameters between these datums have been computed by T.G. Schut in 1991 using the Doppler system derived values of *13 primary stations* in the Netherlands. Geodetic co-ordinates of Sta. Amersfoort "Lieve Vrouwe" Tower (RD-ORIGIN) are in RD1918 and WGS84, resp.:

φ RD.$_O$:	**52° 09' 22".178 N**	λ RD.$_O$:	**5° 23' 15".5 E**	h RD.$_O$:	**0 m**
φ WGS.$_O$:	**52° 09' 18".6200 N**	λ WGS.$_O$:	**5° 23' 13".9327 E**	h WGS.$_O$:	**43.348 m**

9.1.5.1 Transformation Parameters of the Netherlands

The general seven parameter transformation algorithm [9.1.1], (9.07-9.22) is used here. The cartesian co-ordinates $(X, Y, Z)_o$ of a point P on the old datum are to be transformed into its co-ordinates $(X, Y, Z)_n$ on the new datum. X_0, Y_0, Z_0 are cartesian co-ordinates of a defining, Initial Point on the old datum, about which the scale, shift and rotation changes are applied (De Min, 1996; Schut, 1991).

Regular Case I. The geoid is *not* used. In the *general 7-parameter transformation algorithm,* the cartesian co-ordinates X_0, Y_0 and Z_0 of the Initial Point on old datum may be set so that the Initial Point is at the *centre of earth*: $X_0=Y_0=Z_0=0$.

Replacing geoid height values by zero will result in latitude (φ), longitude (λ) and no height.

Exemplified in Case I.

Regular Case I - the following WGS84 to RD1918 S-transformation parameters are given:

$X_{O\,RD}$	=	0	Initial Point (earth centred)
$Y_{O\,RD}$	=	0	Initial Point (earth centred)
$Z_{O\,RD}$	=	0	Initial Point (earth centred)
ΔX	=	- 565.040 m	Bessel-Amersfoort → WGS84 - Case I
ΔY	=	- 49.910 m	
ΔZ	=	- 465.840 m	
ε	=	- 1.9848 10^{-6} rad	
ψ	=	+1.7439 10^{-6} rad	
ω	=	- 9.0587 10^{-6} rad	
Δk	=	- 4.0772 ppm	

Case I, using $\mathbf{C}_7^{(3)}$ transformation algorithm (9.04) in [11.7.8], pp. 266, see example [9.1.5.2].

General Case II. The geoid is used. In the *general 7-parameter transformation algorithm,* the cartesian co-ordinates X_0, Y_0 and Z_0 of the Initial Point on old datum may be set in such a way that the Initial Point is at *Origin "Amersfoort"* of the old Datum Bessel-Amersfoort.
Geodetic co-ordinates of the RD-ORIGIN (φ RD.$_o$, λ RD.$_o$) are converted into X_0, Y_0 and Z_0.

Exemplified in Case II.

General Case II - the following WGS84 to RD1918 S-transformation parameters are given:

$X_{O\ RD}$	=	3 903 453.1482	Initial point (Datum Amersfoort)
$Y_{O\ RD}$	=	368 135.3134	Initial point (Datum Amersfoort)
$Z_{O\ RD}$	=	5 012 970.3051	Initial point (Datum Amersfoort)
ΔX	=	- 593.032 m	Bessel-Amersfoort → WGS84 - Case II
ΔY	=	- 26.000 m	
ΔZ	=	- 478.741 m	
ε	=	- 1.9848 10^{-6} rad	
ψ	=	+1.7439 10^{-6} rad	
ω	=	- 9.0587 10^{-6} rad	
Δk	=	- 4.0772 ppm	

Case II is exemplified using the general $\mathbf{C}_7^{(3)}$ transformation algorithm (9.07) in [11.7.8], page 266, see example [9.1.5.3].

9.1.5.2 Bessel-Amersfoort to WGS84 Transformation - the Netherlands

S-transformation from Bessel-Amersfoort (RD1918) to WGS84 - example (Strang van Hees, 1997).

Ellipsoid old:	Bessel-Amersfoort	Ellipsoid new:	WGS84
Semi-Major axis a_0:	**6 377 397.155**	Semi-Major axis a_n:	**6 378 137**
Recipr. flattening f^{-1}_0:	**299.15281 285**	Recipr. flattening f^{-1}_n:	**298.25722 3563**
Delta X_m:	**565.040**	Omega $_{rad}$:	**+9.0587 D-06**
Delta Y_m:	**49.910**	Epsilon $_{rad}$:	**+1.9848 D-06**
Delta Z_m:	**465.840**	Psi $_{rad}$:	**- 1.7439 D-06**
		Delta k:	**4.0772 D-06**

$C_7^{(3)}$ Curvilinear Geodetic Datum Transformation

DGPS General Case I

Transformation of Bessel-Amersfoort Geodetic Co-ordinates into WGS84 Geodetic Co-ordinates:

Input:	Latitude	Longitude	H	N_0	Output:	Latitude	Longitude	h_n	N_n

Given: Initial point

$X_{RD.O}$:	**0**	$Y_{RD.O}$:	**0**	$Z_{RD.O}$:	**0**

Sta. DUT	RD1918		WGS84
Latitude-old:	**51° 59' 13".3938 N**	Latitude-new I=	51° 59' 09".9145 N
Longitude-old:	**4° 23' 16".9953 E**	Longitude-new I=	4° 23' 15".9533 E
H- old:	**30.809** m	h-new II=	74.312m
Nsep:	**- 0.113** m	N-new II=	43.503 m
h-old=	30.696 m		

Calculated Cartesian Co-ordinates

X_{RD-i}=	3 924 096.8506	Y_{RD-i}=	301 119.8207	Z_{RD-i}=	5 001 429.8963
X_{WGS-i}=	3 924 689.3397	Y_{WGS-i}=	301 145.3379	Z_{WGS-i}=	5 001 908.6872

9.1.5.3 Bessel-Amersfoort to WGS84 Transformation - the Netherlands

S-transformation from Bessel-Amersfoort (RD1918) to WGS84

Ellipsoid old:	Bessel-Amersfoort	Ellipsoid new:	WGS84
Semi-Major axis a_0:	**6 377 397.155**	Semi-Major axis a_n:	**6 378 137**
Recipr. flattening $f^{-1}{}_0$:	**299.15281 285**	Recipr. flattening $f^{-1}{}_n$:	**298.25722 3563**
Delta X_m:	**593.032**	Omega $_{rad}$:	**+9.0587 D-06**
Delta Y_m:	**26.000**	Epsilon $_{rad}$:	**+1.9848 D-06**
Delta Z_m:	**478.741**	Psi $_{rad}$:	**- 1.7439 D-06**
		Delta k:	**4.0772 D-06**

Given: Initial point:

$X_{RD.O}$:	**3 903 453.1482**	$Y_{RD.O}$:	**368 135.3134**	$Z_{RD.O}$:	**5 012 970.3051**
$X_{WGS.O}$=	3 904 046.1802	$Y_{WGS.O}$=	368 161.3134	$Z_{WGS.O}$=	5 013 449.0461

$C_7^{(3)}$ Curvilinear Geodetic Datum Transformation

DGPS General Case II

Transformation of Bessel-Amersfoort Geodetic Co-ordinates into WGS84 Geodetic Co-ordinates:

Input:	Latitude	Longitude	H	N_0	Output:	Latitude	Longitude	h_n	N_n

Given: Initial point

$X_{RD.O}$:	**3 903 453.1482**	$Y_{RD.O}$:	**368 135.3134**	$Z_{RD.O}$:	**5 012 970.3051**

Sta. DUT	RD1918	WGS84	
Latitude-old:	**51° 59' 13"3938 N**	Latitude-new II=	51° 59' 09".9144 N
Longitude-old:	**4° 23' 16".9953 E**	Longitude-new II=	4° 23' 15".9530 E
H:	**30.809** m	h-new II=	74.312 m
Nsep:	**- 0.113** m	N-new II=	43.503 m
h-old=	30.696 m		

Calculated Cartesian Co-ordinates

$X_{RD.i}$ =	3 924 096.8506	$Y_{RD.i}$=	301 119.8207	$Z_{RD.i}$ =	5 001 429.8963
$X_{RD.i-RD.o}$=	20 643.7024	$Y_{RD.i-RD.o}$=	- 67 015.4927	$Z_{RD.i-RD.o}$=	- 11 540.4088
$X_{WGS.i}$=	3 924 689.3396	$Y_{WGS.i}$=	301 145.3376	$Z_{WGS.i}$=	5 001 908.6872

Example from (Strang van Hees, 1997).

Note

Formulae used are those given in [9.1.1], Transformation Equations, (9.01-9.07). *The interested reader may also consult (De Min, 1996) for more information about a new map of geoid heights in The Netherlands.*

9.1.6 Cartesian Co-ordinates

9.1.6.1 Cartesian Co-ordinates - South-West Pacific

3-D SWP-90 Trimble Network - Datum: WGS84 (Schutz et al, 1993).

Ellipsoid old:	WGS84	Ellipsoid new:	WGS84
Semi-Major axis a_0:	**6 378 137**	Semi-Major axis a_n:	**6 378 137**
Recipr. flattening f^{-1}_0:	**298.25722 3563**	Recipr. flattening f^{-1}_n:	**298.25722 3563**
Delta X:	**0**	Omega:	**0**
Delta Y:	**0**	Epsilon:	**0**
Delta Z:	**0**	Psi:	**0**
		Delta k:	**0**

Cartesian Co-ordinates

Conversion of geodetic co-ordinates into X, Y, Z co-ordinates and vice versa

Input: Latitude Longitude H N_0		Output: X Y Z Latitude Longitude h_n	
Sta. Rarotonga			
Latitude-old:	**21° 12' 04".46407 S**	Latitude-new=	21° 12' 04".46407 S
Longitude-old:	**200° 12' 01".40340 E**	Longitude-new=	159° 47' 58".59660 W
H:	**16.49**		
Nsep:	**0**		
h-old=	16.49	h-new=	16.49
X= - 5583136.63305	Y= - 2054237.43604		Z= - 2292187.98612
Sta. Tongatapu			
Latitude-old:	**21° 10' 24".49999 S**	Latitude-new=	21° 10' 24".49999 S
Longitude-old:	**184° 41' 27".91899 E**	Longitude-new=	175° 18' 32".08101 W
H:	**62.290**		
Nsep:	**0**		
h-old=	62.290	h-new=	62.290
X= - 5930280.62598	Y= - 486629.24091		Z= - 2289337.90607
Sta. Espiritu Santo			
Latitude-old:	**15° 26' 57".31117 S**	Latitude-new=	15° 26' 57".31117 S
Longitude-old:	**167° 12' 14".55054 E**	Longitude-new=	167° 12' 14".55054 E
H:	**125.161**		
Nsep:	**0**		
h-old=	125.161	h-new=	125.161
X= - 5996538.76607	Y= 1361935.34014		Z= - 1688098.97032

9.1.6.2 Cartesian Co-ordinates - Switzerland

3-D Netz Turtmann: Terrestr. Netz 1985-1986 - Datum: Berne Geodetic, CH-1903+

Ellipsoid old:	Bessel 1841	Ellipsoid new:	Bessel 1841
Semi-Major axis a_0:	**6 377 397.155**	Semi-Major axis a_n:	**6 377 397.155**
Recipr. flattening f^{-1}_0:	**299.15281 285**	Recipr. flattening f^{-1}_n:	**299.15281 285**
Delta X:	**0**	Omega:	**0**
Delta Y:	**0**	Epsilon:	**0**
Delta Z:	**0**	Psi:	**0**
		Delta k:	**0**

Cartesian Co-ordinates

Conversion of geodetic co-ordinates into X, Y, Z co-ordinates and vice versa

Input: Latitude Longitude H N_0 Output: X Y Z Latitude Longitude h_n

Sta. 7.Turt

Latitude-old:	**51^g 45 35 20036 N**	Latitude-new=	51^g 45 35 20036 N
Longitude-old:	**8^g 55 72 49986 E**	Longitude-new=	8^g 55 72 49986 E
H:	**622.4474**		
Nsep:	**1.3098**		
h-old=	623.7572	h-new=	623.7572

X= 4373700.0722 Y= 591465.9898 Z= 4588963.1846

Sta. 1.Brun

Latitude-old:	**51^g 46 04 18828N**	Latitude-new=	51^g 46 04 18828 N
Longitude-old:	**8^g 51 89 93719 E**	Longitude-new=	8^g 51 89 93719 E
H:	**1008.8876**		
Nsep:	**1.2478**		
h-old=	1010.1354	h-new=	1010.1354

X= 4373824.6037 Y= 588806.6270 Z= 4589719.3096

Sta. 2.Brae

Latitude-old:	**51^g 47 52 29006 N**	Latitude-new=	51^g 47 52 29006N
Longitude-old:	**8^g 54 84 62246 E**	Longitude-new=	8^g 54 84 62246 E
H:	**1506.8276**		
Nsep:	**1.3621**		
h-old=	1508.1897	h-new=	1508.1897

X= 4372830.4541 Y= 590733.7474 Z= 4591102.8446

9.2 Accuracy

The term accuracy refers to the closeness between calculations and their correct or true values, see [2.4.1].

9.2.1 Transformation Round-Trip Errors

The "round-trip errors" are the differences of latitude and longitude in degrees or height in metres between the starting and ending co-ordinates, illustrated in Figure 56, Figure 57, and Figure 58 for a latitudinal distance 0° N to 84° N, a longitudinal distance 0° E to 180° E, and are calculated as follows:

Latitudes, longitudes and heights are converted to the X, Y and Z co-ordinates, which are converted back to latitudes, longitudes and heights (Floyd, 1985).

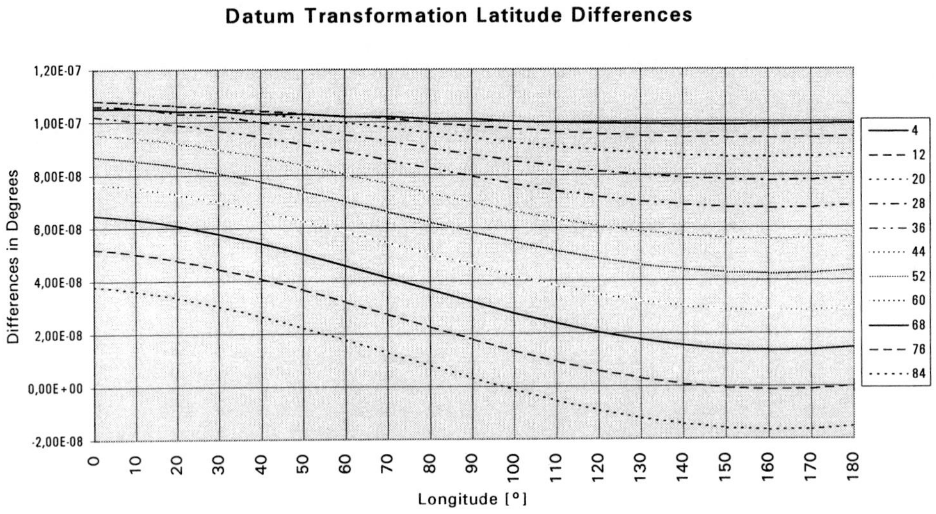

Datum Transformation Latitude Differences

Figure 56: Datum transformation round-trip error of latitude

Datum Transformation Longitude Differences

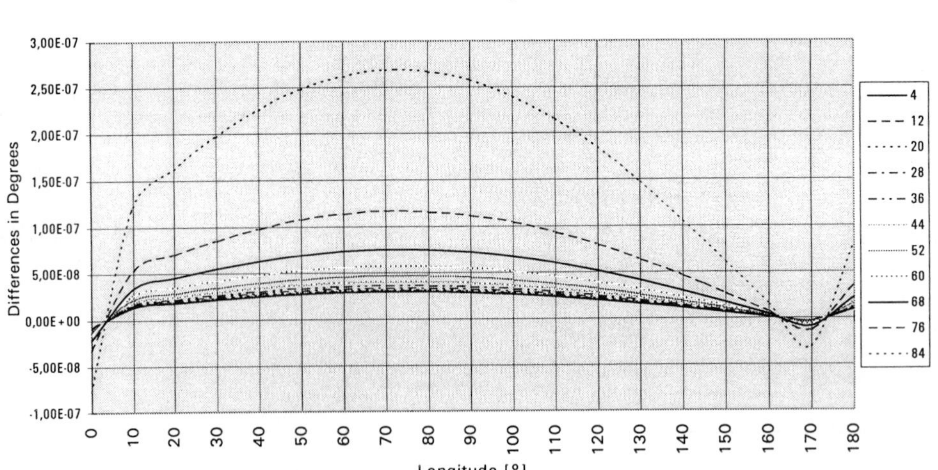

Figure 57: Datum transformation round-trip error of longitude

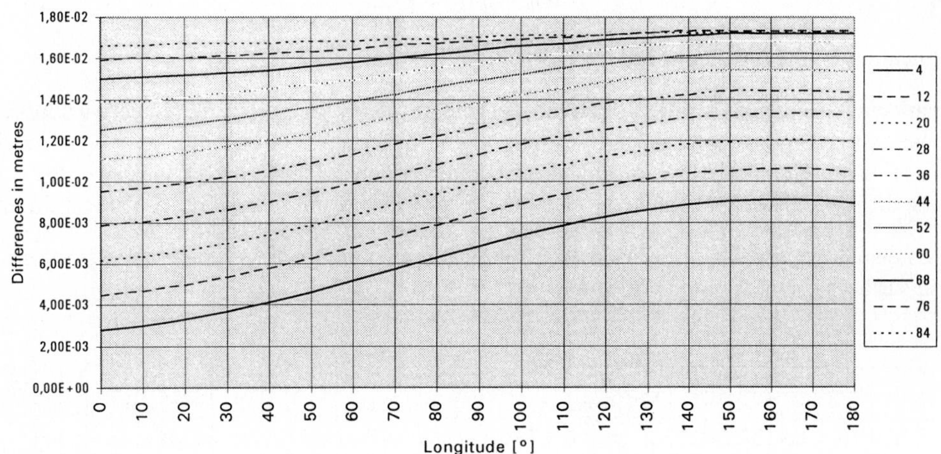

Figure 58: Datum transformation round-trip error of height

Input data:

latitude	=	0° N to 84° N
longitude	=	0° E to 180° E
δX	=	60 m
δY	=	- 75 m
δZ	=	- 375 m
ω	=	2".1
ε	=	0".35
ψ	=	- 0".3
δk	=	- 0.3D-04
H	=	5000 m
Nsep	=	80 m.

9.3 Using Bi-linear Interpolation

Summary

Using bi-linear interpolation to correct a transformation.

Transformation of two datums result in different co-ordinate values. In essence there are two sources for the differences, the first being the conformal projection between differing ellipsoids, the second source is commonly known as distortion. The first source results in differences which are smoothly varying over the Earth, the curvature of the differences being continuous, where the second may result in location-dependent differences.

This section deals with aspects of bi-linear interpolation, which minimises geometrical distortions between the curved grid surfaces of transformed projections.

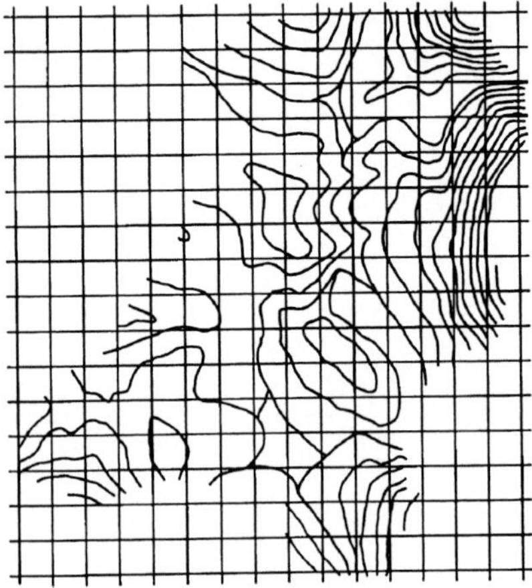

Figure 59: ED50 corrections for ΔE

Figure 60: ED50 corrections for ΔN

Bi-linear interpolation comprises transfer of the weighted mean of the digital number obtained for the four nearest co-ordinated points of any square: ED50 corrections for ΔE (Figure 61).

Practical Example Bi-linear Interpolation

Here it is used to correct a transformation from the Stereographic projection of *The Netherlands*, reference ellipsoid Bessel-Amersfoort (RD1918), into UTM GK projection, European Datum of 1950 (ED50), reference ellipsoid International 1924 and vice versa. RD1918 and ED50 do not exactly map into one another.

The graphs Figure 59 and Figure 60 - inserted for demonstration purposes only (original size is A4, showing corrective data) - show the ΔE, ΔN correction curves for Eastings and Northings, respectively.
The graph values at the grid-intersections were transferred to matrices for application of *bi-linear interpolation* in programs RDED0031.BAS, and RDED0032.BAS [11.7.9] for UTM Zone 31, and UTM Zone 32, respectively.
An appropriate test case is given in [9.3.3].

The approach of RDED003x relies on a two-step process:

- transformation by a two-dimensional polynomial from RD1918 to ED50, and vice versa
- using bi-linear interpolation, based on the grids, which effectively removes the distortions.

Two tabulated data sets are required for the complete computation of a transformation; one for northing-shifts and another for easting-shifts. Thus, two mathematical surfaces must be prepared for a region or country.

Note

Please note that, if needed, bi-linear interpolation can be used for many purposes, such as contour charts of geoidal heights or other physical sizes.

9.3.1 Transformations Between Two Geodetic Datums

In the past, private companies have developed different methodologies for the transformation of data between Bessel-Amersfoort (RD1918) and ED50, which are both *geodetic datums.* RD1918 is used for all national surveys within the Netherlands and ED50 is used for all surveys on the Dutch continental shelf and coastal margins.
The independently derived methods optimise transformations within a particular area of the North Sea. They are often the result of fitting very high-order polynomials to a limited number of accurate co-ordinated data points.

The shifts between Bessel-Amersfoort and ED50 arise from a difference in the reference and co-ordinate systems as well as small differences or local distortions arising due to differences in adjustment and survey methodologies.

A two-dimensional polynomial, prepared by the Topografische Dienst of the Netherlands (TDN), Department of Defence, was fitted to the actual (observed) datum shifts between RD1918 and ED50, using five 1st order stations: Aardenburg, Berkheide, Eierland, Finsterwolde and Ubagsberg. This method often provides results with an accuracy of better than 1 m. Local distortions, implicit within RD1918, are evident and require a correction in eastings and northings, ΔE_i and ΔN_i, respectively (Linden, 1985).

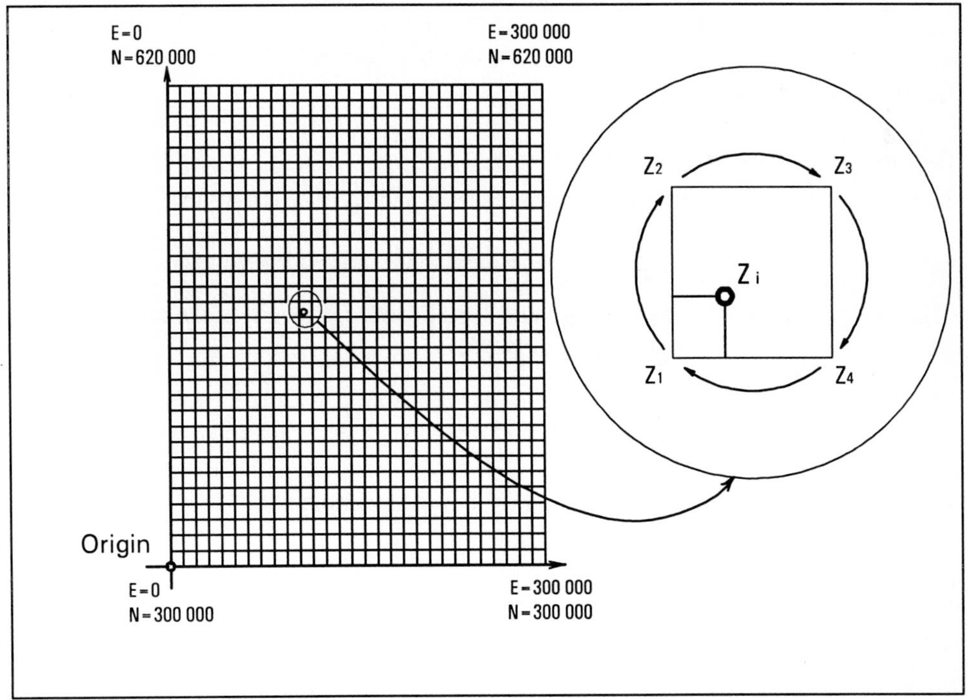

Figure 61: Location of a grid square for interpolation

Contour Charts of Shifts

The corrections ΔE_i, ΔN_i can be depicted as contour charts which show the corrections *in mm* to the transformed data points of RD1918. These contour correction charts, prepared by the TDN, developed from co-ordinate differences by subtracting the transformed Bessel 1841 numerical data from the UTM ED50 numerical data. The difficult aspect of data selection and the computation of two tabulated digital data sets of ΔE_i, ΔN_i-shifts has been prepared by the TDN.

Here, the application of bi-linear interpolation is described as it applies to the correction of transformed positional data between the local geodetic datums RD1918 and ED50. The author has implemented this approach and technique in a computer program, known as RDED0031 (and RDED0032), an acronym standing for the RD1918 to ED50 Conversion, UTM Zone 31 (or UTM Zone 32). RDED003x [11.7.9] was developed to provide a uniform methodology in which a simple interpolation routine provides estimates of ΔE_i, ΔN_i values at random non-nodal points within the area of the five first order stations mentioned above.

Actual application of the processed data was designed in a simple application program for the Hewlett-Packard HP-9845B Desktop (256 KB) computer (Bjork, 1974; Hooijberg, 1984; Olson, 1977).

> *Please note that a discussion in detail of using a two-dimensional polynomial and the RD1918 Stereographic projection / Bessel ellipsoid of 1841 grid is outside the scope of this publication.*

For further reading, see (Alberda, 1978; Heuvelink, 1918; Linden, 1985; Luymes, 1924; Stem, 1989b; Strang van Hees, 1997).

The Practical Aspects of Bi-linear Interpolation

Consider a (demo) matrix of grid locations shown in Figure 59 and Figure 60. Note that an estimate at an unknown grid location can be found for each point within this matrix grid.
Tabulated data grids require bi-linear interpolation to be useful. For any data set of a square grid, a minimum of four values are required. The nodal points Z_1, Z_2, Z_3, Z_4 are known corrections (ΔE_i, ΔN_i, respectively) at grid intersecting points, and the given grid interval (Δ=10 km) is of great importance (Figure 61). The method will determine a smooth surface.

9.3.2 Bi-linear Interpolation Scheme

The polynomial used for the bi-linear interpolation is simple and accurate (Dewhurst, 1990; OS, 1995b). I and J are calculated, positional indices. The *sign convention* is ED50 minus RD1918 (local geodetic datum). Clockwise from the south-west corner of the cell, the PE-indices (and similar for PN-indices) are:

Z_1 = PE (I, J) the (easting) index of the *lower left corner* of the cell in which the unknown point resides.
Z_2 = PE (I, J + 1)
Z_3 = PE (I+1, J + 1)
Z_4 = PE (I+1, J)

The following A, B, C, and D are coefficients of the following polynomial (9.27a, 9.27) and are all functions of the shift values of the surrounding nodal points (Figure 61).

$$A = Z_1 \quad (9.23)$$

$$B = (Z_4 - Z_1) / \Delta \quad (9.24)$$

$$C = (Z_2 - Z_1) / \Delta \quad (9.25)$$

$D = (Z_1 - Z_2 + Z_3 - Z_4) / \Delta^2$, in which Z_1, Z_2, Z_3, Z_4 are known corrections ΔE_i, ΔN_i, respectively at grid points used in the interpolation process. (9.26)

The origin is located in the lower-left corner of a grid square of dimension Δ on a side. The sub-gridding technique is known as bi-linear interpolation (Bjork, 1974; Olson, 1977) and is defined as:

$$Z = A + B * X + C * Y + D * X * Y \tag{9.27a}$$

The polynomial surface, fit to the four surrounding nodal points Z_1, Z_2, Z_3 and Z_4, is recasted as:

$$Z = A + C * Y + (B + D * Y) * X \tag{9.27}$$

and interpolated for point P_i at which place Z is the calculated correction at the unknown point (Figure 61, Figure 62). It is applied *twice*: for eastings and northings.

This equation (9.27) is solved using index location based on the row and column organisation of the grid, as where X and Y represent the co-ordinates of the unknown point.

The grids are organised as matrix rows, minimum easting to maximum easting, and matrix columns, minimum northings to maximum northings. Thus in this case (Figure 61):

- Matrix Rows: Eastings run from 0 m to 300 000 m in 10 km (Δ) steps, and
- Matrix Columns: Northings run from 300 000 to 620 000 m in 10 km (Δ) steps.

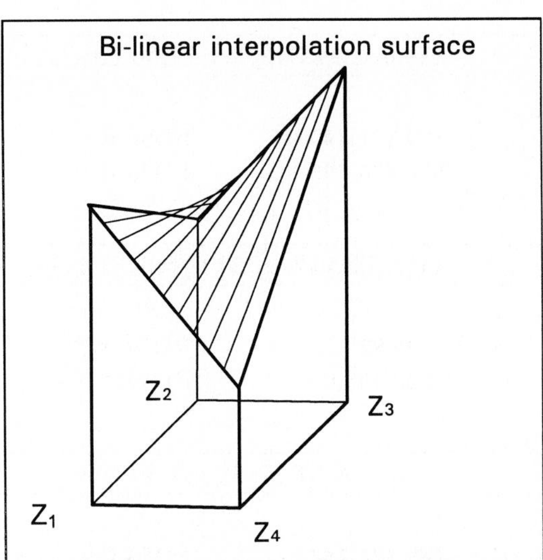

Figure 62: Easting or northing cross sections of the bi-linear interpolation surface are defined by straight-line elements

Interpolated results are calculated to nearest millimetre, using an integer precision matrix.

The RDED003x program [11.7.9] provides for the application of the interpolation method outlined above on all of the tabulated 10 km gridded data sets. The program user merely has to ensure that the proper gridded data set resides in the program or storage area. Eastings and Northings provided by the user will be converted between datums via the program. RDED0031 allows for the conversion of individual points in UTM Zone 31, CM 3°E, and program RDED0032 for individual points in UTM Zone 32, CM 9° E.

9.3.3 Transformation from Bessel to ED50 and Vice Versa - the Netherlands

From old RD1918 to ED50 Zone 31

RD1918-X :	**-101 444.41**	ED50-E =	563 833.073
RD1918-Y :	**- 55 189.93**	ED50-N =	5 722 715.345
RD1918-X :	**- 76 658.12**	ED50-E =	586 955.364
RD1918-Y :	**- 5 087.40**	ED50-N =	5 773 589.130

From ED50 Zone 31 to old RD1918

ED50-E :	**563 833.07**	RD1918-X =	-101 444.409
ED50-N :	**5 722 715.35**	RD1918-Y =	-55 189.924
ED50-E :	**586 955.36**	RD1918-X =	-76 658.122
ED50-N :	**5 773 589.14**	RD1918-Y =	-5 087.391

From new RD1918 to ED50 Zone 31

RD1918-X :	**189 121.443**	ED50-E =	695 920.572
RD1918-Y :	**510 886.095**	ED50-N =	5 830 185.093

From ED50 Zone 31 to new RD1918

ED50-E :	**695 920.57**	RD1918-X =	189 121.443
ED50-N :	**5 830 185.09**	RD1918-Y =	510 886.095

From new RD1918 to ED50 Zone 32

RD1918-X :	**189 121.443**	ED50-E =	289 439.450
RD1918-Y :	**510 886.095**	ED50-N =	5 830 794.479

From ED50 Zone 32 to new RD1918

ED50-E :	**289 439.45**	RD1918-X =	189 121.439
ED50-N :	**5 830 794.48**	RD1918-Y =	510 886.093

9.3.4 Accuracy

Judging the exact accuracy of RDED003x.BAS without some basis to use as truth is not easy. The results of [9.3.3] represent a comparison against original data of first order stations in the Netherlands.

RDED003x can be blamed for being an interpolation method rather than a more traditional geodetic technique. Unfortunately, no transformation technique can satisfactorily accommodate (RD1918) distortions.

Users in the Netherlands are advised to check interpolation error(s) pertaining to application(s). Users are advised to contact the TDN, Emmen / the Netherlands, to ensure having the current data sets.

Note

The Bi-linear interpolation method is documented in:

- *(Bjork, 1974)*
- *(Dewhurst, 1990) for transformation of NAD27 to NAD83*
- *(NIMA, 1991) TR 8350.2-B for transformation of local geodetic datum to WGS84*
- *(Hooijberg, 1984) for transformation of ED50 grid to RD1918 grid*
- *(Olson, 1977)*
- *(OS, 1995b) for transformation of OSGRS80 grid to the National grid.*

10. Miscellaneous Co-ordinate Systems

10.1 Military Applications

Large-scale military maps use conformal projections. A rectangular grid is superimposed on military maps to assist in the location of points. This concept is similar to that of the GK- and LCC systems introduced in the previous sections.

For the military maps, the world-wide grid system adopted by the US DoD NIMA makes use of the Universal Transverse Mercator (UTM) grid and the Universal Polar Stereographic (UPS) grid superimposed on the respective projections.

In some military applications a modified co-ordinate system is used for security reasons. The system is usually a grid system with the co-ordinates offset by moving the *false origin* or by changing the *meridian* in longitude.

The programs GK000000.BAS, LCC00000.BAS, and OM000000.BAS allow such modified systems to be defined by new parameters.

The following application has been in use from the outbreak of World War I at the battlefield in "Nord-Est France" in 1915 until the end of World War II in 1945 [7.2] (Linden, 1985).

10.1.1 Lambert Nord de Guerre - North-East France

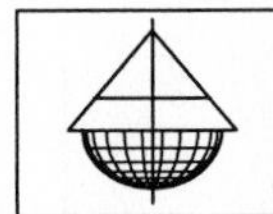

One Parallel and a Scale Factor

Zone Parameters:		*Lambert Nord de Guerre*	L1 = 500000
Reference Ellipsoid:		Plessis |[75]	L2 = .760405965600031
Semi-Major axis	a:	**6 376 523**	L3 = 5453601.53061231
Recipr. Flattening	f^{-1}:	**308.64**	L4 = 5753601.53061231
Standard Parallel	φ_o:	**55^g N**	L5 = 5453601.53061231
Lon. Grid Origin	λ_p:	**6^g E** of Paris	L6 = 11597505.03498878
Scale Factor	k_o:	**0.99950 90800**	L7 = .99950908
False Easting	E_o:	**500 000**	
False Northing	N_b:	**300 000**	

Conversion of Plessis Geographicals to Lambert Grid

The Direct calculation is to convert Geodetic Co-ordinates into LCC-Planar Co-ordinates:

Input:	Latitude	Longitude	Output:	Easting	Northing	Convergence	Scale Factor

Latitude:	**54^g.20 21 757 N**	Easting=	501272.5807
Longitude:	**6^g.01 92 543 E**	Northing=	220199.6558
Convergence=	+0^g.01 46 4108	Scale Factor=	0.999 586 976 611

[75] "Ellipsoide modifiée de Plessis" or Delambre ellipsoid 1810, version II

10.2 Civil Applications

10.2.1 Gauss-Schreiber Grid

Theoretical speculation about the figure of the Earth gave way to serious attempts to determine it by measurement. In 1821 Gauss started measuring an arc of longitude from Göttingen to Altona in the Kingdom of Hannover (Germany) which was accurate enough to connect it to the Dutch triangulation network in 1824. Gauss himself was very much interested in the practical side of his profession. He was closely involved in the work of the Survey, particularly with the observations for determining the shape of the Earth and the determination of this arc.

Although by 1827 the primary triangulation of Gauss covered the whole of the Kingdom of Hannover, the old *projection* would have been unsatisfactory if used to cover the whole of the Kingdom. Therefore the survey committee's recommendations made it necessary to adopt a new projection, a fact recognised at the time by Gauss. The book about the projection method used by Gauss was still unfinished when he died. The arithmetic involved was so laborious that errors were almost unavoidable.

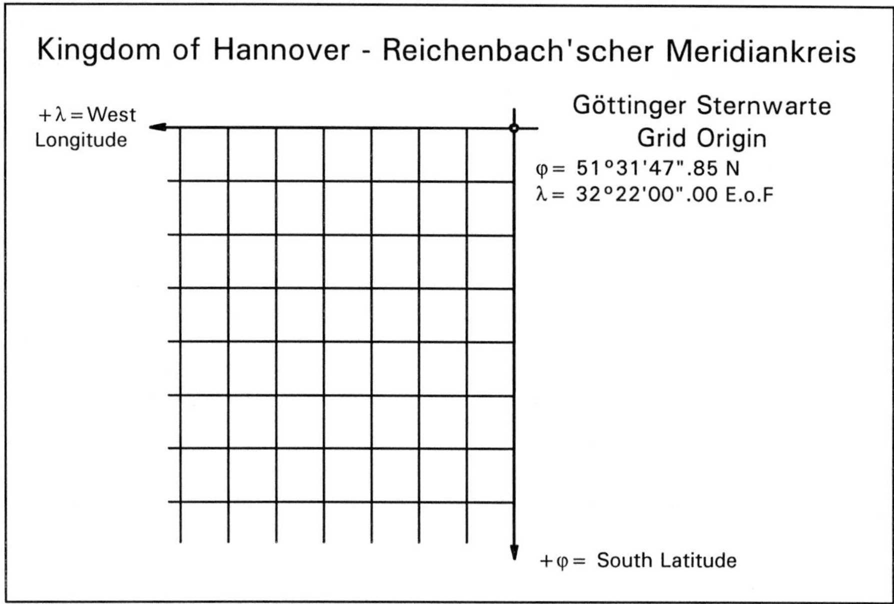

Figure 63: Gauss-Schreiber grid of the "Hannover'schen Landesvermessung 1866

To finish the work of Gauss demanded a certain level of mathematical ability. Commander Oscar Schreiber was a remarkable man, who had many of the qualities needed. He completed the work begun by Gauss between 1860 and 1866. Not only did he remodel the calculating procedure, but he also studied ways of perfecting the methods used and directing the computations.

A large part of Schreiber's volume "Theorie der Projectionsmethode der Hannover'schen Landesvermessung" deals with work and recommendations which was of the greatest international importance in geodesy, because the character of the projection changed to meet the varying demands made on it (Schreiber, 1866).

Note

> *The Gauss-Schreiber Grid 1866 uses the origin of the "Göttinger Sternwarte", and a different quadrant convention: latitude and longitude are reckoned to be positive south and west of the origin, respectively (Figure 63). All distances are measured in "Gauss' metres".*

10.2.1.1 Gauss-Schreiber Co-ordinate System 1866 of Hannover

Example for:	Hannover (Schreiber, 1866)		
Zone Parameters:	Local System	*Using Gauss' metres*	
Reference Ellipsoid:	Walbeck 1817	False Easting:	**0 m_G**
Semi-Major axis:	**6 376 723.661 m_G**	False Northing:	**0 m_G**
Recipr. Flattening:	**302.78**	Parallel of Origin φ_o:	**51° 31' 47".85 N**
Scale Factor:	**1.0000**	Central Meridian λ_o:	**32° 22' 00" E o F** [76]

Conversion of Walbeck 1817 Geographicals to Transverse Mercator Grid

The Direct calculation is to convert Geodetic Co-ordinates into GK-Planar Co-ordinates:

Ellipsoidal and Grid Calculation *(manual)*

Berlin			
Latitude:	**52° 30' 16".70 N**	Easting=	- 20 931.4588
Longitude:	**32° 03' 30".00 E**	Northing=	108 471.8613
Convergence=	- 0° 14' 40".6801	Scale Factor=	1.000 005 378 169
Leipzig			
Latitude:	**51° 20' 20".10 N**	Easting=	- 162 837.8162
Longitude:	**30° 01' 45".00 E**	Northing=	- 18 655.7430
Convergence=	- 1° 49' 32".3279	Scale Factor=	1.000 325 597 329
Breslau			
Latitude:	**51° 06' 56".00 N**	Easting=	163 482.5118
Longitude:	**34° 42' 07".50 E**	Northing=	- 43 500.1570
Convergence=	+1° 49' 05".9520	Scale Factor=	1.000 328 197 281

Scale Factor | Convergence | Arc-to-chord or (t - T) correction

Berlin-Breslau			Breslau-Berlin		
Grid bearing	t_{1-2}	129° 29' 28".6112	Grid bearing	t_{2-1}	309° 29' 28".6112
- (t-T)	δ_{1-2}	+0° 00' 15".6015	- (t-T)	δ_{2-1}	- 0° 00' 39".2556
Proj. Azimuth	T_{1-2}	129° 29' 13".0098	Proj. Azimuth	T_{2-1}	309° 30' 07".8669
+ convergence	γ_1	- 0° 14' 40".6801	+ convergence	γ_2	+1° 49' 05".9520
True Azimuth	α_{1-2}:	129° 14' 32".3296	True Azimuth	α_{2-1}:	311° 19' 13".8188

Note

Results shown here are computed by GK000000.BAS. Due to different quadrant convention, all signs should be reversed to compare with the original calculation, see Figure 63.

The Meridian of Ferro (Canary Islands, now Hierro) was established 20° West of Paris by Royal Act of King Louis the XIII of France in the year 1630. It was also accepted by Germany and Austria. The Meridian of Ferro is accepted as 17° 39' 59".41, rounded to 17° 40' W of Greenwich (Gretschel, 1873).

[76] Meridian is situated east of the Isle of Ferro

10.2.2 Redesigning a Local Co-ordinate System

Reduction of distances to the surface of the ellipsoid.

Using planar co-ordinates, a horizontal distance - especially at some elevation above mean sea level (MSL) - may differ significantly from the grid distance. This can easily be detected by EDM within the measuring capabilities of such an instrument. This difference in the ground / grid distance can be reduced to almost zero by devising a map projection using an elevated surface - an auxiliary ellipsoid - for a local area (Figure 64).

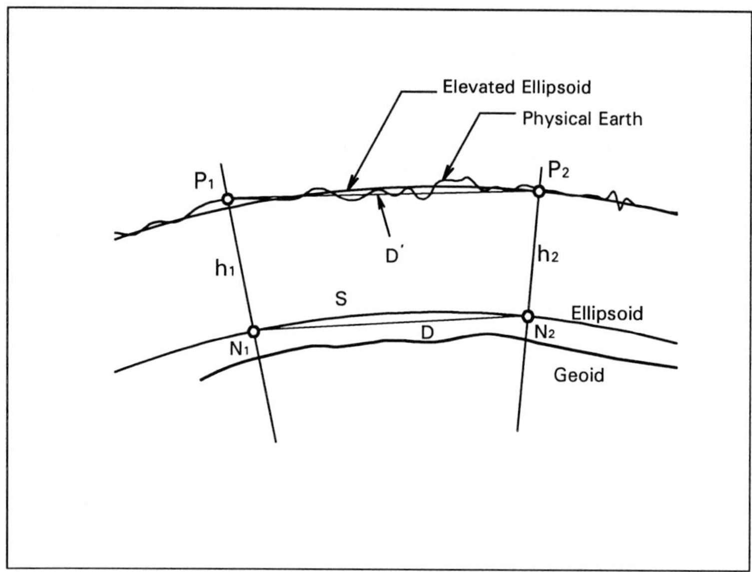

Figure 64: An auxiliary ellipsoid to minimise corrections for distance

Changing a national reference system into a local 2- or 3-D co-ordinate system without the necessity of reducing distances to the surface of the ellipsoid implies a mathematical operation with formulae. This system will project features with the benefits of the national reference system. Thus, grid and ground distance differences are virtually eliminated - existing values of the planar co-ordinates can be retained - without reduction to MSL.

The design of such a projection system is defined by the semi-major axis and e.g. the flattening, including the use of a multiplication factor. An elevation was never a design criterion, but *Michigan* is an exception. Since most of the physical earth in Michigan is within 200 feet of 800 US survey feet above the vertical datum, (Berry, 1971) designed NAD27 projections at a reference surface 800 ft above MSL.

Originally, C&GS selected three Transverse Mercator projections for US State Michigan with TM Zone East, TM Zone Central, and TM Zone West (in 1934). However, when the State passed the Michigan Co-ordinate System law for the NAD27 in 1964, the legislature adopted three Michigan Lambert co-ordinate system projections: LCC North, LCC Central, and LCC South, with *semi-major axis a = 6 378 450. 047 484 48* metres with the multiplication factor of *1.00003 82*. Therefore - except areas exceeding 1 000 ft. in elevation - the elevation factor in Michigan is insignificant for the NAD27 co-ordinates (Bowring, 1979; Floyd, 1985).

Steps for Designing a Local Co-ordinate System.

The steps to define a local co-ordinate system having the geometrical integrity necessary to preserve data-sharing options with the State Geodetic Reference System are given in an article with the following examples in [10.2.2.1; 10.2.2.2] (Burkholder, 1993).

10.2.2.1 Experimental Reference Systems - GK Oregon Tech. - USA

Example for	SPCS83 Oregon Tech.		
Zone Parameters:	*Experimental GK-system*		
Reference Ellipsoid:	GRS80 Modified	False Easting:	**50 000 m**
Semi-Major axis:	**6 379 452**	False Northing:	**20 000 m**
Recipr. Flattening:	**298.25722 21008 827**	Parallel of Origin φ_o:	**42° 12' 00" N**
Scale Factor:	**0.99999 8**	Central Meridian λ_o:	**121° 47' 00" W**

Conversion of NAD83-Modified Geographicals to Transverse Mercator Grid

The Direct calculation is to convert Geodetic Co-ordinates into GK-Planar Co-ordinates:

Input:	Latitude	Longitude	Output:	Easting	Northing	Convergence	Scale Factor

Latitude:	**42° 15' 32".91566 N**	Easting=	50 119.14875
Longitude:	**121° 46' 54".80271 W**	Northing=	26 570.83984
Convergence=	+ 0° 00' 03".4951	Scale Factor=	0.999 998 000 175

Conversion of Transverse Mercator Grid to NAD83-Modified Geographicals

The Inverse process is to convert GK-Planar Co-ordinates into Geodetic Co-ordinates:

Input:	Easting	Northing	Output:	Latitude	Longitude	Convergence	Scale Factor

Easting:	**50 119.14875**	Latitude=	42° 15' 32".91566 N
Northing:	**26 570.83984**	Longitude=	121° 46' 54".80271 W
Convergence=	+ 0° 00' 03".4951	Scale Factor=	0.999 998 000 175

10.2.2.2 Experimental Reference Systems - LCC Oregon Tech. - USA

Example for		SPCS83 Oregon Tech.	
Zone Parameters:		*Experimental Lambert System*	L2 = .6725821382091374
Reference Ellipsoid:		GRS80 Modified	L3 = 7037182.784247107
Semi-Major axis	a:	**6 379 452 m**	L4 = 7037182.784247107
Recipr. Flattening	f^{-1}:	**298.25722 21008 827**	L5 = 7029775.738146574
Lower Parallel	φ_l:	**42° 14' 00" N**	L6 = 12128718.29344483
Upper Parallel	φ_u:	**42° 18' 00" N**	L7 = .9999998313903994
Lat. Grid Origin	φ_b:	**42° 12' 00" N**	F0 = 6.686920927320709D-03
Central Parallel	φ_o=	42° 16' 00".0107306 N	F2 = 5.201458331082286D-05
Lon. Grid Origin	λ_p:	**121° 47' 00" W**	F4 = 5.544580077481781D-07
Scale Factor	k_o=	0.999 999 831 390	F6 = 6.717675115728047D-09
False Easting	E_o:	**20 000 m**	F8 = 8.907661557192931D-11
False Northing	N_b:	**0 m**	

Conversion of NAD83-Modified Geographicals to Lambert Grid

The Direct calculation is to convert Geodetic Co-ordinates into LCC-Planar Co-ordinates:

Input: Latitude Longitude Output: Easting Northing. Convergence Scale Factor

Latitude:	**42° 15' 32".91566 N**	Easting=	20 119.14896
Longitude:	**121° 46' 54".80271 W**	Northing=	6 570.85353
Convergence=	+0° 00' 03".4956	Scale Factor=	0.999 999 839 986

Conversion of Lambert Grid to NAD83-Modified Geographicals

The Inverse process is to convert LCC-Planar Co-ordinates into Geodetic Co-ordinates:

Input: Easting Northing. Output: Latitude Longitude Convergence Scale Factor

Easting:	**20 119.1490**	Latitude=	42° 15' 32".91566 N
Northing:	**6 570.8535**	Longitude=	121° 46' 54".80271 W
Convergence=	+0° 00' 03".4956	Scale Factor=	0.999 999 839 986

10.2.3 Conversion of Co-ordinates Between Projection Zones

Conversions are required each time a survey line crosses from one zone to another, see [5.2], Conversions and Transformations (Field, 1980; Clarke, 1973).

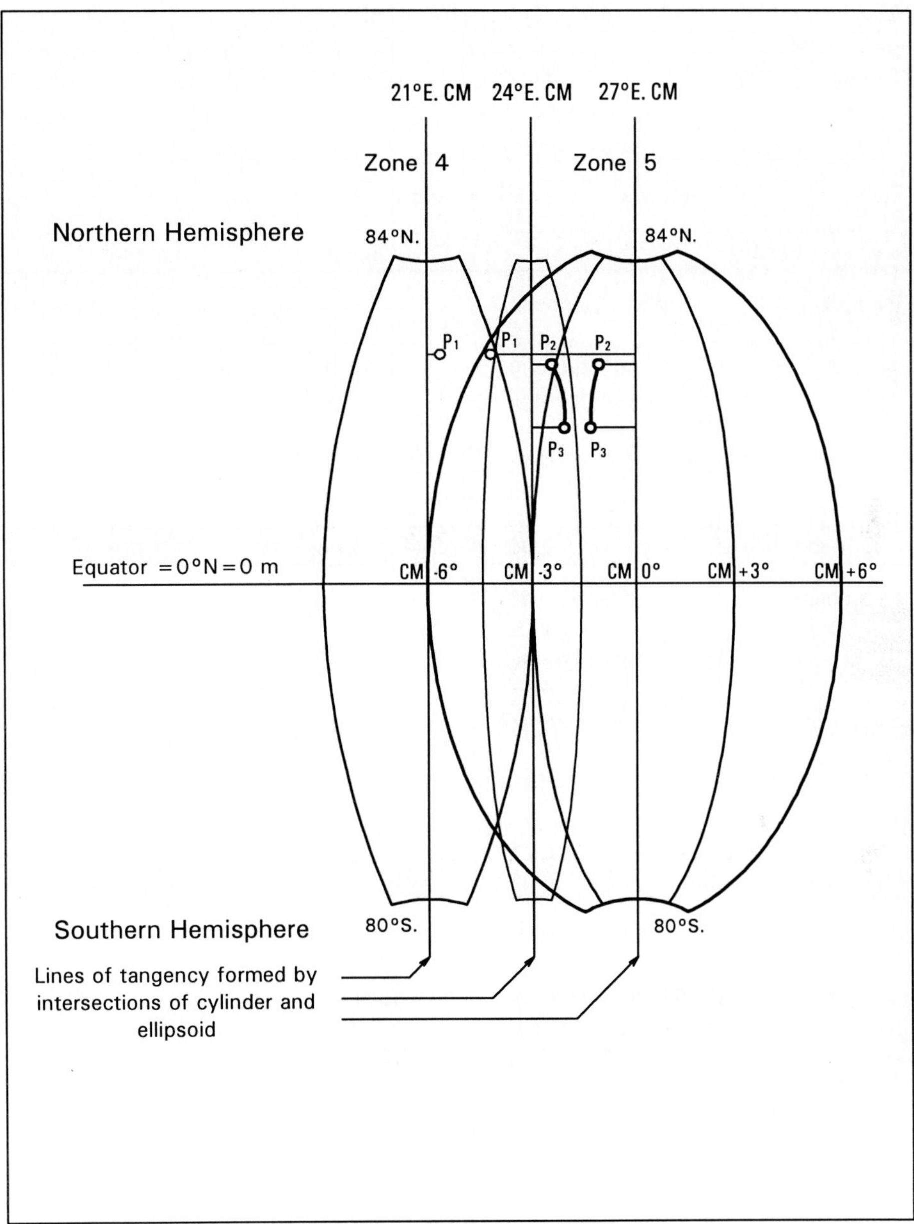

Figure 65: Conversion between 3° - 6° zone systems and 6° - 6° zone systems

To convert plane co-ordinates in one system to plane co-ordinates in another system or zone, it is best to calculate the geodetic co-ordinates from the plane co-ordinates of the first system and then convert those position values to the plane co-ordinates of the adjacent zone or system (Figure 65).

10.2.3.1 CK42 Conversion from 6° Wide Zone into 3° and 6° Wide Zones - Bulgaria

Zone Parameters	CK42, Zone 4 [77]		
Reference Ellipsoid:	Krassovsky 1940	False Easting:	**500 000 m**
Semi-Major axis:	**6 378 245**	False Northing:	**0 m**
Recipr. Flattening:	**298.3**	Parallel of Origin φ_o:	**0° 00' 00" N**
Scale Factor:	**1**	C.M. of zone λ_o:	**21° 00' 00" E**

Conversion from CM 21°E (6°) into CM 24°E (3°) and CM 27°E (6°)

The Inverse calculation is to convert GK-Planar Co-ordinates into Geodetic Co-ordinates:

Input:	Easting	Northing.	Output:	Latitude	Longitude
Zone Parameters		CK42, Zone 4:			
C.M.	λ_o:	21° E			
Easting	E_1:	**501 500.000**	Latitude	φ_1=	47° 50' 00".335083 N
Northing	N_1:	**5 300 000.000**	Longitude	λ_1=	21° 01' 12".128378 E

The Direct calculation is to convert Geodetic Co-ordinates into GK-Planar Co-ordinates of CM 24°

6° zone → 3° zone

Input: Latitude Longitude		Output: Latitude Longitude	Convergence	Scale Factor
Zone Parameters	CK42 - 24°			
C.M. λ_o:	24° E			
Latitude φ_1:	**47° 50' 00".33508 N**	Easting	E_1=	276 910.3034
Longitude λ_1:	**21° 01' 12".12838 E**	Northing	N_1=	5 304 301.6859
Convergence=	- 2° 12' 34".7203	Scale Factor	k_1=	1.000 611 342 397

The Direct calculation is to convert Geodetic Co-ordinates into GK-Planar Co-ordinates of CM 27°

6° zone → 6° zone

Input: Latitude Longitude		Output: Latitude Longitude	Convergence	Scale Factor
Zone Parameters	CK42, Zone 5:			
C.M. λ_o:	27° E			
Latitude φ_1:	**47° 50' 00".33508 N**	Easting	E_1=	52 381.4871
Longitude λ_1:	**21° 01' 12".12838 E**	Northing	N_1=	5 317 343.4834
Convergence=	- 4° 26' 22".7468	Scale Factor	k_1=	1. 002 461 885 478

Note

This procedure does not change the datum. See [9], Spatial Co-ordinate Calculations, for the procedure to use when changing from one datum to another datum.

[77] Examples taken from (Tarczy-Hornoch, 1959)

10.2.3.2 CK42 Conversion from 3° Wide Zone into 6° Wide Zone - Bulgaria

Zone Parameters	CK42 |[78]		
Reference Ellipsoid:	Krassovsky 1940	False Easting:	**500 000 m**
Semi-Major axis:	**6 378 245**	False Northing:	**0 m**
Recipr. Flattening:	**298.3**	Parallel of Origin φ_o:	**0° 00' 00" N**
Scale Factor:	**1**	C.M. of zone λ_o:	**24° 00' 00" E**

Conversion from CM 24°E (3°) into CM 27°E (6°)

The Inverse process is to convert GK-Planar Co-ordinates into Geodetic Co-ordinates:

Input: Easting Northing. Output: Latitude Longitude Conv. Scale Factor (t - T)

Zone Parameters		CK42 - 24°			
C.M.	λ_o:	24° E			
Easting	E_2:	**537 127.9991**	Latitude	φ_2=	42° 24' 17".6282 N
Northing	N_2:	**4 696 793.1726**	Longitude	λ_2=	24° 27' 03".5764 E
Convergence=		+0° 18' 14".8964	Scale Factor	k_2=	1.000 016 952 766
Easting	E_3:	**589 184.1732**	Latitude	φ_3=	42° 19' 53".2714 N
Northing	N_3:	**4 689 104.7299**	Longitude	λ_3=	25° 04' 55".3915 E
Convergence=		+0° 43' 43".4014	Scale Factor	k_3=	1.000 097 819 628
(t - T)	$\delta_{2\text{-}3}$=	+0° 00' 01".0625	(t - T)	$\delta_{3\text{-}2}$=	- 0° 00' 01".4009

The Direct calculation is to convert Geodetic Co-ordinates into GK-Planar Co-ordinates of Zone 5

3° zone → 6° zone

Input: Latitude Longitude Output: Latitude Longitude Conv. Scale Factor (t - T)

Zone Parameters		CK42, Zone 4			
C.M.	λ_o:	27° E			
Latitude	φ_2:	**42° 24' 17".6282 N**	Easting	E_2=	290 147.0589
Longitude	λ_2:	**24° 27' 03".5764 E**	Northing	N_2=	4 699 843.6807
Convergence=		- 1° 43' 10".5146	Scale Factor	k_2=	1.000 541 630 875
Latitude	φ_3:	**42° 19' 53".2714 N**	Easting	E_3=	341 918.5249
Longitude	λ_3:	**25° 04' 55".3915 E**	Northing	N_3=	4 690 319.6963
Convergence=		- 1° 17' 30".6515	Scale Factor	k_3=	1.000 307 345 379
(t - T)	$\delta_{2\text{-}3}$=	- 0° 00' 04".6515	(t - T)	$\delta_{3\text{-}2}$=	+0° 00' 04".2350

GK conversions are also given in:

- [6.3.2.3]: Australia, UTM - Zone 54 → UTM - zone 55
- [6.3.3]: Belgium, UTM - Zone 31 → UTM - Zone 32
- [9.3.3]: The Netherlands, UTM - Zone 31 → UTM - Zone 32.

[78] Examples taken from (Tarczy-Hornoch, 1959)

10.2.4 Conversion of Co-ordinates Between Projection Systems

Conversion of co-ordinates from one grid system into a different grid system using the construction as outlined in [5.2; 10.2.3].

The preferred procedure is to transform the grid co-ordinates from the first grid system to geographic positions. Then transform the geographic positions to grid co-ordinates of the second grid system.

Note

> *This procedure does not change the datum. See [9], Spatial Co-ordinate Calculations, for the procedure to use when changing from one datum to another.*

Example given:

Co-ordinates may be transformed from one grid system to another, for instance, between a Lambert grid and a UTM-grid or between different grid zones. Conversion from:

State Plane Co-ordinate System Lambert "Texas North Central", to UTM - Zone 14 and 15

[10.2.4.1 → 10.2.4.2] (DA, 1958).

10.2.4.1 SPCS27 - Lambert - Texas North Central - USA

Zone Parameters		*SPCS Texas, Zone North Central*	L2 =.5453944178463301
Reference Ellipsoid		Clarke 1866 (NAD27)	L3 = 32691654.35505039
Semi-Major axis	a:	**20 925 832.16 ft** [79]	L4 = 32691654.35505039
Recipr. Flattening	f^{-1}:	**294.97869 82**	L5 = 32187809.30045893
Lower Parallel	φ_l:	**32° 08' 00" N**	L6 = 44846543.95917029
Upper Parallel	φ_u:	**33° 58' 00" N**	L7 = .9998726293324699
Lat. Grid Origin	φ_b:	**31° 40' 00" N**	F0 = 6.761032571864926D-03
Central Parallel	φ_o=	33° 03' 05".8352 N	F2 = 5.317220426784075D-05
Lon. Grid Origin	λ_p:	**97° 30' 00" W**	F4 = 5.73056251095528D-07
Scale Factor	k_o=	0.999 872 629 332	F6 = 7.019628022792133D-09
False Easting	E_o:	**2 000 000 ft**	F8 = 9.412928282121681D-11
False Northing	N_b:	**0 ft**	

Conversion of Lambert Grid to NAD27 Geographicals

The Inverse calculation is to convert LCC- Planar Co-ordinates into Geodetic Co-ordinates

Input: Easting Northing Output: Latitude Longitude Convergence Scale Factor

Easting:	**2 439 603.2551**	Latitude=	34° 15' 34".7420 N
Northing:	**946 451.2651**	Longitude=	96° 02' 43".1579 W
Convergence=	+0° 47' 36".1444	Scale Factor=	1.000 094 894 466

[79] US survey feet

10.2.4.2 UTM - Universal Transverse Mercator Grid - USA

Conversion from Lambert to UTM Zone 14 and Zone 15

Zone Parameters:	UTM, Zone 14 and Zone 15		
Reference Ellipsoid:	Clarke 1866	False Easting:	**500 000 m**
Semi-Major axis:	**6 378 206.4 m**	False Northing:	**0 m**
Recipr. Flattening:	**294.97869 82**	Parallel of Origin $\varphi_{0\,Z14}$:	**0° 00' 00" N**
Scale Factor:	**0.9996**	CM Zone 14 λ_0:	**99° 00' 00" W**
		Parallel of Origin $\varphi_{0\,Z15}$:	**0° 00' 00" N**
		CM Zone 15 λ_0:	**93° 00' 00" W**

Conversion of NAD27 Geographicals to UTM-Grid Zone 14

The Direct calculation is to convert Geodetic Co-ordinates into UTM-Planar Co-ordinates

Input:	Latitude	Longitude	Output:	Easting	Northing	Convergence	Scale Factor

Zone 14: CM 99°W

Latitude:	**34° 15' 34".7420 N**	Easting=	772 075.8149
Longitude:	**96° 02' 43".1579 W**	Northing=	3 794 702.1725
Convergence=	+ 1° 39' 51".6267	Scale Factor=	1. 000 512 588 175

Conversion of NAD27 Geographicals to UTM-Grid Zone 15

The Direct calculation is to convert Geodetic Co-ordinates into UTM-Planar Co-ordinates

Input:	Latitude	Longitude	Output:	Easting	Northing	Convergence	Scale Factor

Zone 15: CM 93°W

Latitude:	**34° 15' 34".7420 N**	Easting=	219 574.6186
Longitude:	**96° 02' 43".1579 W**	Northing=	3 794 948.5090
Convergence=	- 1° 42' 55".6734	Scale Factor=	1.000 569 468 231

A similar conversion is also given in:

[10.2.2.1; 10.2.2.2]:______________ Oregon Tech, USA: transverse Mercator → Lambert.

For further reading, see (Schödlbauer, 1982).

11. Appendix

11.1 The World Geographic Reference System

The World Geographic Reference System (GEOREF) is not a US military grid, but it is a unique reference defining method useful for reporting, plotting of positions, and map storage (Hager, 1990) |[80].
Supposing the map is graduated in latitude and longitude, it provides an identification whatever map projection. The system divides the surface of the earth into quadrangles, each of which is identified by a simple systematic letter code as follows:

- The world is divided into 24 longitudinal belts of 15° width, which are lettered from A to Z inclusive (omitting I and O), extending eastwards from the 180° meridian. The world is divided into 12 latitudinal belts of 15° height, which are lettered from A to M inclusive (omitting I), extending northwards from the South Pole. These 24 x 12=288 quadrangles of 15° are identified by two letters. The first letter is that of the longitude zone and the second letter that of the latitude band. GEOREF of Salisbury Cathedral in the UK is referring to the 15° quadrangle as MK (Figure 66).

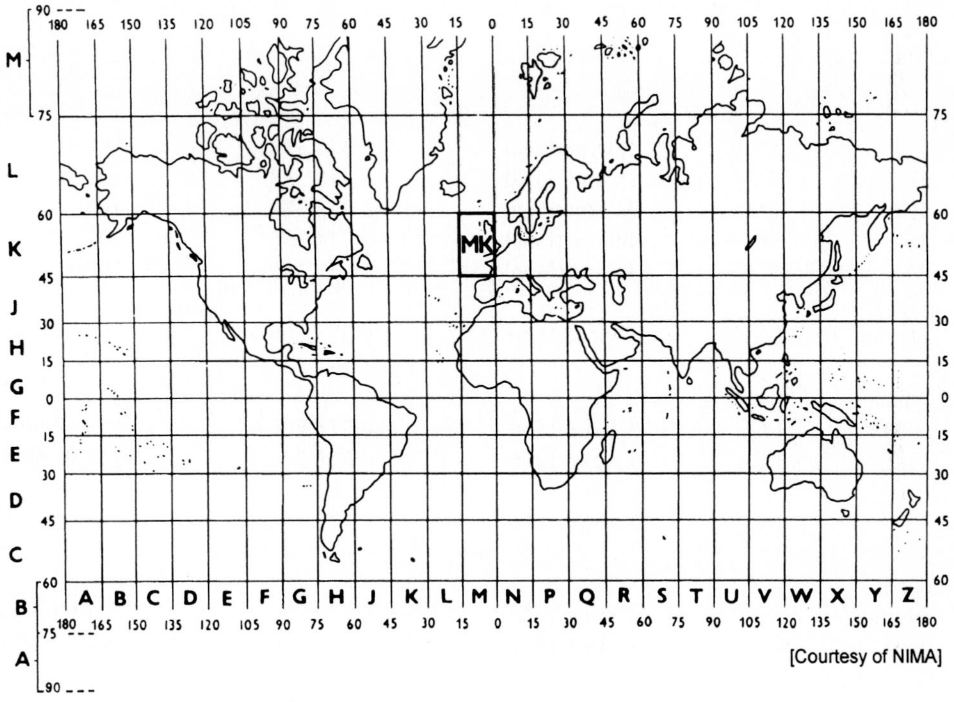

Figure 66: GEOREF system of 15° quadrangle identification letters

[80] Used by the US Army for interallied position reporting, for air defence and strategic air operations

Non-standard Systems in Current Use

Name	Projection	Ellipsoid	Parameters	
			a	f^{-1}
Austria M28	TM	Bessel 1841	6 377 397.155	299.15281285
Austria M31	TM	Bessel 1841	6 377 397.155	299.15281285
Austria M34	TM	Bessel 1841	6 377 397.155	299.15281285
Ceylon |[+]	TM	Everest |[81]	6 974 310.6	300.8017
Costa Rica				
Norte	LCC	Clarke 1866	6 378 206.4	294.9786982
Sud	LCC	Clarke 1866	6 378 206.4	294.9786982
Cuba				
Norte	LCC	Clarke 1866	6 378 206.4	294.9786982
Sud	LCC	Clarke 1866	6 378 206.4	294.9786982
Dominican Rep.	LCC	Clarke 1866	6 378 206.4	294.9786982
Egypt				
I	TM	International	6 378 388	297
II	TM	International	6 378 388	297
III	TM	International	6 378 388	297
IV	TM	International	6 378 388	297
V	TM	International	6 378 388	297
El Salvador	LCC	Clarke 1866	6 378 206.4	294.9786982
Guatemala				
Norte	LCC	Clarke 1866	6 378 206.4	294.9786982
Sud	LCC	Clarke 1866	6 378 206.4	294.9786982
Haiti	LCC	Clarke 1866	6 378 206.4	294.9786982
Honduras				
Norte	LCC	Clarke 1866	6 378 206.4	294.9786982
Sud	LCC	Clarke 1866	6 378 206.4	294.9786982

[81] Everest ellipsoid and units are in indian yards

Non-standard Systems in Current Use

Origin		False Origin [82]		
Latitude	Longitude	Easting	Northing	Scale Factor
0° 00' 00".000 N	10° 20' E / 28° E.o.FERRO	0.000	0.000	
0° 00' 00".000 N	13° 20' E / 31° E.o.FERRO	0.000	0.000	1.0000
0° 00' 00".000 N	16° 20' E / 34° E.o.FERRO	0.000	0.000	1.0000
7° 00' 00".729 N	80° 46' 18".160 E	176 000.000	176 000.000	1.0000
10° 28' 00".000 N	84° 20' 00".000 W	500 000.000	271 820.522	0.99995 696
9° 00' 00".000 N	83° 40' 00".000 W	500 000.000	327 987.436	0.99995 696
22° 21' 00".000 N	81° 00' 00".000 W	500 000.000	280 296.016	0.99993 602
20° 43' 00".000 N	76° 50' 00".000 W	500 000.000	229 228.939	0.99994 848
18° 49' 00".000 N	71° 30' 00".000 W	500 000.000	277 063.657	0.99991 102
0° 00' 00".000 N	25° 30' 00".000 E	300 000.000	0.000	0.99985
28° 30' 00".000 E	25° 30' 00".000 E	300 000.000	0.000	0.99985
31° 30' 00".000 E	25° 30' 00".000 E	300 000.000	0.000	0.99985
34° 30' 00".000 E	25° 30' 00".000 E	300 000.000	0.000	0.99985
37° 30' 00".000 E	25° 30' 00".000 E	300 000.000	0.000	0.99985
13° 47' 00".000 N	89° 00' 00".000 W	500 000.000	295 809.184	0.99996 704
16° 49' 00".000 N	90° 20' 00".000 W	500 000.000	292 209.579	0.99992 226
14° 54' 00".000 N	90° 20' 00".000 W	500 000.000	325 992.681	0.99989 906
18° 49' 00".000 N	71° 30' 00".000 W	500 000.000	277 063.657	0.99991 102
15° 30' 00".000 N	86° 10' 00".000 W	500 000.000	296 917.439	0.99993 273
13° 47' 00".000 N	87° 10' 00".000 W	500 000.000	296 215.903	0.99995 140

(cont'd)

[82] Add 1 000 000 to co-ordinate when Easting and / or Northing co-ordinate becomes negative

Non-standard Systems in Current Use

Name	Projection	Ellipsoid	Parameters	
			a	f^{1}
India				
I	LCC	Everest |81	6 974 310.6	300.8017
II A	LCC	Everest	6 974 310.6	300.8017
II B	LCC	Everest	6 974 310.6	300.8017
III A	LCC	Everest	6 974 310.6	300.8017
III B	LCC	Everest	6 974 310.6	300.8017
IV A	LCC	Everest	6 974 310.6	300.8017
IV B	LCC	Everest	6 974 310.6	300.8017
Levant	LCC	Clarke 1880	6 378 249.2	293.4660213
Nicaragua				
Norte	LCC	Clarke 1866	6 378 206.4	294.9786982
Sud	LCC	Clarke 1866	6 378 206.4	294.9786982
Nigeria				
West belt	TM	Clarke 1880	6 378 249.14533	293.465
Mid belt	TM	Clarke 1880	6 378 249.14533	293.465
East belt	TM	Clarke 1880	6 378 249.14533	293.465
Northwest Africa	LCC	Clarke 1880	6 378 249.2	293.4660213
Palestine	TM	Clarke 1880	6 378 300.79	293.466307656
Panama	LCC	Clarke 1866	6 378 206.4	294.9786982

(Hager, 1990).

Non-standard Systems in Current Use

Origin		False Origin [82]		
Latitude	Longitude	Easting	Northing	Scale Factor
32° 30' 00".000 N	68° 00' 00".000 E	3 000 000.000	1000 000.000	0.99878 6408
26° 00' 00".000 N	74° 00' 00".000 E	3 000 000.000	1000 000.000	0.99878 6408
26° 00' 00".000 N	90° 00' 00".000 E	3 000 000.000	1000 000.000	0.99878 6408
19° 00' 00".000 N	80° 00' 00".000 E	3 000 000.000	1000 000.000	0.99878 6408
19° 00' 00".000 N	100° 00' 00".000 E	3 000 000.000	1000 000.000	0.99878 6408
12° 00' 00".000 N	80° 00' 00".000 E	3 000 000.000	1000 000.000	0.99878 6408
12° 00' 00".000 N	104° 00' 00".000 E	3 000 000.000	1000 000.000	0.99878 6408
34° 39' 00".000 N	37° 21' 00".000 E	300 000.000	300 000.000	0.99962 56
13° 52' 00".000 N	85° 30' 00".000 W	500 000.000	359 891.816	0.99990 314
11° 44' 00".000 N	85° 30' 00".000 W	500 000.000	288 876.327	0.99992 228
0° 00' 00".000 N	4° 30' 00".000 E	230 738.266	0.000	0.99975
0° 00' 00".000 N	8° 30' 00".000 E	670 553.984	0.000	0.99975
0° 00' 00".000 N	12° 30' 00".000 E	1 110 369.702	0.000	0.99975
34° 00' 00".000 N	0° 00' 00".000 E	1 000 000.000	500 000.000	0.99908
31° 44' 02".749 N	35° 12' 43".490 E	170 251.555	126 867.909	1.0000
8° 25' 00".000 N	80° 00' 00".000 W	500 000.000	294 865.303	0.99989 909

Cont'd from [11.1], page 215.

- Each 15° quadrangle is subdivided into fifteen 1° zones of longitude, being lettered from A to Q inclusive (omitting I and O), eastwards from the western meridian of the quadrangle. Each 15° quadrangle is subdivided into fifteen 1° bands of latitude being lettered from A to Q inclusive (omitting I and O), northwards from the southern parallel of the quadrangle.

A 1° quadrangle may be identified by four letters. GEOREF of Salisbury Cathedral is the 1° quadrangle as MK PG (Figure 67).

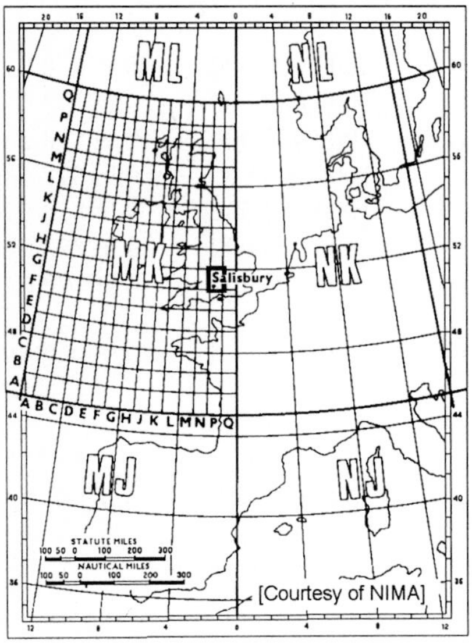

Figure 67: GEOREF 1° identification letters

- Each 1° quadrangle is divided into 60' of longitude, numbered eastwards from its western meridian, and 60' of latitude, numbered northwards from its southern parallel. The location may be west of the prime meridian or south of the equator. The position of a point to an accuracy of 1' in latitude and longitude (i.e. 2 km or less) can now be given by quoting four letters and four numerals. The first two numerals identify a 1' quadrangle by the number of minutes of longitude by which the point lies eastwards of the western meridian of the 1° quadrangle. The last two numerals are the number of minutes of latitude by which the point lies northwards of the southern parallel of the 1° quadrangle. The GEOREF of Salisbury Cathedral is: MK PG 12 04 (Figure 68).

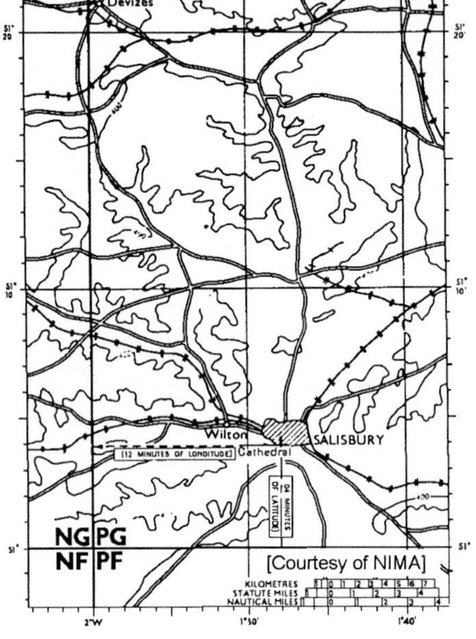

Figure 68: GEOREF map reference

- Each of the 1° quadrangles may be further divided into decimal parts (1/10th and 1/100th) eastwards and northwards. Thus, four letters and six numerals will define a location to 0'.1 and four letters and eight numerals will define a location to 0'.01.

11.2 Some Non-Standard Grid Systems in Current Use

The transverse Mercator projection system is the basis for the standard Gauss-Krüger grid system. Discussed in this book are four world-wide standard Gauss-Krüger (GK) projection and grid systems:

- [6.2.2] the Xi'an-grid system of China
- [6.2.3] the SUR-grid system of the CIS
- [6.2.4] the GK-grid system of Germany
- [6.3] the UTM-grid system of the USA, the US Army standard.

Outside the scope of this book are unique projection systems, such as the stereographic grid system of the Netherlands, the New Zealand grid system NZMG by W.I. Reilly, the oblique Mercator CH-1903 of Switzerland by Max Rosenmund, the Madagascar oblique Mercator or Gauss-Laborde grid system, and the Indonesian Equatorial System, which is a normal Mercator grid system (Grafarend, 1996c).

Dissimilar nonstandard grid systems were originally developed by a country and later conveniently adopted by other countries with or without modifications. Except in specific geographic areas, the systems were devised at different periods of time and have no direct relationship to one another.
The origin of each grid is generally located close to the centre of the grid. The unit of measure is either metres, feet or yards and may bear no relation to the origins of adjacent grids. The preceding pages show some of these nonstandard systems in current use.

11.3 Topographic Mapping of Antarctica

General

A section of the Antarctica continent, Terre Adélie was *first explored* by d'Urville of France in 1840. However, accurate maps of Terre Adélie were produced by France one hundred years later.
The *first topographic survey* of Antarctica was undertaken by Norway, based on aerial photography obtained in the period 1936 - 1937. A similar technology was used by the German Antarctic expedition during 1938-1939. After World War II, the Antarctic, lying almost entirely within the Antarctic Circle, was intensively investigated by the Byrd expedition in 1946-1947.

Australia, France, Norway, and New Zealand made territorial claims within the continent. The ice-free area of the continent was claimed by Argentina, Chile, and the United Kingdom. In 1959 an agreement was reached among twelve States: Argentina, Australia, Belgium, Chile, France, Japan, New Zealand, Norway, Republic of South Africa, United Kingdom, the USA, and the USSR. This resulted in signing of the *Antarctic Treaty* in 1961 by these States.
With time, more states have acceded to the Antarctic Treaty Organisation (ATO), including the People's Republic of China, the Federal Republic of Germany, Italy, and Poland. IMW (International Map of the World) sheet lines were conducted for all mapping of Antarctica.

Since 1980, most imagery has been obtained by means of *Advanced Very High Resolution Radiometer* (AVHRR)-imagery. Right now, scientific activities, as administered by SCAR (Special Committee on Antarctic Research), are concentrated on research in areas of earth-science. Maps and AVHRR-imagery are now prepared by a few states, such as the CIS, FRG, Japan, USA, and the UK.

Universal Transverse Mercator (UTM) and Lambert Conformal Conical (LCC) projections are used north of 80° S, and the Universal Polar Stereographic (UPS) projection for the area south of 80° S. The World Geodetic System of 1984 superseded WGS72 and now forms the reference datum (Böhme, 1993; Harper, 1977; Al-Bayari, 1996).

Antarctica - Argentina

Argentina was involved in the production of charts of Tierra del Fuego, Antártida and Islas del Atlantico Sur on the UPS grid system.

Antarctica - Australia

In 1939 the first map of the whole of Antarctica was produced on an Azimuthal Equidistant projection. Topographic surveys of the Australian possessions between 42°E and 160°E - discontinued by the French territory of Terre Adélie at 140°E - were undertaken by the Australian National Antarctic Research Expedition, and by the Australian Surveying and Land Information Group.

The Australian Heard and Macdonald Isles, in the southern Indian Ocean, have been mapped using the UTM grid. Macquarie Island, to the south of Australia, is projected on the Transverse Mercator (GK), and the Australian Antarctic Territory on UTM, LCC, or UPS grid systems (Rizos, 1990).

Antarctica - Belgium

Topographic survey work was carried out by the Expedition Antarctique de Belge in an area of Queen Maud Land. Maps of the Belgica Mountains region, Princess Astrid Coast and Princess Ragnhild Coast are plotted using the LCC projection and grid system, Queen Maud Land uses the UPS projection and grid system.

Antarctica - Chile

Chile's area of interest - "Territorio Chileno Antarctico" - uses the LCC projection and grid system.

Antarctica - China

The People's Republic of China acceded to the ATO in 1985. The investigation involved the establishment of the Zhong Shan Station on King George Island. The Fildes Peninsula is plotted on the GK projection.

Antarctica - CIS

The areas surveyed by CNIIGAiK (Institute for Geodesy, Air Survey and Cartography) are situated in the Eastern part of Antarctica: Pravda Coast, Queen Maud Land and Enderby Land.

Photomaps based on AVHRR-imagery, and World Map / Karta Mira are plotted on the CK42 projection. The maps of the "Antarktiky" are plotted on a Modified Polyconic projection. General maps on a small scale, such as *General'naja karta Antarktiky*, are on a Polar Stereographic grid, using the Krassovsky System of 1942 from 1956.

Antarctica - France

Antarctica was first explored by d'Urville in 1840. Since 1947, various Expeditions Polaires Françaises (EPF) have been carried out. The EPF organisation is responsible for all mapping of the "Terres Australes et Antarctiques Françaises" in the southern Indian Ocean, the Kerguelen and Crozet archipelagos, de la Possession, the isles of Amsterdam and St. Paul on the UTM grid. The maps of Terre Adélie are on a UPS grid, and Antarctica on a Gauss-Laborde (transverse Mercator) grid system.

Antarctica - Federal Republic of Germany

During the German Antarctic Expedition of 1938-1939, aerial surveys were undertaken to initiate an intended production of mapping. The "Georg von Neumayer" research station was set up to produce maps based on AVHRR-imagery. Topographic glaciological sheets have been published of the Filchner-Ronne Ice Shelf. Projections used: LCC-, and UPS grid system.

Antarctica - Japan

Japanese topographic mapping surveys by the Antarctic Research Expedition are based on AVHRR-imagery of the eastern border of Queen Maud Land, the Ongul Isles, Sor Rondane Mountains region, and a general map of Antarctica. Projections used: LCC-, and UPS grid system (Böhme, 1993).

Antarctica - New Zealand

The survey of the Ross Dependency was undertaken by the New Zealand Geological and Survey Antarctic Expedition. The NZMS Survey Department has produced topographic maps of small isles located near New Zealand, such as the Snares, Bounty, and Antipodes. Campbell is projected on NZMS- and UTM-grid systems, and Auckland Island is on a UTM. The maps of Ross Dependency, Ross Sea Region, Antarctic Regions and Continents are on the UPS grid system.

Antarctica - Norway

Roald Amundsen, who had broken through the Northwest Passage for the first time, succeeded in reaching the South Pole in 1911.

In 1936 aerial photography was undertaken during Lars Christensen's Norsk Antartktis Expedisjon of the "Det Antarktiske Kystland". Surveys by expeditions of Queen Maud Land, Sor Rondane, Maudheimvidda Aust, Peter I Isle, Crown Princess Märtha Coast, and Bouvet Isle provided maps on the LCC-, and UPS grid and projection system.

Antarctic - Poland

All topographic surveys in South Georgia, South Orkney Isles, Graham Land and the area around the Polish "Arctowski Station" at Admiralty Bay on King George Island, were carried out by the Polski Instytut Ekologii and Geofizyczny. Maps are on a LCC grid, using the Krassovsky System of 1942 (Böhme, 1993).

Antarctica - South Africa

The maps of Queen Maud Land and the South African Territory lying south of the African mainland - Prince Edward and Marion Island - are on a LCC grid, using Clarke 1866.

Antarctica - United Kingdom

The British Antarctic Territory, located south of South America, includes South Georgia and South Sandwich Isles - formerly the Falkland Islands -, the South Orkney Isles, the South Shetland Isles and parts of continental Antarctica.

Digitally derived topographic maps by airborne radio sounding and AVHRR-imagery covering the isles of South Orkney, and Eastern Alexander are based on WGS84. The maps of the British Antarctic Territory, South Georgia and South Sandwich Isles are plotted on the LCC-, and UPS grid and projection systems, based on the Clarke 1880 or International 1924 ellipsoid.

Antarctica - USA

The principle survey activities were concentrated on the western coastal zones bordering the Ross Sea, McMurdo Sound, the Filchner Ice Shelf, Enderby Land, Kemp Land and Wilkes Land. In 1991, the photomaps of Antarctica are produced from AVHRR-imagery. Maps are plotted on the LCC-, and UPS grid and projection systems based on the Clarke 1866, International 1924 or WGS84 ellipsoid (Böhme, 1993).

11.4 Spatial Databases

A report "The Use and Value of a Geodetic Reference System" by (Epstein / Duchesneau, 1984) discusses a benefit vs. cost framework for identifying and assessing economic value arising from the use of a geodetic reference system. In the context of economic value, a geodetic reference system itself can be viewed as an information for the production process. Using a geodetic reference system avoids costs, but does not automatically generate any benefits, these occur only if a demand exists.

Most benefit occurs with the use of spatial information based on a geodetic system by civil engineers, computer scientists, geographers, geologists, geophysicists, forest engineers and other users. These users have the need to integrate information produced for geodetic purposes. Activities and decisions, which generate an economic benefit by using universally compatible spatial information, provide an opportunity to measure the benefits obtained. The establishment of *universal multidimensional databases* and information networks is a part of the solution. It is a key element to (Leick, 1990):

- strengthen the world-wide research infrastructure, because fundamental, technological and geodetic research requires good computer-based communication media to work effectively while geographically dispersed, through better and quicker access to relevant information sources
- improve the process and speed of innovation and stimulate the exploitation of results, through the links provided between universities, research centres, standardisation and specialists
- lead to consistent approach and implementation of world-wide common standards. The objective is the provision of a common integrated computer communication infrastructure and associated services. It will allow exchange of data and research results on a world-wide basis.

Traditionally, the national geodetic network of every country used a properly defined datum and ellipsoid. New reference frames will be created and research is underway to arrive at a refined, optimal and universal geoid together with a world-wide geodetic reference frame. To acquire, integrate, process and disseminate spatial data in a digital dataflow without operational limits, several types of instruments are in use, such as:

- Multispectral Scanning Systems (MSS), Remote Sensing (ERTS, LandSat) instruments
- spaceborne photogrammetric cameras, such as the Russian KFA3000, KWR1000
- airborne photogrammetric CCD cameras
- terrestrial photogrammetric tools, such as CycloMedia (Beers, 1995) |[83]
- total-stations.

Spatial data of such instruments are used for integration |[84] to other sets of similar data.

The accuracies of GPS observations exceed those of an existing first-order geodetic network, providing the combined least squares adjustment of GPS, baselines and terrestrial surveys. Because these tools and spatial surveying techniques yield three-dimensional vectors, the spatial graphical and/or alpha-numerically geodetic database is expected to generate benefits. Of course the best universal multidimensional database and information network design - the requirements for the exchange of observation data among co-operating Mapping Agencies - has to be identified. Eventually, spatial measurements, least squares network adjustments and spatial database administration will be performed by using a *multidimensional DataBase Management System* (DBMS) to arrive at the expected benefits.

The outline of the process and dataflow is given in Figure 69.

[83] CycloMedia, the latest terrestrial photogrammetric system for 355° horizontal and 15° vertical scanning and digital image recording, processing, and mapping.

[84] The methods used in classical algorithms as by their nature are unnecessary slow and cumbersome. Using multidimensional calculation techniques, the new algorithms, P4P and twin P4P, support the advanced CCD-sensor instruments, such as mentioned above (Grafarend, 1997a, 1997b).

In many respects the future of topographic and associated disciplines from spatial data is already with us. For further reading, see (Beers, 1995; Burkholder, 1995a, 1995b; Epstein, 1984; Grafarend, 1985, 1997a, 1997b; Jivall, 1995; Leick, 1990).

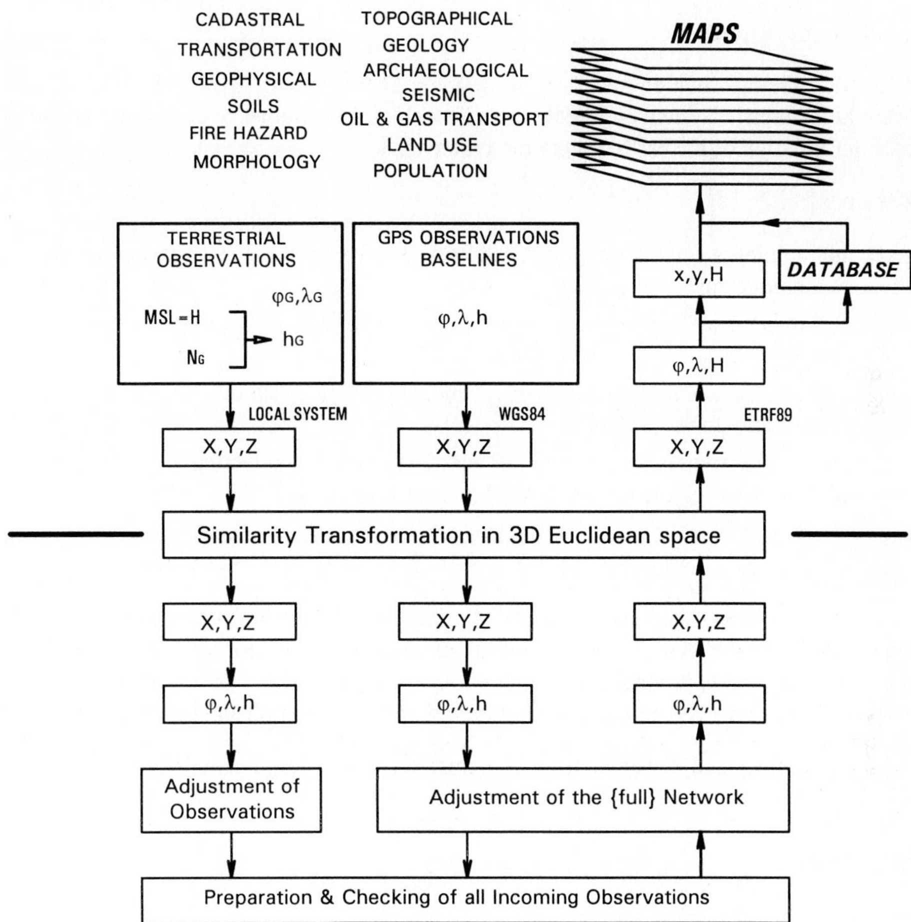

Figure 69: Sketch of spatial dataflow using a multidimensional database

11.5 Organisation of the International Earth Rotation Service

IERS Terms of Reference

The International Earth Rotation Service (IERS) was established in 1987 by IAU and IUGG and it started operation on January 1, 1988. It replaces the International Polar Motion Service (IPMS) and the earth-rotation section of the Bureau International de l'Heure (BIH); the activities of BIH on time are continued at Bureau International des Poids et Mesures (BIPM). IERS is a member of the Federation of Astronomical and Geophysical Data Analysis Services (FAGS).

IERS should provide the information necessary to define a Conventional Terrestrial Reference System, a Conventional Celestial Reference System and relate them as well as their frames to each other and to other reference systems used in termination of the earth orientation parameters.

IERS is responsible for:

- defining and maintaining a conventional terrestrial reference system based on observing stations that use the high-precision techniques in space geodesy
- defining and maintaining a Conventional Celestial Reference System based on extragalactic radio sources, and relating it to other celestial reference systems
- determining the earth orientation parameters connecting these systems, the terrestrial and celestial coordinates of the pole and universal time
- organising operational activities for observation and data analysis, collecting and archiving appropriate data and results, and disseminating the results to meet the needs of users.

IERS consists of a *Central Bureau* and *Co-ordinating Centres* for each of the principal observing techniques, and is supported by many other organisations that contribute to the tasks of observation and data processing.

The Co-ordinating Centres are responsible for developing and organising the activities in each technique to meet the objectives of the service. The Central Bureau combines the various types of data collected by the service, and disseminates the appropriate information on earth-orientation, the terrestrial and celestial reference systems to the user community. It can include sub-bureaux for the accomplishment of specific tasks.

The Central Bureau decides and disseminates the announcements of leap seconds in UTC and values of DUT1 to be transmitted with time signals.

IERS Organisation

The principal centres of IERS are as follows:

IERS Central Bureau

- Observatoire de Paris, Paris, France
- Terrestrial Frame Section, IGN, St Mandé, France
- Earth Orientation Section, Observatoire de Paris, France
- Celestial Frame Section, Observatoire de Paris, France.

IERS Sub-Bureau

- Rapid Service and Predictions, National Earth Orientation, US Naval Observatory, Washington DC
- Atmospheric Angular Momentum, Climate Analysis Centre, NOAA/National Weather Service, Washington, DC.

IERS Co-ordinating Centre

- VLBI Co-ordinating Centre, NGSG, Astronomy and Satellite Branch, Silver Spring, MD, USA
- LLR Co-ordinating Centre, OCA / CERGA, Grasse, France
- GPS Co-ordinating Centre, Jet Propulsion Laboratory, Pasadena, CA, USA
- SLR Co-ordinating Centre, Centre for Space Research, The University of Texas, Austin, TX, USA.

11.5.1 IERS Reference System

Rotation of the Earth and related space-time References. The IERS Reference System is composed of two parts: *IERS Standards* and *IERS Reference Frames.*

IERS Standards

The IERS standards used for a Report are the IERS Standards 1993 |[85], published in IERS Technical Note No 17. These are a set of constants and models used by the IERS Analysis Centres for Lunar Laser Ranging (LLR), Satellite Laser Ranging (SLR), Very Long Baseline Interferometry (VLBI), Global Positioning System (GPS), and by the IERS Central Bureau in the combination of results (IERS, 1993a, 1993b). The values of the constants are adopted from recent analyses; in some cases they differ from the current IAU and IAG conventional ones. The models represent, in general, the state of the art in the field concerned.

IERS Reference Frames

The IERS reference frames consist of the IERS Terrestrial Reference Frame (ITRF) and IERS Celestial Reference Frame (ICRF); both frames are realised through lists of co-ordinates of Fiducial Stations, terrestrial sites or compact extragalactic radio sources.

IERS Terrestrial Reference Frame

The origin, the reference directions and the scale of ITRF are implicitly defined by the co-ordinates adopted for the terrestrial sites. The origin of the ITRF is located at the centre of mass of the Earth with an uncertainty of 10 cm. The unit of length is the metre (SI). The IERS Reference Pole (IRP) and IERS Reference Meridian (IRM) are consistent with the corresponding directions in the BIH Terrestrial System (BTS) within 0".003. The BIH reference pole was adjusted to the Conventional International Origin (CIO) in 1967; it was then kept stable independently until 1987. The uncertainty of the tie of the BIH reference pole with the CIO was 0".03.

IERS Celestial Reference Frame

The origin of the ICRF is at the barycentre of the solar system. The direction of the polar axis is the one given for epoch J2000.0 by the IAU 1976 Precession and the IAU 1980 Theory of Nutation. The origin of right ascensions is in agreement with that of the FK5 within 0".01.

IERS Earth Orientation Parameters

The IERS Earth Orientation Parameters (EOP) are the parameters which describe the rotation of the ITRF to the ICRF, in conjunction with the conventional Precession-Nutation model. They model the unpredictable part of the Earth's rotation.

IERS Co-ordinates of the Pole

X, Y are the co-ordinates of the Celestial Ephemeris Pole (CEP) relative to the IRP, the IERS Reference Pole. The CEP differs from the instantaneous rotation axis by quasi-diurnal terms with amplitudes under 0".01. The X-axis is in the direction of IRM, the IERS Reference Meridian; the Y-axis is in the direction 90° West longitude.

[85] "IERS Standards" is a yearly report published by IERS with detailed values or constants in IERS "Technical Note"

IERS Celestial Pole offsets

$d\psi$, $d\varepsilon$ are offsets in longitude and in obliquity of the Celestial pole with respect to its position defined by the conventional IAU precession / nutation models.

11.6 List of Acronyms and Abbreviations

For information, see: [86]

1PPS One pulse per second signal
AAG Association of American Geographers
AAM Atmospheric Angular Momentum
ACA American Cartographic Association
ACM Association for Computing Machinery
ACS Active Control System
ACSM American Congress on Surveying and Mapping
ADOS African Doppler Survey
AER Atmospheric and Environmental Research Ins.
AFB Air Force Base
AFN Australian Fiducial Network
AGD Australian Geodetic Datum
AGI Année Géophysique Internationale
AGI Association for Geographic Information
AGN Astronomisch Geodätisches Netz
AGRS Active GPS Reference Frame
AGU American Geophysical Union
AHD Australian Height Datum
AIG Association Internationale de Géodésie
AIGA Association Internationale de Géomagnétisme et d'Aéronomie
AIMPA Association Internationale de Météorologie et de Physique Atmosphérique
AIR Associazione Italiana Razzi
AISH Association Internationale des Sciences Hydrologiques
AISPIT Association Internationale de Sismologie et de Physique de l'Intérieur de la Terre
AISPO Association Internationale de Sciences Physiques de l'Océan
AIUB Astronomical Institute, University of Bern
AIVCIT Association Internationale de Volcanologie et de Chimie de l'Intérieur de la Terre
AM Automated Mapping
AM/FM Automated Mapping/Facilities Management
AMG Australian Map Grid
AMPS Advanced Mobile Phone System
AMS American Mathematical Society
AMS Army Map Service
ANN Australian National Network
ANSI American National Standards Institute
APL Applied Physics Laboratory of John Hopkins University
ARGN Australian Regional GPS Network
AROF Ambiguity Resolution On the Fly
ASA American Standards Association
ASCII American Standard Code for Information Interchange/American Standards Committee II
ASPRS American Society for Photogrammetry and Remote Sensing
ATGIC Accurate Time for GPS Integrity Channel
ATS Average Terrestrial System
AVHRR Advanced Very High Resolution Radiometer
AVL Automatic Vehicle Location
AVLN Automatic Vehicle Location and Navigation
B-N Baker-Nunn camera
BASIC Beginner's All-purpose Symbolic Instruction Code
BDG Bearing and Distances to Geographicals
BE Broadcast Ephemeris
BEK Bavarian ErdmessungsKommission
BER Bit Error Rate
BIH Bureau International de l'Heure
BIOS Basic Input-Output System
BIP Band Interleaved by Pixel
BIPM Bureau International des Poids et Mesures
BIT Binary digit
BPI Bits Per Inch
BPS Bits Per Second
BRS Barycentric Reference System
BSI British Standards Institute
BTS BIH Terrestrial System
C Clairaut's Constant
C&GS US Coast and Geodetic Survey
C/A Coarse / Acquisition Code
CAD Computer Aided Drafting
CAE Computer Aided Engineering
CAM Computer Aided Mapping
CCD Charge Coupled Device
CCDS Comité Consultatif pour la Définition de la Seconde
CCIR Consultative Committee on International Radio communications
CCITT Consultative Committee on International Telephone and Telegraph
CDDIS NASA Crustal Dynamics Data Information System
CDP NASA Crustal Dynamics Project
CE US Corps of Engineers
CE-GPS European Complement to GPS
CEP Celestial Ephemeris Pole
CERGA Centre d'Etudes et de Recherches Géodynamiques et Astronomiques
CETEX Committee on Contamination by Extra Terrestrial Exploration
CFA Harvard-Smithsonian Centre For Astrophysics
CGI Commissione Geodetica Italiana
CGPM Conférence Générale des Poids et Mesures
CGSIC Civil GPS Service Interface Committee
CIE Commission International de l'Eclairage
CIGNET Co-operative International GPS Network
CIM Carte du Monde au Millionième
CINA Commission Internationale de Navigation Aérienne

[86] IERS Yearbook, Bollettino di Geodesia e Scienze Affini, (Leick, 1990), and other sources

CIO Conventional International Origin
CIPM Comité International des Poids et Mesures
CIS Commonwealth of Independent States
CIS Conventional Inertial Systems
CISC Complex Instruction Set Computer
CIUS Conseil International des Unions Scientifiques
CK CIS' Geodetic Control Network 1942
CM Central Meridian
CNES Centre National d' Etudes Spatiales, the French space agency
CODE Centre for Orbit Determination in Europe
COGO Co-ordinate Geometry
ConUS Contiguous United States
COSPAR Committee on Space Research
CPS Characters Per Second
CPU Central Processing Unit
CRL Communication Research Laboratory
CRT Cathode Ray Tube
CS Central Synchroniser
CSAGI Comité Spécial de l'Année Géophysique Internationale
CSOC Consolidated Space Operations Centre
CSR Centre for Space Research, University of Texas
CTP Conventional Terrestrial Pole
CTRF Conventional Terrestrial Reference Frame
CTS Conventional Terrestrial System
D/A Digital to Analogue converter
DA Department of the Army
DARPA Defense Advanced Research Projects Agency
DB Database
dB decibel
DBMS Data Base Management System
DDL Data Definition Language
DEM Digital Elevation Matrix/Model
DGFII Deutsches Geodätisches Forschungsinstitut, Abt. 1
DGIWG Digital Geographic Information Working Group, NATO
DGPS Differential GPS
DHDN Das Deutsche Hauptdreiecksnetz
DIGEST Digital Geographic information Exchange Standard
DIN Deutsche Industrial Norms
DLG Digital Line Graph (USGS)
DLG-E Digital Line Graph-Enhanced (USGS)
DMA Defense Mapping Agency, now NIMA
DMAAC Defense Mapping Agency Aerospace Centre
DMAHTC Defense Mapping Agency Hydrographic Topographic Centre
DME Distance Measuring Equipment
DMRS Data Management Retrieval System
DoD Department of Defense of the USA
DoE Department of Energy
DOEDOC German and Austrian Doppler campaign
DORIS Doppler Orbitography and Radio- location Integrated by Satellite
DOSE NASA Dynamics Of Solid Earth
DoT Department of Transportation
DPI Dots Per Inch
DPMS Dahlgren Polar Monitoring Service
DREF FRG Reference Network
DSIF Deep Space Instrumentation Facility, JPL (now DSN)
DSN JPL-Deep Space Network
DTM Digital Terrain Model
DUT Delft University of Technology
DXF Data exchange Format
EBCDIC Extended Binary Coded Decimal Interchange Code
ECEF Earth-Centred, Earth-Fixed
ECI Earth-Centred-Inertial
ECM Earth's Centre of Mass
ECMA European Computer Manufacturers Association
ECMWF European Centre for Medium-range Weather Forecasting
ED European Datum
EDM Electronic Distance Measuring
EDOREF French Doppler campaign
EGM Earth Gravitational Model
EIRP Equivalent Isotropic Radiated Power
ELN European Longitude Network
EMR Energy, Mines and Resources Canada
EOEPR European Organisation for Experimental Photogrammetric Research
EOP BIH Earth Orientation Parameter
EOSAT Earth Observation Satellite company
EPA Environmental Protection Agency
EPM Einpunktmatrixdateien
Eq.E Equation of equinox
ERDAS Earth Resources Data Analysis System
ERP Earth Rotation Parameters
ERTS Earth Resources Technology Satellite
ESA European Space Agency
ESOC European Space Agency Operational Centre
ESRI Environmental Systems Research Institute
ETRF European Terrestrial Reference Frame
ETRS European Terrestrial Reference System
EUREF European Terrestrial Reference Frame
FAA Federal Aviation Administration
FAF Federal Armed Forces
FAGS Fédération des Services Astronomiques et Géophysiques d'Analyse de Données
FAGS Federation of Astronomical and Geophysical Data Analysis Services
FBN Federal Base Network
FBSR Feedback shift register
FCC Federal Communications Commission
FGCC Federal Geodetic Control Committee
FGI Finnish Geodetic Institute
FIG Fédération Internationale des Géomètres
FIPS Federal Information Processing Standard
FK5 Fundamental Katalog Nr 5
FORTRAN Formula Translation
FoV Field of View
FRG Federal Republic of Germany
FRP Federal Radionavigation Plan

FTP File Transfer Protocol
GAOUA Main Astronomical Observatory of the Ukrainian Acad. of Sciences
GAST Greenwich apparent sidereal time
GBD Geographicals to Bearing and Distances
GBIS Geo-Based Information System
GCC Ground Control Centre
GCDB Geographic Co-ordinate Data Base
GDA Geocentric Datum of Australia
GDOP Geometric Dilution Of Precision
GEO Geostationary Satellite
GEOPS Geometric Equipotential Ellipsoid of Revolution
GEOREF World Geographic Reference System
GFZ GeoForschungs Zentrum Potsdam, FRG
GIC GPS Integrity Channel
GIRAS Geographic Information Retrieval and Analysis System
GIS Geographic Information System
GK Gauss-Krüger (Transverse Mercator)
GLONASS Global Navigation Satellite System
GLONASS Globalnaya Navigatsionnaya Sputnikovaya Sistema
GLOSS Global Sea Level Observing System
GM Geocentric Gravitational Constant
GMA Geodetic Model of Australia
GMST Greenwich Mean Sidereal Time
GNIS Geographic Names Information System
GNSS Global Navigation Satellite System (GPS, GLONASS, etc.)
GOTDOC Gotland and Sweden's Doppler campaign
GPS Global Positioning System, mainly NAVSTAR's US DoD program
GPSIC GPS Information Center
GRASS Geographic Resources Analysis Support System
GRGS Groupe de Recherches de Géodésie Spatiale
GRID Global Resource Information Database (UN Environment Program)
GRS Geodetic Reference System
GRUB Geodetic Institute of the University of Bonn
GSFC Goddard Space Flight Centre (Greenbelt)
h Ellipsoidal height
H Levelled Height above MSL
HARN High-Accuracy Reference Network
HDOP Horizontal Dilution Of Precision
HDX Half-Duple X transmission
HEO High Elliptical Orbit
HIS Hue, Intensity, and Saturation
HLS Hue, Luminance, and Saturation
HN Höhen Null, Eastern FRG
HOM Hotine Oblique Mercator
HOW Handover word
I/O input/output
IAG International Association of Geodesy
IAGA International Association of Geomagnetism and Aeronomy
IAGS Inter-American Geodetic Survey
IAH International Association of Hydrogeologists
IAHS International Association of Hydrological Sciences
IAMAP International Association of Meteorology and Atmospheric Physics
IAPSO International Association for the Physical Sciences of the Ocean
IASPEI International Association of Seismology and Physics of the Earth's Interior
IASPO Association Internationale des Sciences Physiques de l'Oceán
IAT International Atomic Time
IAU International Astronomical Union
IAVCEI International Association of Volcanology and Chemistry of the Earth's Interior
ICA International Cartographic Association
ICAM Integrated Computer Automated Mapping
ICAN International Committee of Aeronautical Navigation
ICAO International Civil Aviation Organisation
ICES International Council for the Exploration of the Sea - Conseil International pour l'Exploration de la Mer
ICO Intermediate Circular Orbit
ICRF IERS Celestial Reference Frame
ICSU International Council of Scientific Unions
IDMS Integrated Database Management System
IERS International Earth Rotation Service
IESSG Institute of Engineering Surveying and Space Geodesy - Nottingham University
IfAG Institut für Angewandte Geodäsie, FRG
IGES International Graphic Exchange Specification
IGLD International Great Lakes Datum
IGM Istituto Geografico Militare
IGM Italian Geodetic Network
IGN Institut Géographique National, Paris
IGS International GPS Geodynamics Service
IGSN International Gravity Standardisation Network
IGU International Geographical Union
IGY International Geophysical Year
IHB International Hydrographic Bureau, Monaco
ILO International Labour Organisation
ILS Instrument Landing System
ILS International Latitude Service
IMO International Maritime Organisation
IMU Inertial Measurement Unit
IMU International Mathematical Union
IMW International Map of the World on the millionth scale
INMARSAT - International Maritime Satellite organisation
INS Inertial Navigation System
IOC Initial operational capability
ION Institute Of Navigation
IOS International Organisation Standardisation
IPMS International Polar Motion Service
IPX Internetwork Protocol exchange
IPY International Polar Year
IR Infra-Red
IRIS International Radio Interferometric Surveying

IRM IERS Reference Meridian
IRP IERS Reference Pole
IRU Inertial Reference Units
ISA Industry Standard Architecture
ISDN Integrated Services Digital Network
ISG Integrated Survey Grid
ISO International Standards Organisation
ISS Inertial Surveying Systems
ITC International Training Centre for Aerospace Survey and Earth Sciences (NL)
ITO International Trade Organisation
ITRF IERS Terrestrial Reference Frame
ITS Instantaneous Terrestrial System
ITU International Telecommunication Union - Union Internationale des Télécommunications
IUB International Union of Biochemistry
IUBS International Union of Biological Sciences
IUCr International Union of Crystallography
IUFRO International Union of Forestry Research Organisations
IUGG International Union of Geodesy and Geophysics
IUHS International Union of the History of Science Union Internationale d'Histoire des Sciences
IUP International Union of Physiology
IUPAC International Union of Pure and Applied Chemistry
IUPAP International Union of Pure and Applied Physics
IUTAM International Union of Theoretical and Applied Mechanics
IVHS Intelligent Vehicle-Highway System
JD Julian date
JDH Japan Hydrographic Department
JMA Japan Meteorological Agency
JPL Jet Propulsion Laboratory (California Institute of Technology)
JPO Joint Program Office
KONMAC German GPS-campaign
L+T Federal Office of Topography, Berne
L1 L1 carrier (1575.42 MHz)
L2 L2 carrier (1227.6 MHz)
LAN Local Area Network
LAN Longitude of the Ascending Node
LANDSAT US Satellite System
LASER Light Amplification by Stimulated Emission of Radiation
LCC Lambert Conformal Conic
LEO Low Elliptical Orbit
LIS Land Information System
LLR Lunar Laser Ranging
LORAN Long Range Navigation system
LPAC Astronomical Council of USSR
LPF Low Pass Filter
m International metre
MC&G Mapping, Charting, and Geodesy
MCP Multiplex and Cross-connect Processor
MEO Medium Elliptical Orbit
MERIT Monitoring Earth Rotation and Intercomparing Techniques
MGPS Multibase GPS
MIT Massachusetts Institute of Technology
MITLL MIT - Lincoln Laboratories
MJD Modified Julian Day
MLS Microwave Landing System
MODEM Modulator-Demodulator
MRES Multiple Regression Equations
MSL Mean Sea Level
MSS Multispectral Scanner
MST Mean sidereal time
MSTI Mean sidereal time corrected for polar motion
N Geoid undulations
N.g.d.F. Nivellement general de France
N.g.d.M. Nivellement general de Madagascar
NAD27 North American Datum of 1927
NAD83 North American Datum of 1983
NAL National Aerospace Laboratory, Japan
NANU Notice advisory to NAVSTAR users
NAOMZ National Astronomical Observatory, Mizusawa branch
NAP Normaal Amsterdams Peil
NASA National Aeronautics and Space Administration
NATO North Atlantic Treaty Organisation
NAVD88 North American Vertical Datum of 1988
NAVNET Navy VLBI Network
NAVOCEANO - Naval Oceanographic Office
NAVSTAR Navigation Satellite Timing And Ranging
NBS National Bureau of Standards
NCDCDS National Committee for Digital Cartographic Data Standards
NCO Numerically controlled oscillator
NEDOC Netherlands' Doppler campaign
NEOS National Earth Orientation Service
NEP North Ecliptic Pole
NEREF Netherlands Terrestrial Reference Frame
NFCE Navigation Field Control Equipment
NGI National Geografisch Instituut, Brussels
NGRS National Geodetic Reference System
NGS National Geodetic Survey, Silver Springs
NGS National Geographic Society
NGVD29 National Geodetic Vertical Datum of 1929
NH Normal Höhen Punkt, Eastern FRG
NHTSA National Highway Traffic Safety Administration
NIMA National Imagery and Mapping Agency, formerly DMA
NIST National Institute of Standards and Technology
NMC National Meteorological Centre
NN Normal Null, Western FRG
NNSS US Navy Navigation Satellite System (Transit or simply Doppler)
NOAA US National Oceanic and Atmospheric Administration
NORAD North American Air Defence
NorFA Nordic Academy For Advanced Studies

NPL National Physical Laboratory
NRC National Research Council
NRL Naval Research Laboratory
NSWC US Naval Surface Warfare Center
NSWL US Naval Surface Warfare Laboratory
NZMG New Zealand Grid System
OACI Organisation de l'Aviation Civile Internationale
OCA Observatoire de la Côte d'Azur
OCS GPS operational control system
ODGPS Ordinary Differential GPS
OEEPE Organisation Européenne d'Etudes Photogrammétriques Expérimentales
OEM Original Equipment Manufacturer
OM Oblique Mercator
OMM Organisation Météorologique Mondiale World Meteorological Organisation
ONERS Office National d' Etudes et de Recherches Spatiales
ONU Organisation des Nations Unies
OP Observatoire de Paris
Organisation Météorologique Mondiale
OS Operating System
OSGB Ordnance Survey of Great Britain
OSGB(SN) Ordnance Survey of Great Britain Scientific Network
OSI Ordnance Survey of Ireland
OSNI Ordnance Survey of Northern Ireland
OTF On-the-fly ambiguity resolution
P-code Precision code (10.23 MHz)
P. du N. Pierre du Niton (Switzerland)
PAGEOS Passive Geodetic Earth Orbiting Satellite
PAIGH Pan American Institute of Geography and History
PCMCIA Personal Computer Memory Card International Association
PDN Packet Data Network
PDOP Position Dilution Of Precision
PE Precise Ephemeris
PIN Parcel/Personal Identification Number
PIOSA Pan Indian Ocean Science Association
PLL Phase lock loop
PM Prime Meridian
PMR US Navy Pacific Missile Range
PPP Point-to-Point Protocol
PPS Precise Positioning Service
PRARE Precise Range and Range Rate Equipment
PRC Pseudo Range Corrections
PRME Pseudo Range Monitoring Error
PRN Pseudorandom noise
PSAD Provisional South American Datum
PSK Phase-shift keying
PSMSL Permanent Service for Mean Sea Level
QBE Query By Example
QBF Query By Forms
QC Quality Control
QOS Quantum-Optical Stations
QPSK Quadri-phase-shift keying
R&D Research and Development
RADAR Radio Detection And Ranging
RBV Return Beam Vidicon
RD Rijks Driehoeksmeting
RDB Relational Database
RDMS Relational Database Management System
RDOP Relative dilution of precision
RETDOC Doppler campaign for RETrig
RETrig Réseau Europeén de Triangulation
RGB Red-Green-Blue
RINEX Receiver Independent Exchange (format)
RIRT Russian Institute of Radionavigation and Time
RISC Reduced Instruction Set Computing
RMS Root Mean Square
RMSE Root Mean Square Error
RPG Report Program Generator language
RPM Revolutions Per Minute
RS Remote Sensing
RSC Radio Source Co-ordinates
RSO Rectified Skew Orthomorphic
RT-DGPS Real Time-DGPS
RTCA Radio Technical Commission for Aeronautics
RTCM Radio Technical Commission for Maritime Services
RTK Real Time Kinematic
RTRF Regional Terrestrial Reference Frame
S/N Signal-to-Noise ratio
SA Selective Availability of GPS
SAD South American Datum
SAO Smithsonian Astrophysical Observatory
SCAR Special Committee on Antarctic Research
SCG Special Committee for Inter-Union Co-operation in Geophysics
SCOR Scientific Committee on Oceanic Research
SCSI Small Computer Systems Interface
SDTS Spatial Data Transfer Standard, Federal Geographic Data Committee
SECOR Sequential Collation of Range
SGS Soviet Geocentric System
SHA Shanghai Observatory
SI Système International
SIF Standard Interchange Format
SIFET Società Italiana di Fotogrammetria e Topografia
SIGEM Società Italiana di Geofisica e Meteorologia
SIO Scripps Institution of Oceanography
SIP Société Internationale de Photogrammétrie
SLIP Serial Line Internet Protocol
SLR Satellite Laser Ranging
SNR Signal-to-Noise Ratio
SPANDOC Spanish Doppler campaign
SPARC Scaleable Processor Architecture
SPCS State Plane Co-ordinate System
SPOT Système Probatoire d'Observation de la Terre
SPS Standard Positioning Service
SRP Solar radiation pressure
SSB Single Side Band
SSC Sets of Stations Co-ordinates
SSG Special Study Group

STADAN Satellite Tracking and Data Acquisition Network (GSFC)
STN Staatliches Trigonometrisches Netz
SV Space Vehicle (Navstar)
SVN Space vehicle launch number
SWEREF Swedish Reference Network
t Grid azimuth
T Projected geodetic azimuth
TAI International Atomic Time
Temps Atomique International
TCP/IP Transmission Control Protocol / Internet Protocol
TD Tokyo Datum
TDMA Time Division Multiplexing Architecture
TDN Topografische Dienst Nederland
TDOP Time Dilution of Precision
TDT Terrestrial Dynamical Time
TEC Total electron content
TELEX Teleprinter Exchange service
TGBM Tide Gauge Bench Marks
TIFF Tagged Image File Format
TIGER file Topological Integrated Geographic Encoding and Reference files
TLM Telemetry word
TM Registered Trademark
TM Thematic Mapper (on LANDSAT satellite)
TM Transverse Mercator (Gauss-Krüger in Europe)
TOP Technical and Office Protocol
TOPEX Ocean Topographical Experiment
TOPOCOM
US Army Topographic Command (now NIMA)
TOW Time of week
TPC Telecommunications Process Controller
TR Technical Report
TRANSIT Time Ranging and Sequential
TT Terrestrial Time
TYRGEONET -
Tyrrhenian Geodetic Network
UAI Union Astronomique Internationale
UDL Unified Data Language
UDN User Densification Network
UELN Unified European Levelling Network
UERE User Equivalent Range Error
UGGI Union Géodésique et Géophysique Internationale
UGI Union Géographique Internationale
UHF Ultra High Frequency
UIC User Identification Code
UICPA Union Internationale de Chimie Pure et Appliquée
UICPA Union Internationale de Mécanique Théorique et Appliquée

UPLN Unified (East) European Precise Levelling Network
UTC Universal co-ordinated time
UTI Universal time corrected for polar motion
VDOP Vertical dilution of precision
VLBI Very long baseline interferometry
VSAT Very Small Aperture Terminal
WAADGPS -
Wide Area Augmented Differential GPS
WAAS Wide Area Augmentation System
WADGPS Wide Area Differential GPS
WAN Wide Area Network
WGO Working Group of Oceanography (of CSAGI)
WGS US DoD: World Geodetic Reference System
WGS84-G730 -
WGS84 referring to GPS week 730
WHO World Health Organisation
WMO World Meteorological Organisation
WSMR White Sands Missile Range (New Mexico)
WTUSM Wuhan Technical University of Surveying and Mapping
WVR Water vapor radiometer
XNS/ITP Xerox Network Systems' Internet Transport Protocol
Y-code Encrypted P-code
YMCK Yellow-Magenta-Cyan-Black
ZEN Zentral Europaïsches Netz
ZIPE Zentralinstitut für Physik der Erde

11.7 Programs

11.7.1 ELLIDATA.BAS

ELLIDATA.BAS - Date: 05-16-1997

ELLIPSOID CONSTANTS	
Ref. Ellips =	**Bessel 1841**
A =	**6377397.155**
Fl =	**299.15281285**
Scale Factor :	**1**
PI4 =	.7853981633974483
RD =	.0174532925199433
f =	3.342773181616099D-03
1-f (w1) =	.9966572268183839
c =	6398786.848070608
A/B =	1.00335398479203
N =	1.674184800834699D-03
CONT	**<press F5>**
e^2 =	6.674372230688467D-03
e =	8.169683121571159D-02
1-e^2 =	.9933256277693116
SQR(1-e^2) =	.9966572268183839
e^'2 =	6.719218798046065D-03
1+e^'2 =	1.006719218798046
SQR(1+e^'2) =	1.00335398479203
A =	6377397.155
B =	6356078.962821751
r =	6366742.520235041
E1 =	- 6.039696610366733D-02
E2 =	2.535744451207499D-04
E3 =	- 1.313917859616868D-06
F1 =	6.009450686939833D-02
F2 =	3.503304296398412D-04
F3 =	2.834308525744958D-06
1+e^'2 =	1.006719218798046
SQR(1+e^'2) =	1.00335398479203
Lat (D.DDD) :	**50**
N =	6389923.081696984 = ν
M =	6372232.366902328 = ρ
R mean =	6381071.593643641
R polar c =	6398786.848070608
R (azimuth) =	6382601.810046898
R ab mean =	6366738.058910876
r geo-mean =	6366729.136256292
r eq.vol.sph =	6370283.172953796
Gm =	5540279.541931841
Gm :	**5540279.5419**
Lat =	49.99999999859768

□

```
1000 ' ELLIDATA.BAS
1010 CLS
1020 PRINT "ELLIDATA.BAS ,A ";"- Date:
";DATE$
1030  ' ELLIDATA.BAS ,A - Date: 05-16-1997
1040 PRINT     "--------------------------------------"
1050 PRINT "          ELLIPSOID CONSTANTS "
1060  '               M. Hooijberg 1976-1997
1070 PRINT     "--------------------------------------"
1080 DEFDBL A-Z
1090 DEFINT I
1100 PI4=ATN(1)
1110 RD=PI4/45
1120  ' input ellipsoid data
1130 INPUT "Ref. Ellips :", ___ P$
1140 PRINT "Ref. Ellips =";___ P$
1150 INPUT "Major axis :", ___ A
1160 PRINT "A =";__________ A
1170 INPUT "1/f :", _________ FL
1180 PRINT "Fl ="; _________ FL
1190 INPUT "Scale Factor :",__ K0
1200 PRINT "K0 =";_________ K0
1210  ' calculate ellipsoid data
1220 PRINT "PI4 ="; ________ PI4
1230 PRINT "RD ="; ________ RD
1240 F=1/FL
1250 PRINT "f ="; __________ F
1260 W1=1-F
1270 PRINT "1-f (w1) =";_____ W1
1280 C=A/W1
1290 PRINT "c ="; __________ C
1300 AB=1/W1
1310 PRINT "A/B =";A ______ B
1320 N=F/(2-F)
1330 PRINT "N =";__________ N
1340 STOP
1350 EC2=(2-F)/FL
1360 PRINT "e^2 ="; ________ EC2
1370 EC=SQR(EC2)
1380 PRINT "e ="; __________ EC
1390 VE2=1-EC2
1400 PRINT "1-e^2 =";_______ VE2
1410 PRINT "SQR(1-e^2) ="; _ SQR(VE2)
1420 E12=EC2/W1/W1
1430 PRINT "e^ '2 =";_______ E12
1440 PRINT "1+e^ '2 =";_____ 1+E12
1450 PRINT "SQR(1+e^ '2) ="; SQR(1+E12)
1460 PRINT "A =";__________ A
1470 B=A-A/FL
1480 PRINT "B =";__________ B
1490  ' --- forward constants ---
1500 R0=A*(1+N^2/4)/(1+N)
1510 PRINT "r =";___________R0
1520 E1=-N*(36+N*(45+39*N))
1530 PRINT "E1 ="; _________E1
1540 E3=-280*N^3
1550 E2=90*N^2-E3
1560 PRINT "E2 ="; _________E2
1570 PRINT "E3 ="; _________E3
1580  ' --- inverse constants ---
1590 F1=N*(36+N*(-63+93*N))
1600 PRINT "F1 ="; _________F1
1610 F3=604*N^3
1620 F2=126*N^2-F3
1630 PRINT "F2 ="; _________F2
1640 PRINT "F3 ="; _________F3
1650 PRINT "1+e^ '2 =";1+E12
1660 PRINT "SQR(1+e^ '2) ="; SQR(1+E12)
1670 INPUT "Lat (D.DDD) :", _M0
1680 S =M0*RD
1690 S2=SIN(S)^2
1700 C2=COS(S)^2
1710 W=SQR(1-EC2*S2)
1720 V=SQR(1+E12*C2)
1730 N=A/W
1740 PRINT "N ="; __________N;"v"
1750 M=C/V^3
1760 PRINT "M =";__________M;"p"
1770 PRINT "R mean ="; _____SQR(M*N)
1780 PRINT "r polar c =";_____A^2/B
1790 PRINT "R (azimuth) ="; _M*N/(M*S2+N*C2)
1800 PRINT "R ab mean ="; ___(A+B)/2
1810 PRINT "r geo-mean =";___SQR(A*B)
1820 PRINT "r eq.vol.sph ="; __A*(1-F/3-F^2/9)
1830  'calculate meridional arc
1840  'sub meridional arc forwards
1850 C0=COS(S)
1860 C1=C0*C0
1870 G0=R0*K0*(S+SQR(1-C1)*C0/12*(E1+C1*
(E2+C1*E3)))
1880 PRINT "Gm =";_________G0
1890  'sub meridional arc inverse
1900 INPUT "Gm : ",_________G1
1910 PRINT "Gm =";_________G1
1920 W0=G1/R0/K0
1930 C0=COS(W0)
1940 C1=C0*C0
1950 S=W0+SQR(1-C1)*C0/12*(F1+C1*
(F2+C1*F3))
1960 M0=S/RD  'raddms
1970 PRINT "Lat =";_________M0
1980 STOP : GOTO 1130
```

□

11.7.2 REFGRS00.BAS

REFGRS00.BAS - Date: 06-08-1996

GEODETIC REFERENCE SYSTEM

Ellipsoid :	**WGS84**
Ellipsoid =	WGS84
Major axis :	**6378137**
Major axis =	6378137
GM*10^8 M^3 S^-2 :	**3986005D8**
GM*10^8 M^3 S^-2 =	398600500000000
J2*10^-8 :	**108262.9989051944D-8**
J2*10^-8 =	1.082629989051944D-03
W*10^-11 rad S^-1 :	**7292115D-11**
W*10^-11 rad S^-1 =	.00007292115
e2 <estimate> :	**.00669438**
e2 <estimate> =	.00669438
CONT	**<press F5>**
e2 <temp> =	6.694379990109238D-03
e2 <temp> =	6.694379990108909D-03
e2 <temp> =	6.69437999013992D-03
e2 <temp> =	6.694379990156965D-03
e2 <temp> =	6.694379990130334D-03
e2 <temp> =	6.694379990158905D-03
e2 <temp> =	6.694379990152921D-03
e2 <temp> =	6.694379990137185D-03
e2 <temp> =	6.694379990164976D-03
e2 <temp> =	6.694379990137433D-03
e2 <temp> =	6.69437999012346D-03
e2 <temp> =	6.694379990152814D-03
e2 <temp> =	6.694379990137089D-03
e2 <temp> =	6.694379990138545D-03
e2 <temp> =	6.69437999013488D-03
e2 <temp> =	6.694379990131632D-03
e2 <temp> =	6.694379990128753D-03
e2 <temp> =	6.69437999014707D-03
e2 <temp> =	.006694379990132
e2 <mean> :	**.00669437999014**
1/F =	3.352810664746825D-03
F =	298.2572235630584
	<next case>

□

```
1000 'REFGRS00.BAS

1010 CLS
1020 PRINT "REFGRS00.BAS - Date: ";DATE$
1030 'REFGRS00.BAS - Date: 06-08-1996
1040 PRINT "-------------------------------------------"
1050 PRINT    "GEODETIC REFERENCE SYSTEM"
1060 '                  M. Hooijberg 1996
1070 PRINT "-------------------------------------------"
1080 DEFDBL A-Z
1090 INPUT "Ellipsoid          : ",  P$
1100 PRINT "Ellipsoid          = ";  P$
1110 INPUT "Major axis         : ",  A
1120 PRINT "Major axis         = ";  A
1130 INPUT "GM*10^8 M^3 S^-2  : ",      GM
1140 PRINT "GM*10^8 M^3 S^-2  = ";      GM
1150 INPUT "J2*10^-8           : ",  J2
1160 PRINT "J2*10^-8           = ";  J2
1170 INPUT "W*10^-11 rad S^-1 : ",      W
1180 PRINT "W*10^-11 rad S^-1 = ";      W
1190 INPUT "e2 <estimate>     : ", W3
1200 PRINT "e2 <estimate>     = ";      W3
1210 RD=ATN(1 )/45
1220 STOP

1230 COUNT=0
1240 E1=SQR(W3)/SQR(1 -W3)
1250 COUNT=COUNT+1
1260 Q2=(1 +3 /E1^2)*ATN(E1)-3 /E1
1270 E2=3 *J2+4 /15 *W^2*A^3/GM*W3^1.5/Q2
1280 PRINT "e2 <temp>         = ";E2
1290 IF COUNT=20  THEN 1300 ELSE W3=(W3+E2)/2  : GOTO 1240
1300 INPUT "e2 <mean> : ",EE2
1310 PRINT "1/F       = ";1 -SQR(1 -EE2)
1320 PRINT "F         = ";1 /(1 -SQR(1 -EE2))
1330 END
```
□

11.7.3 BDG00000.BAS

BDG00000.BAS - Date: 01-27-1997

DIRECT LONG LINE CALCULATION

ellipsoid: **BESSEL 1841**
ellipsoid= BESSEL 1841
major axis: **6377397.155**
major axis= 6377397.155
1/flattening: **299.15281285**
1/flattening= 299.15281285

Latitude-1 (D,M,S): **45,0,0**
N or S : **N**
Latitude -1= 45 deg 0 min 0 sec N
Longitude-1 (D,M,S): **10,0,0**
E or W : **E**
Longitude-1= 10 deg 0 min 0 sec E
True Bearing 1-2 (D,M,S): **29,3,15.45943**
True Bearing1-2= 29 deg 3 min 15.45943 sec
True Distance: **1320284.369475**
True Distance 1-2= 1320284.369475
No of Integrations: **1000**
No of Integrations: 1000

1 - f= .9966572268183839
C = 6398786.848070608
e'2= 6.719218798046065D-03
Time = *17:39:02*
C= .3486350841733274
S= 1.659248455199551D-04
A= 1.001105713918906
R= .5980706182687509
do-loop :::::::::::::::↓:::::::::::::::::::: ::::::::::::::::::↓:::: line 2140 - 2270

C= .2528748305691283
S= 1.749993990064363D-04
A= 1.00138029751993
R= .5353980803046009

Latitude = i= 500= 50 deg 7 min 7.7994 63544846148 sec N
Longitude= i= 500= 14 deg 29 min 19.1779 3210855908 sec E
::::::::::::::::::↓:::::::::::::::::::: ::::::::::::::::::↓::::::::::::::::::::

C= .3490658492482405
S= 1.658806495262553D-04
A= 1.001104667508712
R= .5983533471896338
D= -3.783037031223165
Time = *17:39:52*

Latitude -2= 55 deg 0 min 1.019798133938821D-04 sec N

Longitude-2= ______________________ 19 deg 59 min 59.9997 6266646158 sec E
True Bearing 2-1= __________________ 216 deg 45 min 7.4002 70836825484 sec
Latitude (D,M,S): __________________ **<next case>**

□

```
1000 'BDG00000.BAS
1010 CLS
1020 PRINT "BDG00000.BAS,A";"- Date: ";DATE$
1030 ' BDG00000.BAS,A - Date: 01-27-1997
1040 PRINT "-------------------------------------------"
1050 PRINT " DIRECT LONG LINE CALCULATION
1060 '              M. Hooijberg 1981-1997
1070 PRINT "-------------------------------------------"
1080 DEFDBL A-Z
1090 DEFINT I
1100 PI=4*ATN(1)
1110 R0=ATN(1)/45
1120 DEF FNASN(X)=ATN(X/SQR(1-X*X))
1130 DEF FNACS(X)=PI/2-ATN(X/SQR(1-X*X))
1140 READ
LA$,LO$,D$,M$,S$,T$,DI$,BR$,P1$,P2$,
P3$,P4$,P5$,A$

1150 ' ellipsoid data
1160 INPUT "ellipsoid:      ",  P$
1170 PRINT "ellipsoid=      ";  P$
1180 INPUT "major axis:     ",  A0
1190 PRINT "major axis=     ";  A0
1200 INPUT "1/flattening:   ",  FL
1210 PRINT "1/flattening=   ";  FL

1220 ' input data
1230 INPUT "Latitude-1 (D,M,S):      ",M1,M2,M3
1240 INPUT "N or S :",NS$
1250 I2=1
1260 IF NS$="N" OR NS$="n" THEN 1300
1270 IF NS$="S" OR NS$="s" THEN 1290
1280 GOTO 1240
1290 I2=-1
1300 PRINT
LA$;P1$;A$;M1;D$;M2;M$;M3;S$;NS$
1310 GOSUB 2720 ' call dmsrad
1320 B=M0
1330 INPUT "Longitude-1 (D,M,S):     ",M1,M2,M3
1340 INPUT "E or W :",EW$
1350 I2=1
1360 IF EW$="E" OR EW$="e" THEN 1400
1370 IF EW$="W" OR EW$="w" THEN 1390
1380 GOTO 1340
1390 I2=-1
1400 PRINT
LO$;P1$;A$;M1;D$;M2;M$;M3;S$;EW$
1410 GOSUB 2720 ' call dmsrad
1420 C=M0
1430 INPUT "True Bearing 1-2
(D,M,S):",M1,M2,M3
1440 PRINT T$;BR$;P3$;A$;M1;D$;M2;M$;M3;S$
1450 I2=1
1460 GOSUB 2720 ' call dmsrad
1470 D=M0
1480 INPUT "True Distance:      ",E
1490 PRINT T$;DI$;P3$;A$;E
1500 INPUT "No of Integrations: ",Z
1510 PRINT "No of Integrations: ";Z
1520 IF Z=0 THEN 1500

1530 ' start integration
1540 ' IZ=Z/2 ' intermediate station
1550 E=E/Z
1560 ' E10=E ' dist.counter
1570 ' E11=0 ' check do-loop
1580 F0=1/FL
1590 H=1-F0
1600 F=A0/H
1610 G=(2-F0)/FL/H/H
1620 ' PRINT "1-f= ";     H
1630 ' PRINT "C  = ";     F
1640 ' PRINT "e'2= ";     G
1650 V=-1
1660 U=PI/2
1670 T=U
1680 IF T<=D THEN 1710
1690 V=0
1700 GOTO 1850
1710 T=2*T
1720 IF T<=D THEN 1770
1730 V=1
1740 D=D-T
1750 E=-E
1760 GOTO 1850
1770 T=T+U
1780 IF T<=D THEN 1820
1790 V=2
1800 T=T-U
1810 GOTO 1740
1820 T=T+U
1830 D=D-T
1840 V=0
1850 J=COS(B)
1860 A=SQR(J*J*G+1)
1870 ' PRINT "A= ";A
1880 R=SIN(D)
1890 H=F*R*J/A
1900 ' PRINT "Time = ";TIME$

1910 ' repeat integration
1920 IF ABS(R)>.99999999999 THEN 2540 ' error
1930 U=SQR(1-R*R)
```

```
1940 S=U*E*A^3/F
1950 J=COS(S/2+B)
1960 A=SQR(J*J*G+1)
1970 R=A*H/F/J
1980 ' PRINT "S= ";S
1990 ' PRINT "A= ";A
2000 ' PRINT "R= ";R
2010 C=C+A*E*R/F/J
2020 ' PRINT "C= ";C
2030 IF ABS(R)>.99999999999 THEN 2540 'error
2040 U=SQR(1-R*R)
2050 S=U*E*A^3/F
2060 ' PRINT "S= ";S
2070 B=B+S
2080 J=COS(B)
2090 A=SQR(J*J*G+1)
2100 R=A*H/F/J
2110 ' PRINT "A= ";A
2120 ' PRINT "R= ";R
2130  GOTO 2280 'skip i/s->

2140 'compute intermediate stations
2150 M0=B
2160 GOSUB 2760 'call raddms
2170 B1=M1:B2=M2:B3=M3
2180 IF I2=>0 THEN NS$=" N"
2190 IF I2<0 THEN NS$=" S"
2200 'display i/s
2210 IF Z=IZ THEN PRINT
LA$;A$;P5$;Z;A$:B1;D$;B2;M$;B3;S$;NS$
2220 M0=C
2230 GOSUB 2760 'call raddms
2240 C1=M1:C2=M2:C3=M3
2250 IF I2=>0 THEN EW$=" E"
2260 IF I2<0 THEN EW$=" W"
2270 IF Z=IZ THEN PRINT
LO$;A$;P5$;Z;A$;C1;D$;C2;M$;C3;S$;EW$

2280 ' NUM=Z/100 'do-loop counter
2290 ' IF NUM-INT(NUM)=0 THEN PRINT "I= ",Z;
2300 ' E11=E11+E10 'do-loop check
2310 ' PRINT "Dist= ";E11
2320 Z=Z-1
2330 IF Z>=1 THEN 1910 'next integration ->
2340 W3=FNASN(R)
2350 D=W3*(V=2)+(W3+2*PI)*(V=1)+
(W3+PI)*(V=0)
2360 ' PRINT "D= ";D

2370 'cont.computation
2380 ' PRINT "Time = ";TIME$
2390 M0=B
2400 GOSUB 2760 'call raddms
2410 NS$="N"
2420 IF I2=-1 THEN NS$="S"
2430 PRINT
LA$;P2$;A$;M1;D$;M2;M$;M3;S$;NS$
2440 M0=C
2450 GOSUB 2760 'call raddms
2460 EW$="E"
2470 IF I2=-1 THEN EW$="W"
2480 PRINT
LO$;P2$;A$;M1;D$;M2;M$;M3;S$;EW$
2490 M0=D
2500 GOSUB 2760 'call raddms
2510 PRINT T$;BR$;P4$;A$;M1;D$;M2;M$;M3;S$
2520 GOTO 1220 'next case
2530 END

2540 'sub sequence error
2550 PRINT "error!" : GOTO 1220 'again
2560 END
2570 J=COS(B)
2580 A=SQR(J*J*G+1)
2590 ' PRINT "A= ";A
2600  RETURN
2610 IF ABS(R)>.99999999999 THEN 2540 'error
2620 U=SQR(1-R*R)
2630 S=U*E*A^3/F
2640 ' PRINT "S= ";S
2650  RETURN
2660 R=A*H/F/J
2670 ' PRINT "R= ";R
2680  RETURN
2690 C=C+A*E*R/F/J
2700 ' PRINT "C= ";C
2710  RETURN

2720 'sub dmsrad
2730
M0=(ABS(M1)+(ABS(M2)+ABS(M3)/60)/60)*R0
2740 M0=M0*I2
2750  RETURN

2760 'sub raddms
2770 I2=1
2780 IF M0<0 THEN I2=-1
2790 W1=ABS(M0/R0)
2800 M1=INT(W1)
2810 W2=(W1-M1)*60
2820 M2=INT(W2)
2830 M3=(W2-M2)*60
2840  RETURN

2850 DATA "Latitude ","Longitude"," deg"," min","
sec ","True ","Distance "
```

```
2860 DATA "Bearing ","-1","-2","1-2","2-1","
i=","="
2870 END
```

□

11.7.4 GBD00000.BAS

GBD00000.BAS - Date: 01-27-1997

INVERSE LONG LINE CALCULATION

ellipsoid: ____________	**BESSEL 1841**
ellipsoid= ____________	BESSEL 1841
major axis: ____________	**6377397.155**
major axis= ____________	6377397.155
1/flattening: ____________	**299.15281285**
1/flattening= ____________	299.15281285
Latitude-1 (D,M,S): ____________	**45,0,0**
N or S : ____________	**N**
Latitude -1 = ____________	45 deg 0 min 0 sec N
Longitude-1 (D,M,S): ____________	**10,0,0**
E or W : ____________	**E**
Longitude -1 = ____________	10 deg 0 min 0 sec E
Latitude-2 (D,M,S): ____________	**55,0,0**
N or S : ____________	**N**
Latitude -2 = ____________	55 deg 0 min 0 sec N
Longitude-2 (D,M,S): ____________	**20,0,0**
E or W : ____________	**E**
Longitude -2 = ____________	20 deg 0 min 0 sec E
Integration no: ____________	**1000**
Integration no: ____________	1000
F0= ____________	3.342773181616099D-03
C= ____________	6398786.848070608
e2= ____________	6.674372230688467D-03
e'2= ____________	6.719218798046065D-03
Clairaut = ____________	2193687.104605701
Clairaut = ____________	2193647.010496139
Time = ____________	*17:42:00*
R= ____________	.4856815900450406
C= ____________	.1746748966018393
R= ____________	.4857254755149233
do-loop ::::::::::::::::↓:::::::::::::::::::::: ______	:: only ::↓:: if available: line 2310 - 2450
R= ____________	.535285363003062
R= ____________	.535341252721878
C= ____________	.2528741276467541
Latitude = i= 500 = ____________	50 deg 7 min 7.6747 96042902017 sec N
Longitude= i= 500 = ____________	14 deg 29 min 19.0329 4396123817 sec E
::::::::::::::::::::↓:::::::::::::::::::::______	::::::::::::::::::::↓:::::::::::::::::::::
R= ____________	.5982810732549452
C= ____________	.3490641264926239
R= ____________	.5983517765990244
Time = ____________	*17:42:28*
::::::::::::::::::::↓:::::::::::::::::::::______	::::::::::::::::::::↓:::::::::::::::::::::

V= ______________________ 1.493562555661956
True Distance 1-2 = ______________ 1320284.3694 74921
True Bearing 1-2 =______________ 29 deg 3 min 15.4594 3149107359 sec
True Bearing 2-1 =______________ 216 deg 45 min 7.4001 64316707958 sec

□

```
1000 'GBD00000.BAS
1010 CLS
1020 PRINT "GBD00000.BAS,A";" - Date:";DATE$
1030 ' GBD00000.BAS,A- Date: 01-27-1997
1040 PRINT "-------------------------------------------"
1050 PRINT " INVERSE LONG LINE CALCULATION
1060 '                M. Hooijberg 1981-1997
1070 PRINT "-------------------------------------------"
1080 DEFDBL A-Z
1090 DEFINT I
1100 PI=4*ATN(1)
1110 R0=ATN(1)/45
1120 DEF FNASN(X)=ATN(X/SQR(1-X*X))
1130 DEF FNACS(X)=PI/2-ATN(X/SQR(1-X*X))
1140 READ
LA$,LO$,D$,M$,S$,T$,DI$,BR$,P1$,P2$,
P3$,P4$,P5$,A$

1150 ' ellipsoid data
1160 INPUT "ellipsoid: ",       P$
1170 PRINT "ellipsoid= ";       P$
1180 INPUT "major axis: ",      A0
1190 PRINT "major axis= ";      A0
1200 INPUT "1/flattening: ",    FL
1210 PRINT "1/flattening= ";    FL

1220 ' input data
1230 INPUT "Latitude-1 (D,M,S): ",M1,M2,M3
1240 INPUT "N or S :",NS$
1250 I2=1
1260 IF NS$="N" OR NS$="n" THEN 1300
1270 IF NS$="S" OR NS$="s" THEN 1290
1280 GOTO 1240
1290 I2=-1
1300 PRINT
LA$;P1$;A$;M1;D$;M2;M$;M3;S$;NS$
1310 GOSUB 3340 ' call dmsrad
1320 B=M0
1330 INPUT "Longitude-1 (D,M,S): ",M1,M2,M3
1340 INPUT "E or W :",EW$
1350 I2=1
1360 IF EW$="E" OR EW$="e" THEN 1400
1370 IF EW$="W" OR EW$="w" THEN 1390
1380 GOTO 1340
1390 I2=-1
1400 PRINT
LO$;P1$;A$;M1;D$;M2;M$;M3;S$;EW$
1410 GOSUB 3340 ' call dmsrad
1420 C=M0
1430 INPUT " Latitude-2 (D,M,S): ",M1,M2,M3
1440 INPUT "N or S :",NS$
1450 I2=1
1460 IF NS$="N" OR NS$="n" THEN 1500
1470 IF NS$="S" OR NS$="s" THEN 1490
1480 GOTO 1440
1490 I2=-1
1500 PRINT
LA$;P2$;A$;M1;D$;M2;M$;M3;S$;NS$
1510 GOSUB 3340 ' call dmsrad
1520 D=M0
1530 INPUT "Longitude-2 (D,M,S): ",M1,M2,M3
1540 INPUT "E or W :",EW$
1550 I2=1
1560 IF EW$="E" OR EW$="e" THEN 1600
1570 IF EW$="W" OR EW$="w" THEN 1590
1580 GOTO 1540
1590 I2=-1
1600 PRINT
LO$;P2$;A$;M1;D$;M2;M$;M3;S$;EW$
1610 GOSUB 3340 ' call dmsrad
1620 E=M0
1630 INPUT "Integration no: ",Z
1640 PRINT "Integration no: ";Z
1650 IF Z=0 THEN 1630

1660 ' start integration
1670 ' IZ=Z/2 ' intermediate station
1680 L=B
1690 K=B
1700 M=C
1710 N=D
1720 O=E
1730 F0=1/FL
1740 H=1-F0
1750 F=A0/H
1760 V=(2-F0)/FL
1770 G=V/H/H
1780 ' PRINT "F0= ";       F0
1790 ' PRINT "C= ";        F
1800 ' PRINT "e2= ";       V
1810 ' PRINT "e' 2= ";     G
1820 GOSUB 3080
1830 W=A
1840 X=J
1850 K=D
1860 GOSUB 3080
1870 IF D>B THEN 1880 ELSE 3060 ' error
1880 Y=E-C
1890 IF B=0 THEN B=1D-18
1900 IF Y<>0 THEN 1930
1910 P=Y
1920 GOTO 1940
1930 P=ATN(1/((TAN(D)/(1+G)/TAN(B)+
A*V/W*X/J-COS(Y))*SIN(B)/SIN(Y)))
1940 R=SIN(P)
```

```
1950 ' compute clairaut constant
1960 H=F*R*X/W
1970 ' PRINT "Clairaut =";H
1980 K=(B+D)*.5
1990 GOSUB 3080
2000 E=A*H/F/J
2010 IF ABS(E)>.99999999999 THEN 3060 ' error
2020 U=SQR(1-E*E)
2030 V=F/A
2040 W=V/A/A
2050 X=V*W/(E*E*W+U*U*V)
2060 E=X*FNACS(SIN(B)*SIN(D)+COS(B)*
COS(D)*COS(Y))
2070 Q=E
2080 D=P
2090 E=E/Z
2100 ' E10=E ' dist.counter
2110 ' E11=0 ' check do-loop
2120 K=B
2130 GOSUB 3080
2140 IF Y=0 THEN Y=1D-18
2150 D=D-(A^4-A*A)/12/F^2*E*E*Z*Z*SIN(D+D)
2160 R=SIN(D)

2170 ' compute clairaut constant
2180 H=F*R*J/A
2190 ' PRINT "Clairaut =";H
2200 I1=-1
2210 P=D/R0
2220 ' PRINT "Time =";TIME$
2230 ' repeat integration
2240 GOSUB 3110
2250 K=B+S/2
2260 GOSUB 3080
2270 GOSUB 3150
2280 GOSUB 3180
2290 GOSUB 3110
2300 B=B+S
2310 ' intermediate stations, see lines 2140-2270 as
used in BDG00000.BAS
2450 '< LAT = B-RAD, LON = C-RAD >
2460 K=B
2470 GOSUB 3080
2480 GOSUB 3150
2490 ' E11=E11+E10 ' check do-loop
2500 ' PRINT "Dist =";E11
2510 ' NUM=Z/100 ' do-loop counter
2520 ' IF NUM-INT(NUM)=0 THEN PRINT "I= ",Z
2530 Z=Z-1
2540 IF Z>=1 THEN 2240 ' next integration->
2550 ' cont.computation
2560 ' PRINT "Time =";TIME$
2570 D=FNASN(R)
2580 H=N
2590 R=O
2600 E=Q
2610 K=B
2620 GOSUB 3080
2630 S=H-B
2640 IF S<0 THEN I1=0
2650 W=F*S/A^3
2660 S=R-C
2670 IF S<0 THEN I1=1
2680 X=F*J*S/A
2690 IF X=0 THEN U=0 ELSE U=ATN(W/X)
2700 G=1
2710 H=P
2720 IF I1=0 THEN 2860
2730 IF I1=1 THEN 2800
2740 GOSUB 3210
2750 GOSUB 3260
2760 GOSUB 3300
2770 E=E+F
2780 H=H+G
2790 GOTO 2910
2800 GOSUB 3210
2810 GOSUB 3300
2820 E=E-F
2830 H=H-G
2840 GOTO 2910
2850 IF I1=1 THEN 2880
2860 GOSUB 3210
2870 GOTO 2760
2880 GOSUB 3210
2890 GOSUB 3260
2900 GOTO 2810
2910 K=(A*A-1)/COS(B)^2
2920 PRINT T$;DI$;P3$;A$;E
2930 H=H*R0
2940 M0=H
2950 GOSUB 3380 ' call raddms
2960 PRINT T$;BR$;P3$;A$;M1;D$;M2;M$;M3;S$
2970 S=SQR(1+K*COS(N)^2)
2980 R=SQR(1+K*COS(L)^2)
2990 R=PI+FNASN(S/R*COS(L)/COS(N)*SIN(H))
3000 M0=R
3010 GOSUB 3380 ' call raddms
3020 PRINT T$;BR$;P4$;A$;M1;D$;M2;M$;M3;S$
3030 GOTO 1230 ' next case
3040 END

3050 ' sub sequence error
3060 PRINT "error!"
```

```
3070 GOTO 1230
3080 J=COS(K)
3090 A=SQR(J*J*G+1)
3100  RETURN
3110 IF ABS(R)>.99999999999 THEN 3060 'error
3120 U=SQR(1-R*R)
3130 S=U*E*A^3/F
3140  RETURN
3150 R=A*H/F/J
3160 ' PRINT "R=";R
3170  RETURN
3180 C=C+A*E*R/F/J
3190 ' PRINT "C=";C
3200  RETURN
3210 V=-(U+D)*(D<>0)
3220 IF V<>0 THEN 3240
3230 V=FNACS(V)
3240 ' PRINT "V=";V
3250  RETURN
3260 IF V<=PI/2 THEN 3290
3270 V=PI-V
3280 G=-1
3290  RETURN
3300 Y=SQR(W*W+X*X)
3310 G=G*Y*COS(V)/E/R0
3320 F=Y*SIN(V)
3330  RETURN

3340 'sub dmsrad
3350
M0=(ABS(M1)+(ABS(M2)+ABS(M3)/60)/60)*R0
3360 M0=M0*I2
3370  RETURN

3380 'sub raddms
3390 IF M0<0 THEN M0=PI+PI+M0
3400 W1=ABS(M0/R0)
3410 M1=INT(W1)
3420 W2=(W1-M1)*60
3430 M2=INT(W2)
3440 M3=(W2-M2)*60
3450  RETURN

3460 DATA " Latitude "," Longitude"," deg"," min","
sec ","True"," Distance"
3470 DATA " Bearing "," -1 "," -2 "," 1-2 "," 2-1 ","
i=","= "
3480 END
```

□

11.7.5 GK000000.BAS

GK000000.BAS - Date: 06-08-1996

TRANSVERSE MERCATOR - GAUSS-KRÜGER

Ellipsoid : ______________ **ANS1966**
Ellipsoid= ______________ ANS1966
Major axis : ______________ **6378160**
Major axis = ______________ 6378160
1 / flattening : ______________ **298.25**
1 / flattening = ______________ 298.25
Scale Factor : ______________ **0.9996**
Scale Factor= ______________ .9996
False Easting : ______________ **5D5**
False Easting = ______________ 500000
False Northing : ______________ **1D7**
False Northing = ______________ 10000000

Northern or Southern Hemisphere (N/S) : ___ **S**
Latitude of Origin (D,M,S) : ______________ **0,0,0**
Latitude of Origin= ______________ 0 deg, 0 min, 0 sec. S
Longitude Central Meridian (D,M,S) :______ **141,0,0**
East or West (E/W) : ______________ **E**
Longitude CM= ______________ 141 deg, 0 min, 0 sec. E

Lat/lon Eas/Nor [t-T] .. (L,E,T) ??: ________ **L**
Latitude (D,M,S): ______________ **37,39,15.5571**
Latitude = ______________ 37 deg, 39 min, 15.5571 sec. S
Longitude (D,M,S) : ______________ **143,55,30.6330**
East or West (E/W): ______________ **E**
Longitude= ______________ 143 deg, 55 min, 30.633 sec. E
Easting = ______________ 758053.0895581127
Northing = ______________ 5828496.973545677
Convergence = ______________ (-) 1 deg, 47 min, 16.67165343029014 sec.
Scale Factor= ______________ 1.000420299350839

Lat/lon Eas/Nor [t-T"] .. (L,E,T) ?? ________ **E**
Easting : ______________ **758053.0896**
Easting = ______________ 758053.0896
Northing: ______________ **5828496.9735**
Northing = ______________ 5828496.9735
Latitude = ______________ 37 deg, 39 min, 15.55710272309291 sec. S
Longitude= ______________ 143 deg, 55 min, 30.63300197287617 sec. E
Convergence = ______________ (-) 1 deg, 47 min, 16.67168152189118 sec.
Scale Factor= ______________ 1.000420299202572

Lat/lon Eas/Nor [t-T"] .. (L,E,T) ?? ______	**T**
Easting-1 :______	**758053.0896**
Northing-1:______	**5828496.9735**
Easting-2 :______	**800817.4065**
Northing-2:______	**5793905.6504**
Easting -1 =______	758053.0896
Northing -1 = ______	5828496.9735
Easting -2 =______	800817.4065
Northing -2 = ______	5793905.6504
[t-T"]=______	(+) 0 deg, 0 min, 23.92434769977297 sec.
Brg =______	128 deg, 58 min, 7.687415911653943 sec.
Dist = ______	55003.14930740069
Easting -1 =______	800817.4065
Northing -1 = ______	5793905.6504
Easting -2 =______	758053.0896
Northing -2 = ______	5828496.9735
[t-T"]=______	(-) 0 deg, 0 min, 25.17509095475578 sec.
Brg =______	308 deg, 58 min, 7.687415911653943 sec.
Dist = ______	55003.14930740069

□

```
1000 ' GK000000.BAS
1010 CLS
1020 PRINT "GK000000.BAS - Date";DATE$
1030 'GK000000.BAS - Date: 06-29-1995
1040 PRINT "------------------------------------------"
1050 PRINT      "TRANSVERSE MERCATOR
                    GAUSS-KRUEGER
1060 '          M. Hooijberg 1978-1996
1070 PRINT "------------------------------------------"
1080 READ
F$,E$,N$,CM$,SC$,CO$,LT$,LG$,D$,M$,S$,A$,O$
,TT$,BR$,DI$,S1$,S2$
1090 DEFDBL A-Z
1100 DEFINT I
1110 PI=4 *ATN(1 )
1120 RD=ATN(1 )/45

1130 ' ellipsoid data
1140 INPUT "Ellipsoid       : ",P$
1150 PRINT "Ellipsoid:      = ";P$
1160 INPUT "Major axis      : ",A
1170 PRINT "Major axis      = ";A
1180 INPUT "1 / flattening : ",FL
1190 PRINT "1 / flattening = ";FL
1200 INPUT "Scale Factor    : ",K0
1210 PRINT SC$;A$;K0
1220 INPUT "False Easting   : ",E0
1230 PRINT F$;E$;A$;E0
1240 INPUT "False Northing  : ",N0
1250 PRINT F$;N$;A$;N0

1260 ' ellipsoid parameters
1270 N1= 1 /FL/(2 -1 /FL)
1280 N2=N1*N1
1290 E2=(2 -1 /FL)/FL
1300 E3=E2/(1 -E2)
1310 C0=A/SQR(1 -E2)

1320 ' compute meridional arc constants
        [method Hooijberg]| 87
1330 RL=A*(1 +N2/4 )/(1 +N1)
1340 U2=C0*(((((-86625 /8 *E3+11025 )/64*E3-175
)/4 *E3+45 )/16 *E3-3 )/4 *E3)
1350 U4=C0*((((-17325 /4 *E3+3675 )/256 *E3-175
/12)/1 *E3+15 )/32 *E3*E3)
1360 U6=C0*(-1493 /2 +735 *E3)/2048 *E3^3
1370 U8=C0*((-3465 /4 *E3+315 )/1024 *E3^4)
1380 V2=((((16384 *E3-11025 )/64 *E3+175 )/4
*E3-45 )/16 *E3+3 )/4 *E3
1390 V4=(((-20464721 /120 *E3+19413 )/8 *E3-
1477 )/32 *E3+21 )/32 *E3*E3
1400 V6=((4737141 /28 *E3-17121 )/32 *E3+151
)/192 *E3^3
1410 V8=(-427277 /35 *E3+1097 )/1024 *E3^4

1420 ' zone parameters
1430 INPUT "Northern or Southern Hemisphere
(N/S) : ",NS$
1440 I3=1
1450 IF NS$="N" OR NS$="n" THEN 1490
1460 IF NS$="S" OR NS$="s" THEN 1480
1470 GOTO 1430
1480 I3=-1
1490 INPUT "Latitude of Origin (D,M,S) :
",M1,M2,M3
1500 I2=1
1510 IF NS$="S" OR NS$="s" THEN I2=-1
1520 PRINT LT$;O$;A$;M1;D$;M2;M$;M3;S$;NS$
1530 GOSUB 2640 'call dmsrad
1540 LTC=M0
1550 INPUT "Longitude Central Meridian (D,M,S) :
",M1,M2,M3
1560 INPUT "East or West (E/W) : ",EW$
1570 I2=1
1580 IF EW$="E" OR EW$="e" THEN 1620
1590 IF EW$="W" OR EW$="w" THEN 1610
1600 GOTO 1560
1610 I2=-1
1620 PRINT
LG$;CM$;A$;M1;D$;M2;M$;M3;S$;EW$
1630 GOSUB 2640 'call dmsrad
1640 LGC=M0
1650 LT=LTC
1660 GOSUB 2770 'compute meridional arc - for-
ward
1670 S0=S

1680 ' transverse mercator projection conversion
1690 INPUT "Lat/lon Eas/Nor [t-T"] .. (L,E,T) ??
",G$
1700 IF G$="L" OR G$="l" THEN 1740
1710 IF G$="E" OR G$="e" THEN 1960
1720 IF G$="T" OR G$="t" THEN 2180
1730 GOTO 1690

1740 ' forward computation - Lat/Lon to Eas/Nor
1750 INPUT "Latitude (D,M,S):",M1,M2,M3
1760 PRINT LT$;A$;M1;D$;M2;M$;M3;S$;NS$
1770 GOSUB 2640 'call dmsrad
1780 LT=ABS(M0)
1790 INPUT "Longitude (D,M,S) :",M1,M2,M3
1800 INPUT "East or West (E/W):",EW$
```

[87] See example (Hooijberg, 1996), [2.4]

```
1810 I2=1
1820 IF EW$="E" OR EW$="e" THEN 1860
1830 IF EW$="W" OR EW$="w" THEN 1850
1840 GOTO 1800
1850 I2=-1
1860 PRINT LG$;A$;M1;D$;M2;M$;M3;S$;EW$
1870 GOSUB 2640 'call dmsrad
1880 LG=M0
1890 GOSUB 2850 'call radian to tmerc
1900 I1=1
1910 IF E0>E THEN I1=-1
1920 PRINT E$;A$;E
1930 PRINT N$;A$;N
1940 GOTO 2110 ' compute convergence and scale
factor ->

1950 ' inverse computation - Eas/Nor to Lat/Lon
1960  INPUT "Easting : ",E
1970 PRINT E$;A$;E
1980  INPUT "Northing: ",N
1990 PRINT N$;A$;N
2000 GOSUB 3100 'call forward - tmerc to radians
2010 M0=LT
2020 GOSUB 2680 'call raddms
2030 NS$="N"
2040 IF I3=-1 THEN NS$="S"
2050 PRINT LT$;A$;M1;D$;M2;M$;M3;S$;NS$
2060 M0=LG
2070 GOSUB 2680 'call raddms
2080 EW$="E"
2090 IF I2=-1 THEN EW$="W"
2100 PRINT LG$;A$;M1;D$;M2;M$;M3;S$;EW$
2110 M0=LDA
2120 GOSUB 2680 'call raddms
2130 NI$="(+)"
2140 IF I1*I3=-1 THEN NI$="(-)"
2150 PRINT CO$;A$;NI$;M1;D$;M2;M$;M3;S$
2160 PRINT SC$;A$;K
2170 GOTO 1690

2180 ' compute (t-T) correction
2190 INPUT "Easting-1 :",E
2200 INPUT "Northing-1 :",N
2210 INPUT "Easting-2 :",EB
2220 INPUT "Northing-2 :",NB
2230 I4=1

2240 ' 2nd computation
2250 PRINT E$;S1$;A$;E
2260 PRINT N$;S1$;A$;N
2270 PRINT E$;S2$;A$;EB
2280 PRINT N$;S2$;A$;NB
2290 NM=(NB+N)/2
2300 W=(NM-N0+S0)/(K0*RL)
2310 V0 =N1*(3 +N1/4 *(-21 +N1*(31 -657 /16
*N1)))
2320 LATF=W+V0*SIN(W)*COS(W)
2330 SF=SIN(LATF)
2340 SFF=SF*SF
2350 CFF=1 -SFF
2360 NF=E3*CFF
2370 F=(1 -E2*SFF)*(1 +NF)/(K0*A)^2
2380 E4=2*(E-E0)+(EB-E0)
2390 DE=EB-E
2400 DN=NB-N
2410 IF DN=0  THEN DN=1D-34
2420 D12=-DN/6 *E4*F*(1 -E4/27 *E4*F)
2430 NI$="(+) "
2440 IF D12<0 THEN NI$="(-) "
2450 M0=D12
2460 GOSUB 2680 'call raddms
2470 PRINT TT$;A$;NI$;M1;D$;M2;M$;M3;S$
2480 M0=ATN(DE/DN)
2490 IF DE>0 AND DN>0 THEN 2520
2500 IF DN<0 THEN M0=PI+M0
2510 IF DN>0 THEN M0=PI+PI+M0
2520 GOSUB 2680 'call raddms
2530 PRINT BR$;A$;M1;D$;M2;M$;M3;S$
2540 PRINT DI$;A$;SQR(DE*DE+DN*DN)
2550 IF I4=-1 THEN 1690 ' next case
2560 I4=-1
2570 W1=E
2580 W2=N
2590 E=EB
2600 N=NB
2610 EB=W1
2620 NB=W2
2630 GOTO 2240 '2nd computation ->

2640 'sub dmsrad
2650
M0=(ABS(M1)+(ABS(M2)+ABS(M3)/60)/60)*RD
2660 M0=M0*I2
2670   RETURN

2680 'sub raddms
2690 I2=1
2700 IF M0<0 THEN I2=-1
2710 W1=ABS(M0/RD)
2720 M1=INT(W1)
2730 W2=(W1-M1)*60
2740 M2=INT(W2)
2750 M3=(W2-M2)*60
2760   RETURN

2770 '  sub forward - meridional arc
```

```
2780 SL1=SIN(LT)
2790 SLL=SL1*SL1
2800 CL=1-SLL
2810 CL1=SQR(CL)
2820
W=((((U8*CL+U6)*CL+U4)*CL+U2)*CL1*SL1)+R
L*LT
2830 S =K0*W
2840  RETURN

2850 ' sub forward computation - radian to tmerc
2860 L=(LGC-LG)*COS(LT)
2870 LL=L*L
2880 T=TAN(LT)
2890 TT=T*T
2900 GOSUB 2770 ' call meridional arc
2910 NN=E3*CL
2920 R=K0*A/SQR(1 -E2*SLL)
2930 A1=-R
2940 A2=R*T
2950 A3=1 -TT+NN
2960 A4=5 -TT+NN*(9 +4*NN)
2970 A5=5 +TT*(-18 +TT)+NN*(14 -58 *TT)
2980 A6=61 +TT*(-58 +TT)+NN*(270 -330 *TT)
2990 A7=61 +TT*(-479 +TT*(179 -TT))
3000 E =E0+A1*L*(1 +LL/6 *(A3+LL/20
*(A5+A7/42*LL)))
3010 N =N0+(S-S0+A2/2 *LL*(1 +LL/12
*(A4+A6/30 *LL)))*I3
3020 C1=-T
3030 C2=1 +NN
3040 C3=1 +NN*(3 +2 *NN)
3050 C4=5 -4 *TT+NN*(9 -24*TT)
3060 C5=2 -TT
3070 LDA =C1*L*(1 +LL/3 *(C3+C5/5 *LL))
3080 K =K0*(1 +C2/2 *LL*(1 +C4/12 *LL))
3090  RETURN

3100 ' sub inverse computation - meridional arc
3110 W=((N-N0)*I3+S0)/(K0*RL*RD)
3120 CW1=COS(W*RD)
3130 CW=CW1*CW1
3140
LTF=(((V8*CW+V6)*CW+V4)*CW+V2)*CW1*SQ
R(1 -CW)+W*RD

3150 ' sub inverse computation - tmerc to radian
3160 SF=SIN(LTF)
3170 SFF=SF*SF
3180 CFF=1-SFF
3190 RF=K0*A/SQR(1 -E2*SFF)
3200 I1=1
3210 IF E0>E THEN I1=-1
3220 Q=(E-E0)/RF
3230 QQ=Q*Q
3240 TF=TAN(LTF)
3250 TFF=TF*TF
3260 NF=E3*CFF
3270 B2=TF*(1 +NF)
3280 B3=1 +2 *TFF+NF
3290 B4=5 +3 *TFF+NF*((1 -9 *TFF)-4 *NF)
3300 B5=5 +TFF*4 *(7 +6 *TFF)+NF*(6 +8 *TFF)
3310 B6=61 +TFF*45 *(2 +TFF)+NF*(46
+TFF*18*(-14 -5 *TFF))
3320 B7=61 +TFF*(662 +TFF*40 *(33 +18 *TFF))
3330 LT =LTF-B2/2 *QQ*(1 +QQ/12 *(-B4+B6/30
*QQ))
3340 LG =LGC+Q/SQR(CFF)*(1 +QQ/6 *(-
B3+QQ/20 *(B5-B7/42 *QQ)))
3350 D1=TF
3360 D2=1 +NF
3370 D3=1 +TFF-NF*(1+2 *NF)
3380 D4=1 +5 *NF
3390 D5=2 +TFF*(5 +3 *TFF)
3400 LDA =D1*Q*(1 +QQ/3 *(-D3+D5/5 *QQ))
3410 K =K0*(1 +D2/2 *QQ*(1 +D4/12 *QQ))
3420  RETURN

3430 DATA "False ","Easting ","Northing ","
CM","Scale Factor"
3440 DATA "Convergence ","Latitude
","Longitude","deg, ","min, ","sec. "
3450 DATA "= "," of Origin","[t-T"]","Brg ","Dist
","-1 ","-2 "
3460 END
```

□

11.7.6 LCC00000.BAS

LCC00000.BAS - Date : 02-06-1997

LAMBERT CONFORMAL CONICAL

Ellipsoid :	**GRS80**
Ellipsoid=	GRS80
Major axis :	**6378137**
Major axis =	6378137
1 / flattening :	**298.2572221008827**
1 / flattening =	298.2572221008827
Northern or Southern Hemisphere (N/S):	**N**
Two Standard Parallels (1/2) :	**2**
Lat. Lower Parallel (D,M,S) :	**30,7,0**
Lat. Lower Parallel =	30 deg, 7 min, 0 sec N
Lat. Upper Parallel (D,M,S) :	**31,53,0**
Lat. Upper Parallel =	31 deg, 53 min, 0 sec N
Lat. Grid Origin (D,M,S) :	**29,40,0**
Lat. Grid Origin =	29 deg, 40 min, 0 sec N
Lat.Central Parallel=	31 deg, 0 min, 5.007015736016918 sec N
Lon. Grid Origin (D,M,S) :	**100,20,0**
East or West (E/W):	**W**
Lon. Grid Origin =	100 deg, 20 min, 0 sec W
Scale Factor=	.9998817436292895
False Easting :	**7D5**
False Easting =	700000
False Northing :	**3D6**
False Northing =	3000000
CONT	**<press F5>**
L1 =	700000
L2 =	.5150588822350693 =sin φ_o
L3 =	10770561.10342463 =R_b m
L4 =	13770561.10342463
L5 =	10622600.32497674 =R_o m
L6 =	14219009.88127395 =K m
L7 =	.9998817436292895 =k_0
F0 =	6.686920927320709D-03
F2 =	5.201458331082286D-05
F4 =	5.544580077481781D-07
F6 =	6.717675115728047D-09
F8 =	8.907661557192931D-11
C1 =	6378137
C2 =	298.2572221008827
C3 =	6.694380022903416D-03
C4 =	6.739496775481622D-03
C5 =	1.679220394629406D-03

Lat/lon or Eas/nor or [t-T] (L/E/T) : **L**
Latitude (D,M,S): **31,0,5.00701**
Latitude = 31 deg, 0 min, 5.00701 sec N
Longitude (D,M,S): **100,20,0**
East or West (E/W): **W**
Longitude= 100 deg, 20 min, 0 sec W
Easting = 700000
Northing= 3147960.778271255 (=N_o)
Convergence= (+) 0 deg, 0 min, 0 sec
Scale Factor= .9998817436292896

Lat/lon or Eas/nor or [t-T] (L/E/T) : **L**
Latitude (D,M,S): **32,0,0**
Latitude = 32 deg, 0 min, 0 sec N
Longitude (D,M,S): **106,30,0**
East or West (E/W): **W**
Longitude= 106 deg, 30 min, 0 sec W
Easting = 117571.2278474049
Northing= 3274824.816898567
Convergence= (-) 3 deg, 10 min, 34.30718561853494 sec
Scale Factor= 1.00003342420777

Lat/lon or Eas/nor or [t-T] (L/E/T) : **E**
Easting : **117571.2278**
Easting = 117571.2278
Northing: **3274824.8169**
Northing= 3274824.8169
Latitude = 31 deg, 59 min, 59.99999996186251 sec N
Longitude= 106 deg, 30 min, 1.806215976785097D-06 sec W
Convergence= (-) 3 deg, 10 min, 34.30718654884497 sec
Scale Factor= 1.000033424207769

Lat/lon or Eas/nor or [t-T] (L/E/T) : **L**
Latitude (D,M,S): **31,54,15**
Latitude = 31 deg, 54 min, 15 sec N
Longitude (D,M,S): **106,23,16**
East or West (E/W): **W**
Longitude= 106 deg, 23 min, 16 sec W
Easting = 127581.7268708659
Northing= 3263631.550902191
Convergence= (-) 3 deg, 7 min, 6.223397195567944 sec
Scale Factor= 1.00000566280668

Northing-O: **3147960.7782**
Easting -1 : **117571.2278**
Northing-1: **3274824.8169**
Easting -2 : **127581.7269**
Northing-2: 3263631.5509
Northing-O = 3147960.7782
Easting -1 = 117571.2278
Northing-1 = 3274824.8169
Easting -2 = 127581.7269
Northing-2 = 3263631.5509

P1-P2= ____________________	126864.0387 115670.7727
Q1-Q2=____________________	-582428.7722 -572418.2731
R11-R22=__________________	10495736.28627674 10506929.55227674
U1 = ______________________	110703.9873499133
LTC = _____________________	.5410763428155457
LAT3 = ____________________	.5579182182283617
D12 = _____________________	1.410793166643919D-05
[t-T"]=____________________	(+) 0 deg, 0 min, 2.909969791725355 sec
Bearing = __________________	138 deg, 11 min, 33.74291243395447 sec
Dist. =____________________	15016.63397628966
CONT ______________________	**<press F5>**
P1-P2= ____________________	115670.7727 126864.0387
Q1-Q2=____________________	-572418.2731 -582428.7722
R11-R22=__________________	10506929.55227674 10495736.28627674
U1 = ______________________	100078.0784809841
MO = ______________________	6351602.541949948
LTC = _____________________	.5410763428155457
LAT3 = ____________________	.5574201184945228
D12 = _____________________	-1.369278565618012D-05
[t-T"]=____________________	(-) 0 deg, 0 min, 2.824339780355013 sec
Bearing = __________________	318 deg, 11 min, 33.74291243396726 sec
Dist. =____________________	15016.63397628966
Lat/lon or Eas/nor or [t-T] (L/E/T) : ______	***<next case>***

□

```
1000 ' LCC00000.BAS
1010 CLS
1020 PRINT "LCC00000.BAS ,A ";" - Date:
";DATE$
1030 ' LCC00000.BAS,A - Date: 02-06-1997
1040 PRINT "------------------------------------------"
1050 PRINT " LAMBERT CONFORMAL CONICAL"
1060 '              M. Hooijberg 1978-1997
1070 PRINT "------------------------------------------"
1080 READ F$,E$,N$,ST$,PL$,GO$,PU$,CP$,CM$,
CO$,SC$,LA$,LO$,G$,D$,M$,S$,TT$
1090 READ A$,BR$,DI$,S0$,S1$,S2$
1100 DEFDBL A-Z
1110 DEFINT I
1120 DEF FNASN(X)=ATN(X/SQR(1-X*X))
1130 PI=4*ATN(1)
1140 RD=ATN(1)/45

1150 ' ellipsoid data
1160 INPUT "ellipsoid   : ",P$
1170 PRINT "ellipsoid   = ";P$
1180 INPUT "major axis  : ",A
1190 PRINT "major axis  = ";A
1200 INPUT "1/flattening : ",FL
1210 PRINT "1/flattening = ";FL
1220 F=1/FL

1230 ' ellipsoid parameters
1240 E2=(2-1/FL)/FL
1250 E1=SQR(E2)
1260 E3=E2/(1-E2)
1270 N=F/(2-F)

1280 ' compute isometric latitude constants
1290 ' Berry/Burkholder/version - recasted by
        Hooijberg 1994
1300 F0=E2*(1+E2/6*(-1+E2*(1/5+E2/84*
(31/5+5*E2))))
1310 F2=E2*E2*(7/6+E2/5*(-9/2+E2/7*(13-
101/36*E2)))
1320 F4=E2^3*(28/15+1/56*E2*(-467/3+117*E2))
1330 F6=E2^4/45*(4279/28-344*E2)
1340 F8=E2^5*2087/315

1350 ' zone parameters
1360 INPUT "Northern or Southern Hemisphere
(N/S):",NS$
1370 I3=1 ' if north
1380 IF NS$="N" OR NS$="n" THEN 1420
1390 IF NS$="S" OR NS$="s" THEN 1410
1400 GOTO 1360
1410 I3=-1 ' if south

1420 ' compute parameters for one or two parallels
1430 INPUT "Two Standard Parallels (1/2) : ",V$
1440 IF V$="2" THEN 1600
1450 IF V$="1" THEN 1460 ELSE 1430

1460 ' zone parameters for one standard parallel
1470 INPUT "Lat.Std Parallel   (D,M,S)
:",M1,M2,M3
1480 I2=1
1490 PRINT ST$;A$;M1;D$;M2;M$;M3;S$;NS$
1500 GOSUB 3850 ' call dmsrad
1510 W5=M0
1520 LTC=M0
1530 GOSUB 3810 ' call isometric
1540 QO=QW
1550 WO=WW
1560 SS=SIN(M0)
1570 INPUT "Scale Factor: ",K0
1580 K=A*K0*COS(LTC)*EXP(QO*SS)/(WO*SS)
1590 GOTO 1990 ' cont' d -----------------[1]->

1600 ' zone parameters for two standard parallels
1610 INPUT "Lat. Lower Parallel (D,M,S)
:",M1,M2,M3
1620 I2=1
1630 PRINT PL$;A$;M1;D$;M2;M$;M3;S$;NS$
1640 GOSUB 3850 ' call dmsrad
1650 W5=M0
1660 LTL=M0
1670 GOSUB 3810 ' call isometric
1680 QL=QW
1690 WL=WW
1700 INPUT "Lat. Upper Parallel (D,M,S)
:",M1,M2,M3
1710 PRINT PU$;A$;M1;D$;M2;M$;M3;S$;NS$
1720 GOSUB 3850 ' call dmsrad
1730 W5=M0
1740 LTU=M0
1750 GOSUB 3810 ' call isometric
1760 QU=QW
1770 WU=WW
1780 INPUT "Lat. Grid Origin   (D,M,S)
:",M1,M2,M3
1790 PRINT GO$;A$;M1;D$;M2;M$;M3;S$;NS$
1800 GOSUB 3850 ' call dmsrad
1810 W5=M0
1820 LTC=M0
1830 GOSUB 3810 ' call isometric
1840 QO=QW
1850 WO=WW
```

```
1860
SS=(LOG(WU*COS(LTL)/(WL*COS(LTU))))/(QU-QL)
1870 '@SS=.7716421928' [belgium ...]
1880 W5=FNASN(SS)
1890 M0=W5
1900 GOSUB 3890 ' call raddms
1910 PRINT CP$;A$;M1;D$;M2;M$;M3;S$;NS$
1920 GOSUB 3810 ' call isometric
1930 Q0=QW
1940 W0=WW
1950 K=A*COS(LTL)*EXP(QL*SS)/(WL*SS)
1960 '@K=11565915.812935 ' [belgium ...]
1970 R0=K/EXP(Q0*SS)
1980 K0=W0*TAN(W5)*R0/A

1990 ' cont'd for 1 and 2 parallels
2000 RB=K/EXP(QO*SS)
2010 IF R0=0 THEN R0 = RB ' if one parallel
2020 INPUT "Lon. Grid Origin    (D,M,S)
:",M1,M2,M3
2030 INPUT "East or West (E/W):",EW$
2040 I2=1
2050 IF EW$="E" OR EW$="e" THEN 2090
2060 IF EW$="W" OR EW$="w" THEN 2080
2070 GOTO 2030
2080 I2=-1
2090 PRINT CM$;A$;M1;D$;M2;M$;M3;S$;EW$
2100 GOSUB 3850 ' call dmsrad
2110 LGC=M0
2120 PRINT SC$;A$;K0
2130 INPUT "False Easting          : ",E0
2140 PRINT "False Easting  = ";E0
2150 INPUT "False Northing         : ",N0
2160 PRINT "False Northing = ";N0
2170 RC=N0*I3+RB
2180 '@AB=ATN((150000-149256.456)/(5400000-165373.012)) ' [belgium ... ]
2190 '@PRINT "AB = ";AB ' [belgium ...]
2200   STOP

2210 ' compute basic zone constants
2220 PRINT "L1 =";E0
2230 PRINT "L2 =";SS
2240 PRINT "L3 =";RB
2250 PRINT "L4 =";RC
2260 PRINT "L5 =";R0
2270 PRINT "L6 =";K
2280 PRINT "L7 =";K0
2290 PRINT "F0 =";F0
2300 PRINT "F2 =";F2
2310 PRINT "F4 =";F4
2320 PRINT "F6 =";F6
2330 PRINT "F8 =";F8
2340 PRINT "C1 =";A
2350 PRINT "C2 =";FL
2360 PRINT "C3 =";E2
2370 PRINT "C4 =";E3
2380 PRINT "C5 =";N

2390 ' lambert conformal conic projection
2400 INPUT "Lat/lon or Eas/nor or [t-T]  (L/E/T) :
",V$
2410 IF V$="E" OR V$="e" THEN 2850
2420 IF V$="L" OR V$="l" THEN 2460
2430 IF V$="T" OR V$="t" THEN 3230
2440 GOTO 2400

2450 ' compute forward conversion - Lat/Lon to
Eas/Nor
2460 INPUT "Latitude  (D,M,S): ",M1,M2,M3
2470 I2=1
2480 PRINT LA$;A$;M1;D$;M2;M$;M3;S$;NS$
2490 GOSUB 3850 ' call dmsrad
2500 W5=M0
2510 LAT=M0
2520 GOSUB 3810 ' call isometric
2530 QI=QW
2540 WI=WW
2550 INPUT "Longitude (D,M,S): ",M1,M2,M3
2560 INPUT "East or West (E/W):",EW$
2570 I2=1
2580 IF EW$="E" OR EW$="e" THEN 2620
2590 IF EW$="W" OR EW$="w" THEN 2610
2600 GOTO 2560
2610 I2=-1
2620 PRINT LO$;A$;M1;D$;M2;M$;M3;S$;EW$
2630 GOSUB 3850 ' call dmsrad
2640 LON=M0
2650 RI=K/EXP(QI*SS)
2660 LDA=(LGC-LON)*SS
2670 N =RC-RI*COS(LDA)
2680 '@N =RC-RI*COS(LDA+AB)' [belgium ...]
2690 E =E0-RI*SIN(LDA)
2700 '@E =E0-RI*SIN(LDA+AB) ' [belgium ...]
2710 N =N*I3 ' if south
2720 PRINT E$;A$;E
2730 PRINT N$;A$;N
2740 I1=1
2750 IF E-E0<0 THEN I1=-1
2760 NI$=" (+)"
2770 IF I1*I3=-1 THEN NI$=" (-)"
2780 M0=LDA
2790 GOSUB 3890 ' call raddms
```

```
2800 PRINT CO$;A$;NI$;M1;D$;M2;M$;M3;S$
2810 KI=WI*RI*SS/(A*COS(LAT))
2820 PRINT SC$;A$;KI
2830 GOTO 2400 ' next case -->

2840 ' compute inverse conversion - Eas/Nor to Lat/Lon
2850 INPUT "Easting :",E
2860 PRINT E$;A$;E
2870 INPUT "Northing :",N
2880 PRINT N$;A$;N
2890 N=N*I3 ' if south
2900 ND=RC-N
2910 IF ND=0 THEN ND=1D-34
2920 ED=E-E0
2930 LDA1=ATN(ED/ND)
2940 '@LDA1=ATN(ED/ND)+AB ' [Belgium ...]
2950 LON1=LGC+LDA1/SS
2960 QD=SQR(ND*ND+ED*ED)
2970 Q=LOG(K/QD)/SS
2980 X=2*ATN((EXP(Q)-1)/(EXP(Q)+1))
2990 CX=COS(X)
3000 CC=CX*CX
3010 LAT1=X+SQR(1-CC)*CX*(F0+CC*(F2+CC*(F4+CC*(F6+F8*CC))))
3020 M0=LAT1
3030 GOSUB 3890 ' call raddms
3040 PRINT LA$;A$;M1;D$;M2;M$;M3;S$;NS$
3050 W5=M0
3060 GOSUB 3810 ' call isometric
3070 WK=WW
3080 M0=LON1
3090 GOSUB 3890 ' call raddms
3100 EW$=" E"
3110 IF I2=-1 THEN EW$=" W"
3120 PRINT LO$;A$;M1;D$;M2;M$;M3;S$;EW$
3130 I1=1
3140 IF ED<0 THEN I1=-1
3150 NI$=" (+)"
3160 IF I1*I3=-1 THEN NI$=" (-)"
3170 M0=LDA1
3180 GOSUB 3890 ' call raddms
3190 PRINT CO$;A$;NI$;M1;D$;M2;M$;M3;S$
3200 K1=WK*QD*ABS(SS)/(A*COS(LAT1))
3210 PRINT SC$;A$;K1
3220 GOTO 2400 ' next case

3230 ' compute [t-T"] correction
3240 I4=1
3250 INPUT "Northing-O :",N0
3260 INPUT "Easting -1 :",E
3270 INPUT "Northing-1 :",N
3280 INPUT "Easting -2 :",EB
3290 INPUT "Northing-2 :",NB
3300 PRINT N$;S0$;A$;N0
3310 PRINT E$;S1$;A$;E ' <-2nd calc
3320 PRINT N$;S1$;A$;N
3330 PRINT E$;S2$;A$;EB
3340 PRINT N$;S2$;A$;NB
3350 P1=N-N0
3360 P2=NB-N0
3370 PRINT "P1-P2= ";P1;P2
3380 Q1=E-E0
3390 Q2=EB-E0
3400 PRINT "Q1-Q2= ";Q1;Q2
3410 R11=R0-P1
3420 R22=R0-P2
3430 PRINT "R11-R22= ";R11;R22
3440 DE=EB-E
3450 DN=NB-N
3460 U1=P1-Q1*Q1/(2*R11)
3470 PRINT "U1  = ";U1
3480 MO=K0*A*(1-E2)/(1-E2*SS*SS)^1.5
3490 PRINT "MO  = ";MO
3500 LTC=FNASN(SS)
3510 PRINT "LTC  = ";LTC
3520 LAT3=(LTC+(U1+DN/3)/MO)
3530 PRINT "LAT3 = ";LAT3
3540 D12=(SIN(LAT3)/SS-1)*(Q2/R22-Q1/R11)/2
3550 PRINT "D12  = ";D12
3560 M0=D12
3570 NI$=" (+)"
3580 IF D12<0 THEN NI$=" (-)"
3590 GOSUB 3890 ' call raddms
3600 PRINT TT$;A$;NI$;M1;D$;M2;M$;M3;S$
3610 IF DN=0 THEN DN=1D-34
3620 M0=ATN(DE/DN)
3630 IF DE>0 AND DN>0 THEN 3660
3640 IF DN<0 THEN M0=PI+M0
3650 IF DN>0 THEN M0=PI+PI+M0
3660 GOSUB 3890 ' call raddms
3670 PRINT BR$;A$;M1;D$;M2;M$;M3;S$
3680 PRINT DI$;A$;SQR(DE*DE+DN*DN)
3690 IF I4=-1 THEN 2400 ' next case -->
3700  STOP

3710 I4=-1
3720 W1=E
3730 W2=N
3740 E=EB
3750 N=NB
3760 EB=W1
3770 NB=W2
3780 GOTO 3350 ' 2nd calc ->
3790 END
```

```
3800 ' sub isometric
3810 E4=SIN(W5)
3820 QW=(LOG((1+E4)/(1-E4))-
E1*LOG((1+E1*E4)/(1-E1*E4)))/2
3830 WW=SQR(1-E2*E4*E4)
3840  RETURN

3850 ' sub dmsrad
3860
M0=(ABS(M1)+(ABS(M2)+ABS(M3)/60)/60)*RD
3870 M0=M0*I2
3880  RETURN

3890 ' sub raddms
3900 I2=1
3910 IF M0<0 THEN I2=-1
3920 W1=ABS(M0/RD)
3930 M1=INT(W1)
3940 W2=(W1-M1)*60
3950 M2=INT(W2)
3960 M3=(W2-M2)*60
3970  RETURN

3980 DATA "False ","Easting  ","Northing ",
"Lat.Standard Parallel","Lat. Lower Parallel ","Lat.
Grid Origin   "
3990 DATA "Lat. Upper Parallel ","Lat.Central Paral-
lel","Lon. Grid Origin   ","Convergence"
4000 DATA "Scale Factor","Latitude
","Longitude","gon, ","deg, ","min, ","sec "
4010 DATA "[t-T"]","= ","Bearing ","Dist. ","-O ","-
1 ","-2 "
4020 END
```

□

11.7.7 OM000000.BAS

OM000000.BAS - Date: 01-27-1997

OBLIQUE MERCATOR	
Northern or Southern Hemisphere (N/S) : ___	**N**
Eastern or Western Hemisphere (E/W) : ____	**E**
ellipsoid : ________	**EVEREST**
ellipsoid =________	EVEREST, N + E HemiSphere
major axis :________	**6377276.3458**
major axis = ________	6377276.3458
1/flattening : ________	**300.8017**
1/flattening = ________	300.8017
grid scale constant:________	**0.99984**
grid scale constant= ________	.99984
Latitude Centre: ________	**4,0,0**
Lat. Centre= ________	4 deg, 0 min, 0 sec. N
Longitude Centre:________	**115,0,0**
Lon. Centre= ________	115 deg, 0 min, 0 sec. E
Skew Angle True Origin (D,M,S):________	**53,18,56.95370000004016**
Pos. or Neg. (P/N) : ________	**P**
Skew Ang.Tr.Or.= ________	(+) 53 deg, 18 min, 56.95370000004016 sec.
Skew Angle False Origin (D,M,S): ________	**53,7,48.36847496152333**
Skew Ang.Fs.Or.= ________	(+) 53 deg, 7 min, 48.36847496152333 sec.
Atan (True Origin) = ________	1.342376315926919
Sin (False Origin) = ________	.8
Cos (False Origin) =________	.6
False Easting : ________	**0**
False Northing : ________	0
False Easting =________	**0**
False Northing =________	0
Conversion Factor : ________	**0.0497099547**
Conversion Factor =________	.0497099547
a = ________	6377276.843658787
b = ________	1.003303209179641
c = ________	2.991328941527593D-06
d = ________	6355263.713924926
f =________	.7999999999286045
g = ________	.600000000095194
h = ________	1.333333333002799
i =________	.9998400780551293
lda = ________	109.6855202029758
ellipsoid =________	EVEREST, N / E HemiSphere
Lat/lon - Eas/nor - t-T (L/E/T) : ________	**L**
Latitude (D,M,S):________	**5,23,14.1129**
Latitude =________	5 deg, 23 min, 14.1129 sec. N
Longitude(D,M,S):________	**115,48,19.8196**
Longitude = ________	115 deg, 48 min, 19.8196 sec. E

Easting =__________ 33765.15704590832
Northing = __________ 29655.00568624834
Convergence=__________ (+) 0 deg, 14 min, 36.83789009239697 sec.
Scale Factor= __________ .9999001313432248

Lat/lon - Eas/nor - t-T (L/E/T) : __________ **E**
Easting : __________ **33765.157046**
Easting =__________ 33765.157046
Northing : __________ **29655.005686**
Northing = __________ 29655.005686
Latitude = __________ 5 deg, 23 min, 14.11289983558926 sec. N
Longitude = __________ 115 deg, 48 min, 19.81960005922801 sec. E

Lat/lon - Eas/nor - t-T (L/E/T) : __________ **T**
Easting-1 :__________ **33765.15705**
Easting =__________ 33765.15705
Northing-1:__________ **29655.00569**
Northing = __________ 29655.00569
Easting-2 :__________ **34253.65353**
Easting =__________ 34253.65353
Northing-2: __________ **31480.41517**
Northing = __________ 31480.41517

Bearing = __________ 14 deg, 58 min, 54.57521830636043 sec.
Distance =__________ 1889.642447831404
(t-T) =__________ 5.918108039075119

Bearing = __________ 194 deg, 58 min, 54.57521830635244 sec.
Distance =__________ 1889.642447831404
t-T) = __________ -6.515566519456238

Lat/lon - Eas/nor - t-T (L/E/T) : __________ **<next case>**

□

```
1000 ' "OM000000.BAS
1010 CLS
1020 PRINT "OM000000 - Date: ";DATE$
1030 ' OM000000 - Date: 01-27-1997
1040 PRINT "--------------------------------------------"
1050 PRINT "        OBLIQUE MERCATOR "
1060 '              M. Hooijberg 1978-1997
1070 PRINT "--------------------------------------------"
1080 READ
F$,E$,N$,SC$,AC$,LA$,OC$,LO$,CO$,
BR$,DI$,ST$,SF$,D$,M$,S$,A$,TT$
1090 DEFDBL A-Z
1100 DEFINT I
1110 PI=4*ATN(1)
1120 RD=ATN(1)/45
1130 DEF FNASN(X)=ATN(X/SQR(1-X*X))

1140 INPUT "Northern or Southern Hemisphere
(N/S) : ",NS$
1150 I3=1
1160 IF NS$="N" OR NS$="n" THEN 1200
1170 IF NS$="S" OR NS$="s" THEN 1190
1180 GOTO 1140
1190 I3=-1
1200 INPUT "Eastern or Western Hemisphere
(E/W) : ",EW$
1210 IF EW$="E" OR EW$="e" OR EW$="W" OR
EW$="w" THEN 1240
1220 GOTO 1200

1230 ' input ellipsoid data
1240 INPUT "ellipsoid   : ",     P$
1250 PRINT "ellipsoid   = ";     P$;
1260 PRINT ", ";NS$;" / ";EW$;" HemiSphere"
1270 INPUT "major axis  : ",     A
1280 PRINT "major axis  = ";     A
1290 INPUT "1/flattening : ",    FL
1300 PRINT "1/flattening = ";    FL
1310 INPUT "grid scale constant: ",      K0
1320 PRINT "grid scale constant= ";      K0

1330 ' calculate Oblique Mercator zone
1340 INPUT "Latitude Centre:  ",   M1,M2,M3
1350 I2=1
1360 PRINT AC$;A$;M1;D$;M2;M$;M3;S$;NS$
1370 GOSUB 2630 ' call dmsrad
1380 LATC=M0
1390 INPUT "Longitude Centre: ",   M1,M2,M3
1400 PRINT OC$;A$;M1;D$;M2;M$;M3;S$;EW$
1410  I2=1
1420 IF EW$="W" OR EW$="w" THEN I2=-1
1430 GOSUB 2630 ' call dmsrad
1440 LONC=M0
1450 INPUT "Skew Angle True Origin (D,M,S): "
,M1,M2,M3
1460 INPUT "Pos. or Neg. (P/N) : " ,PN$
1470 I2=1
1480 PI$="(+)"
1490 IF PN$="P" OR PN$="p" THEN 1530
1500 IF PN$="N" OR PN$="n" THEN 1510 ELSE
1460
1510 PI$="(-)"
1520 I2=-1
1530 PRINT ST$;A$;PI$;M1;D$;M2;M$;M3;S$
1540 GOSUB 2630 ' call dmsrad
1550 T0=TAN(M0)
1560 AT0=M0
1570 INPUT "Skew Angle False Origin (D,M,S): "
,M1,M2,M3
1580 PRINT SF$;A$;PI$;M1;D$;M2;M$;M3;S$
1590 GOSUB 2630 ' call dmsrad
1600 S0=SIN(M0)
1610 C0=COS(M0)
1620 TT0=M0

1630 PRINT "Atan (True Origin) =";T0
1640 PRINT "Sin (False Origin) =";S0
1650 PRINT "Cos (False Origin) =";C0
1660 INPUT "False Easting  :",E0
1670 INPUT "False Northing :",N0
1680 PRINT "False Easting  =";E0
1690 PRINT "False Northing =";N0
1700 INPUT "Conversion Factor :",M
1710 PRINT "Conversion Factor =";M

1720 ' compute ellipsoid parameter
1730 E2=(2-1/FL)/FL

1740 ' compute isometric latitude constants
1750 ' Berry/Burkholder/version-recasted by
          Hooijberg 1994
1760 F0=E2*(1+E2/6*(-1+E2*(1/5+E2/84*
(31/5+5*E2))))
1770 F2=E2*E2*(7/6+E2/5*(-9/2+E2/7*
(13-101/36*E2)))
1780 F4=E2^3*(28/15+1/56*E2*(-467/3+117*E2))
1790 F6=E2^4/45*(4279/28-344*E2)
1800 F8=E2^5*2087/315
1810 SC=SIN(LATC)
1820 CC=COS(LATC)
1830 E3=E2/(1-E2)
1840 E1=SQR(E2)
1850 B=SQR(1+E3*CC^4)
1860 W3=1-E2*SC*SC
1870 W4=SQR(W3)
```

```
1880 AA=A*B*SQR(1-E2)/(W3)
1890 QC=.5*(LOG((1+SC)/(1-SC))-E1*LOG
((1+E1*SC)/(1-E1*SC)))
1900 W1=B*SQR(1-E2)/(W4*CC)
1910 C=LOG(W1+SQR(W1*W1-1))-B*QC
1920 D=K0*AA/B
1930 E=A*CC/W4
1940 F=SIN(AT0)*E/AA
1950 K=FNASN(F)
1960 G=COS(K)
1970 H=TAN(K)
1980 GI=K0*AA/A
1990 K=-TAN(K)

2000 ' compute OM constants
2010 PRINT "a =";  AA
2020 PRINT "b =";  B
2030 PRINT "c =";  C
2040 PRINT "d =";  D
2050 PRINT "f =";  F
2060 PRINT "g =";  G
2070 PRINT "h =";  H
2080 PRINT "i =";  GI
2090 Z=            QC*B+C
2100 Z=LONC-FNASN((EXP(Z)-EXP(-Z))/2*H)/B
2110 PRINT "lda =";Z/RD

2120 ' Oblique Mercator Projection
2130 PRINT "--------------------------------------"
2140 PRINT "ellipsoid =";P$;
2150 PRINT ", ";NS$;" / ";EW$;"  HemiSphere"
2160 PRINT ""
2170 INPUT "Lat/lon - Eas/nor - t-T  (L/E/T) : ",V$
2180 IF V$="L" OR V$="l" THEN 2230
2190 IF V$="E" OR V$="e" THEN 2490
2200 IF V$="T" OR V$="t" THEN 3130
2210 GOTO 2160 '  return

2220 ' compute forward conversion Lat/Lon to
Eas/Nor
2230 INPUT "Latitude (D,M,S): ",M1,M2,M3
2240 I2=1
2250 PRINT LA$;A$;M1;D$;M2;M$;M3;S$;NS$
2260 GOSUB 2630 ' call dmsrad
2270 LAT=M0
2280 INPUT "Longitude(D,M,S): ",M1,M2,M3
2290 I2=1
2300 IF EW$="W" OR EW$="w" THEN I2=-1
2310 PRINT LO$;A$;M1;D$;M2;M$;M3;S$;EW$
2320 GOSUB 2630 ' call dmsrad
2330 LON=M0
2340 GOSUB 2760 ' call radian to omerc
2350 NA=NA*I3
2360 PRINT E$;A$;EA
2370 PRINT N$;A$;NA

2380 ' compute convergence and scale factor
2390 M0=LDA
2400 GOSUB 2670 ' call raddms
2410 I4=+1
2420 NI$="(+)"
2430 IF LON-LONC<0 THEN I4=-1
2440 IF I3*I4=-1 THEN NI$="(-)"
2450 PRINT CO$;A$;NI$;M1;D$;M2;M$;M3;S$
2460 PRINT SC$;A$;K
2470 GOTO 2160 ' next case

2480 ' compute inverse conversion Eas/Nor to
Lat/Lon
2490 INPUT "Easting  : " ,EA
2500 PRINT E$;A$;EA ' B
2510 INPUT "Northing : ",NA
2520 PRINT N$;A$;NA
2530 IF NS$="S" OR NS$="s" THEN NA=-NA
2540 GOSUB 2960 ' call omerc to radian
2550 M0=LAT1
2560 GOSUB 2670 ' call raddms
2570 PRINT LA$;A$;M1;D$;M2;M$;M3;S$;NS$
2580 M0=LON1
2590 GOSUB 2670 ' call raddms
2600 PRINT LO$;A$;M1;D$;M2;M$;M3;S$;EW$
2610 GOTO 2160 '  RETURN
2620 END

2630 ' sub dmsrad
2640
M0=(ABS(M1)+(ABS(M2)+ABS(M3)/60)/60)*RD
2650 M0=M0*I2
2660   RETURN

2670 ' sub raddms
2680 I2=1
2690 IF M0<0 THEN I2=-1
2700 W1=ABS(M0/RD)
2710 M1=INT(W1)
2720 W2=(W1-M1)*60
2730 M2=INT(W2)
2740 M3=(W2-M2)*60
2750   RETURN

2760 ' sub radian to omerc
2770 LO1=Z-LON
2780 SL=SIN(LAT)
2790 CL=COS(LAT)
2800 QL=.5*(LOG((1+SL)/(1-SL))-E1*LOG
((1+E1*SL)/(1-E1*SL)))
```

```
2810 W1=B*QL+C
2820 J=(EXP(W1)-EXP(-W1))/2
2830 K=(EXP(W1)+EXP(-W1))/2
2840 L=B*LO1
2850 CW=COS(L)
2860 SW=SIN(L)
2870 U=D*ATN((J*G-F*SW)/CW)
2880 V=D/2*LOG((K-F*J-G*SW)/(K+F*J+G*SW))
2890 EA=(U*S0+V*C0+E0)*M
2900 NA=(U*C0-V*S0+N0)*M
2910 W1=F-J*G*SW
2920 W2=K*G*CW
2930 LDA= ATN(W1/W2)-TT0
2940 K= SQR(1-E2*SL^2)*GI*COS(U/D)/(CL*CW)
2950   RETURN

2960 ' sub omerc to radian
2970 SM=S0/M
2980 CM=C0/M
2990 U=(SM*EA+CM*NA-E0*S0-N0*C0)/D
3000 V=(CM*EA-SM*NA+N0*S0-E0*C0)/D
3010 S=(EXP(V)+EXP(-V))/2
3020 R=(EXP(V)-EXP(-V))/2
3030 SU=SIN(U)
3040 CU=COS(U)
3050 Q =(.5*LOG((S-R*F+G*SU)/(S+R*F-G*SU))-
C)/B
3060 X=2*ATN((EXP(Q)-1)/(EXP(Q)+1))

3070 ' isorad
3080 W2=COS(X)
3090 W1=W2*W2
3100 LAT1=X+SQR(1-W1)*W2*(F0+W1*
(F2+W1*(F4+W1*(F6+F8*W1))))
3110 LON1=Z+1/B*ATN((R*G+SU*F)/CU)
3120   RETURN

3130 ' compute (t-T)" correction
3140 INPUT "Easting-1  : ",EA
3150 PRINT E$;A$;EA
3160 INPUT "Northing-1 : ",NA
3170 PRINT N$;A$;NA
3180 INPUT "Easting-2  : ",EB
3190 PRINT E$;A$;EB
3200 INPUT "Northing-2 : ",NB
3210 PRINT N$;A$;NB
3220 I5=0
3230 EC=EB-EA
3240 NC=NB-NA
3250 IF NC=0 THEN NC=1D-34
3260 M0=ATN(EC/NC)
3270 IF EC>0 AND NC>0 THEN 3300
3280 IF NC<0 THEN M0=PI+M0
3290 IF NC>0 THEN M0=PI+PI+M0
3300 GOSUB 2680 ' call raddms
3310 PRINT BR$;A$;M1;D$;M2;M$;M3;S$
3320 PRINT DI$;A$;SQR(EC*EC+NC*NC)
3330 NA=NA*I3
3340 NB=NB*I3
3350 W3=((2*EA+EB)*C0-(2*NA+NB)*S0-
N0*M*S0)*((EB-EA)*S0+(NB-NA)*C0)
3360 W3=-(W3*(B/M)^2)/((K0*A)^2*RD/600)
3370 PRINT TT$;A$;W3*I3
3380 NA=NA*I3
3390 NB=NB*I3
3400 W1=EA
3410 W2=NA
3420 EA=EB
3430 NA=NB
3440 EB=W1
3450 NB=W2
3460 I5=I5+1
3470 IF I5=1  THEN 3230 ELSE 2160
3480 GOTO 3350

3490 DATA "False ","Easting  ","Northing ","Scale
Factor","Lat. Centre"
3500 DATA "Latitude ","Lon. Cen-
tre","Longitude","Convergence ","Bearing    "
3510 DATA "Distance   ","Skew Ang.Tr.Or.","Skew
Ang.Fs.Or."," deg, "," min, "
3520 DATA "sec. ","= ","(t-T) "
3530 END
```

□

11.7.8 TRM00000.BAS

TRM00000.BAS - Date: 06-09-1996

DATUM S-TRANSFORMATION - REGULAR CASE - ITALY

Ellipsoid - old:	**WGS84**
Major axis - old:	**6378137**
1 / flattening :	**298.257223563**
Ellipsoid - old=	WGS84
Major axis - old=	6378137
1 / flattening - old =	298.257223563
Ellipsoid - new:	**INTERNATIONAL**
Major axis - new:	**6378388**
1 / flattening - new:	**297**
Ellipsoid - new=	INTERNATIONAL
Major axis - new=	6378388
1 / flattening - new=	297
Delta X :	**169.5**
Delta X =	169.5
Delta Y :	**79**
Delta Y =	79
Delta Z :	**12.9**
Delta Z =	12.9
Omega - [z] :	**1.2**
Omega =	1.2
Epsil - [x] :	**0.6**
Epsilon =	.6
Psi - [y] :	**-1.5**
Psi =	-1.5
Delta K :	**2.8D-6**
Delta K =	.0000028
Ppm or Sec (P/S) :	**S**
omega =	5.817764173314432D-06
epsilon =	2.908882086657216D-06
psi =	-7.27220521664304D-06
General or Regular (G/R)? :	**R** { *REGULAR CASE* }
Latitude (D,M,S) :	**43,46,39.4948**
North or South (N/S):	**N**
Latitude - old =	43 deg, 46 min, 39.4948 sec. N
Longitude (D,M,S) :	**11,15,37.0789**
East or West (E/W) :	**E**
Longitude - old =	11 deg, 15 min, 37.0789 sec. E
Height-old [MSL] :	**114.20**
h-old =	114.2
N-old [geoid sep] :	**0**
N-old =	0
h old [H+N] =	114.2

X0 = 4523893.6827600[illegible]	Y0 = 900703.8401914619	Z0 = 4390364.837328862
XN = 4524113.01748306[illegible]	YN = 900771.8142313181	ZN = 4390354.511526449

iteration= ______________ 2.775557561562891D-17

Latitude - new = ______________ 43 deg, 46 min, 37.09090302502972 sec. N
Longitude- new = ______________ 11 deg, 15 min, 38.14475374126989 sec. E
h- new = ______________ 64.61674582608976
N- new = ______________ -49.58325417391024

TRM00000.BAS - Date: 06-09-1996

DATUM S-TRANSFORMATION - GENERAL CASE - 1 - ITALY

Ellipsoid - old: ______________ **WGS84**
Major axis - old: ______________ **6378137**
Reciprocal flattening : ______________ **298.257223563**
Ellipsoid - old= ______________ WGS84
Major axis - old= ______________ 6378137
1 / flattening - old= ______________ 298.257223563

Ellipsoid - new : ______________ **ED50**
Major axis - new: ______________ **6378388**
1 / flattening : ______________ **297**
Ellipsoid - new= ______________ ED50
Major axis - new= ______________ 6378388
1 / flattening - new= ______________ 297

Delta X : ______________ **169.5**
Delta X = ______________ 169.5
Delta Y : ______________ **79.0**
Delta Y = ______________ 79
Delta Z : ______________ **12.9**
Delta Z = ______________ 12.9
Omega - [z] : ______________ **1.2**
Omega = ______________ 1.2
Epsil - [x] : ______________ **0.6**
Epsilon = ______________ .6
Psi - [y] : ______________ **-1.5**
Psi = ______________ -1.5
Delta K : ______________ **2.8D-6**
Delta K = ______________ .0000028
Ppm or Sec (P/S) : ______________ **S**
omega = ______________ 5.817764173314432D-06
epsilon = ______________ 2.908882086657216D-06
psi = ______________ -7.27220521664304D-06
General or Regular (G/R)? : ______________ **G** { *GENERAL CASE* }

X origin :	**0**	Y origin :	**0**	Z origin :	**0**
XO =	0	YO =	0	ZO =	0 Origin

Latitude (D,M,S) : ______________ **43,46,39.4948**
North or South (N/S): ______________ **N**
Latitude - old= ______________ 43 deg, 46 min, 39.4948 sec. N
Longitude (D,M,S) : ______________ **11,15,37.0789**
East or West (E/W) : ______________ **E**

Longitude - old= ______ 11 deg, 15 min, 37.0789 sec. E
Height - old [MSL] : ______ **114.20**
h-old = ______ 114.2
N-old [geoid sep] : ______ **0**
N-old = ______ 0
h old [H+N] = ______ 114.2
X0 = 4523893.682760074 Y0 = 900703.8401914619 Z0 = 4390364.837328862
XT = 4523893.682760074 YT = 900703.8401914619 ZT = 4390364.837328862 c-o
XN = 4524113.017378991 YN = 900771.8142692522 ZN = 4390354.511625901

Latitude- = ______ 43 deg, 46 min, 37.0909 0747343936 sec. N
Longitude- new = ______ 11 deg, 15 min, 38.1447 5631350991 sec. E
h- new = ______ 64.616 7462853482
N- new = ______ -49.583 2537146518

□

DGPS - DATUM S-TRANSFORMATION - GENERAL CASE - 2A - NETHERLANDS

Ellipsoid - old: ______ **BESSEL-AMERSFOORT**
Major axis - old: ______ **6377397.155**
1 / flattening: ______ **299.15281285**
Ellipsoid - old= ______ BESSEL-AMERSFOORT
Major axis - old= ______ 6377397.155
1 / flattening old = ______ 299.15281285

Ellipsoid - new= ______ **WGS 1984**
Major axis - new= ______ **6378137**
1 / flattening = ______ **298.257223563**
Delta X = ______ **593.032**
Delta Y = ______ **26**
Delta Z = ______ **478.741**
Omega = ______ .0000090587
Epsilon = ______ .0000019848
Psi = ______ -.0000017439
Delta K = ______ .0000040772
XO= 0 YO= 0 ZO= 0 Origin rd
Latitude - old = ______ 52 deg, 9 min, 22.178 sec. N
Longitude- old = ______ 5 deg, 23 min, 15.5 sec. E
H-old= ______ 0
N-old= ______ 0
h old [H+N]= ______ 0
X0 = 3903453.1482 16324 Y0 = 368135.3134 486838 Z0 = 5012970.3050 96434

[the triplet X_0, Y_0, Z_0 is used in following calculation]

□

TRM00000.BAS - Date: 06-09-1996

DGPS - DATUM S-TRANSFORMATION - GENERAL CASE - 2B - NETHERLANDS

Ellipsoid - old: ______ **BESSEL-AMERSFOORT**
Major axis - old: ______ **6377397.155**
1 / flattening: ______ **299.15281285**
Ellipsoid - old= ______ BESSEL-AMERSFOORT
Major axis old= ______ 6377397.155
1 / flattening old = ______ 299.15281285

Ellipsoid - new: ______ **WGS84**
Major axis - new: ______ **6378137**
1 / flattening: ______ **298.257223563**
Ellipsoid - new= ______ WGS84
Major axis - new = ______ 6378137
1 / flattening new = ______ 298.257223563
Delta X : ______ **593.032**
Delta X = ______ 593.032
Delta Y : ______ **26**
Delta Y = ______ 26
Delta Z ______ **478.741**
Delta Z = ______ 478.741
Omega - [z] : ______ **9.0587D-6**
Omega = ______ .0000090587
Epsil - [x] : ______ **1.9848D-6**
Epsilon = ______ .0000019848
Psi - [y] : ______ **-1.7439D-6**
Psi = ______ -.0000017439
Delta K : ______ **4.0772D-6**
Delta K = ______ .0000040772
Ppm or Sec (P/S) : ______ **P**
omega = ______ .0000090587
epsilon = ______ .0000019848
psi = ______ -.0000017439
General or Regular (G/R)? : ______ **G** { *GENERAL CASE* }

X origin : **3903453.1482**	Y origin : **368135.3134**	Z origin : **5012970.3051** (←2A)
XO = 3903453.1482	YO = 368135.3134	ZO = 5012970.3051 Origin

Latitude (D,M,S) : ______ **51,59,13.3938**
North or South (N/S): ______ **N**
Latitude- old = ______ 51 deg, 59 min, 13.3938 sec. N
Longitude (D,M,S) : ______ **4,23,16.9953**
East or West (E/W) : ______ **E**
Longitude- old = ______ 4 deg, 23 min, 16.9953 sec. E
Height-old [MSL] : ______ **30.809**
h- old = ______ 30.809
N-old [geoid sep] : ______ **-.113**
N- old = ______ -.113
h old [H+N] = ______ 30.696

X0 = 3924096.8506 21058	Y0 = 301119.8206 968471	Z0 = 5001429.8962 66938
XT = 20643.7024 2105756	YT = -67015.4927 0315286	ZT = -11540.4088 3306204 c-o
XN = 3924689.3395 90998	YN = 301145.3375 507697	ZN = 5001908.6872 2618

iteration= ____________________ 0
Latitude - new=____________________ 51 deg, 59 min, 9.9144 31085273492 sec. N
Longitude - new =____________________ 4 deg, 23 min, 15.9530 3321545168 sec. E
h - new =____________________ 74.312 37253069412
N - new = ____________________ 43.503 37253069412

□

Sta. 0021 NEREF93 (Husti, 1996; Salzmann, 1996):

Latitude (D,M,S) : ____________________ **51,59,13.0288**
North or South (N/S):____________________ **N**
Latitude- old = ____________________ 51 deg, 59 min, 13.0288 sec. N
Longitude (D,M,S) : ____________________ **4,23,15.9269**
East or West (E/W) :____________________ **E**
Longitude- old =____________________ 4 deg, 23 min, 15.9269 sec. E
Height-old [MSL] :____________________ **30.638**
h- old =____________________ 30.638
N-old [geoid sep] : ____________________ **-.113**
N- old = ____________________ -.113
h old [H+N] =____________________ 30.525
X0 = 3924107.1664 39985 Y0 = 301100.1667 213726 Z0 = 5001422.8148 57224
XT = 20654.0182 3998493 YT = -67035.1466 7862742 ZT = -11547.4902 42776 c-o
XN = 3924699.6552 61597 YN = 301125.6833 876589 ZN = 5001901.6058 08614
iteration = ____________________ 1.387778780781446D-17
Latitude - new = ____________________ 51 deg, 59 min, 9.5494 76597312037 sec. N
Longitude - new =____________________ 4 deg, 23 min, 14.8847 8720778484 sec. E
h-new =____________________ 74.141 52013149578
N-new = ____________________ 43.503 52013149578

□

```
1000 ' TRM00000.BAS
1010 CLS
1020 PRINT "TRM00000.BAS - Date: ";DATE$
1030 'TRM00000.BAS - Date: 06-08-1996
1040 PRINT "-------------------------------------------"
1050 PRINT "   DATUM S-TRANSFORMATION "
1060 '                M. Hooijberg 1996
1070 PRINT "-------------------------------------------"
1080 READ P$,AA$,FL$,DX$,DY$,DZ$,DK$,LA$,LO$,D$,M$,S$
1090 READ OM$,EP$,PS$,HT$,NG$,OL$,NW$,A$
1100 DEFDBL A-Z
1110 DEFINT I-J
1120 PI=4 *ATN(1 )
1130 R =ATN(1 )/45
1140 R1=R/3600

1150 ' ellipsoid data
1160 INPUT "Ellipsoid - old        : ",PO$
1170 INPUT "Major axis - old    : ",A0
1180 INPUT "1 / flattening : ",FL0
1190 PRINT P$;OL$;A$;PO$
1200 PRINT AA$;OL$;A$;A0
1210 PRINT FL$;OL$;A$;FL0
1220 INPUT "Ellipsoid - new        : ",PN$
1230 INPUT "Major axis - new    : ",AN
1240 INPUT "1 / flattening   : ",FLN
1250 PRINT P$;NW$;A$;PN$
1260 PRINT AA$;NW$;A$;AN
1270 PRINT FL$;NW$;A$;FLN

1280 ' transformation parameters
1290 INPUT "Delta X      : ",DX
1300 PRINT DX$;A$;DX
1310 INPUT "Delta Y      : ",DY
1320 PRINT DY$;A$;DY
1330 INPUT "Delta Z      : ",DZ
1340 PRINT DZ$;A$;DZ
1350 INPUT "Omega - [z]   : ",OG
1360 PRINT OM$;A$;OG
1370 INPUT "Epsil - [x]   : ",EP
1380 PRINT EP$;A$;EP
1390 INPUT "Psi  - [y]   : ",PS
1400 PRINT PS$;A$;PS
1410 INPUT "Delta K      : ",DK
1420 PRINT DK$;A$;DK
1430 INPUT "Ppm or Sec (P/S) : ",PP$
1440 IF PP$="S" OR PP$="s" THEN 1470
1450 IF PP$="P" OR PP$="p" THEN 1500
1460 GOTO 1430
1470 OG=OG*R1
1480 EP=EP*R1
1490 PS=PS*R1
1500 PRINT "omega   = ";OG ' <-----
1510 PRINT "epsilon  = ";EP
1520 PRINT "psi     = ";PS
1530 INPUT "General or Regular (G/R)? : ",FM$
1540 IF FM$="R" OR FM$="r" THEN 1630
1550 IF FM$="G" OR FM$="g" THEN 1570
1560 GOTO 1530
1570 INPUT "X origin : ",XO  '<-----
1580 INPUT "Y origin : ",YO
1590 INPUT "Z origin : ",ZO
1600 PRINT "XO = ";XO;" ";
1610 PRINT "YO = ";YO;" ";
1620 PRINT "ZO = ";ZO; "Origin"

1630 INPUT "Latitude  (D,M,S)  : ",M1,M2,M3
1640 I2=1
1650 INPUT "North or South (N/S): ",NS$
1660 IF NS$="N" OR NS$="n" THEN 1700
1670 IF NS$="S" OR NS$="s" THEN 1690
1680 GOTO 1650
1690 I2=-1
1700 PRINT LA$;OL$;A$;M1;D$;M2;M$;M3;S$;NS$
1710 GOSUB 2560 'call dmsrad
1720 LT0=M0
1730 INPUT "Longitude (D,M,S) : ",M1,M2,M3
1740 INPUT " East or West (E/W) : ",EW$
1750 I2=1
1760 IF EW$="E" OR EW$="e" THEN 1800
1770 IF EW$="W" OR EW$="w" THEN 1790
1780 GOTO 1740
1790 I2=-1
1800 PRINT LO$;OL$;A$;M1;D$;M2;M$;M3;S$;EW$
1810 GOSUB 2560 'call dmsrad
1820 LG0=M0

1830 ' compute ellipsoidal constants
1840 BN=AN*(1 -1 /FLN)
1850 E20=(2 -1 /FL0)/FL0
1860 E2N=(2 -1 /FLN)/FLN
1870 E3N=E2N/(1 -E2N)
1880 INPUT "Height-old [MSL]  : ",H
1890 PRINT HT$;OL$;A$;H
1900 INPUT "N-old [geoid sep] : ",NS0
1910 PRINT NG$;OL$;A$;NS0
1920 RN0=A0/SQR(1 -E20*SIN(LT0)^2)
1930 H0=H+NS0
1940 PRINT "h old [H+N] = ";H0

1950 ' compute old xyz system
```

```
1960 X0=(RN0+H0)*COS(LT0)*COS(LG0)
1970 Y0=(RN0+H0)*COS(LT0)*SIN(LG0)
1980 Z0=(RN0*(1 -E20)+H0)*SIN(LT0)
1990 PRINT "X0 = ";X0;" ";
2000 PRINT "Y0 = ";Y0;" ";
2010 PRINT "Z0 = ";Z0
2020 IF FM$="R" OR FM$="r" THEN 2120

2030 ' compute new - general - xyz system
2040 XT=X0-XO: YT=Y0-YO: ZT=Z0-ZO
2050 PRINT "XT = ";XT;" ";
2060 PRINT "YT = ";YT;" ";
2070 PRINT "ZT = ";ZT;"c-o"
2080 XN=X0+DX+(DK*XT+OG*YT-PS*ZT)
2090 YN=Y0+DY+(-OG*XT+DK*YT+EP*ZT)
2100 ZN=Z0+DZ+(PS*XT-EP*YT+DK*ZT)
2110 GOTO 2160

2120 ' compute new - regular - xyz system
2130 XN=DX+(X0+OG*Y0-PS*Z0)*(1 +DK)' -dkx
2140 YN=DY+(Y0-OG*X0+EP*Z0)*(1 +DK)' -dky
2150 ZN=DZ+(Z0+PS*X0-EP*Y0)*(1 +DK)'  -dkz
2160 PRINT "XN = ";XN;" ";
2170 PRINT "YN = ";YN;" ";
2180 PRINT "ZN = ";ZN

2190 ' algorithm bowring 1976
2200 P=SQR(XN*XN+YN*YN)
2210 IF P=<1D-30 THEN P=1D-30
2220 TH=ATN(AN*ZN/(BN*P))
2230 W1=ZN+E3N*BN*SIN(TH)^3
2240 W2=P-E2N*AN*COS(TH)^3
2250 LTN=ATN(W1/W2)

2260 ' algorithm Schuhr, 1996
2270 LTT=LTN ' <---
2280 TH=ATN((BN/AN)*TAN(LTN))
2290 W1=ZN+E3N*BN*SIN(TH)^3
2300 W2=P-E2N*AN*COS(TH)^3
2310 LTN=ATN(W1/W2)
2320 PRINT "iteration=";ABS(LTT-LTN)
2330 IF (ABS(LTT-LTN))>.00000000001  GOTO
2260
2340 M0=LTN
2350 GOSUB 2600 'call raddms
2360 NS$="N"
2370 IF I2=-1 THEN NS$="S"
2380 PRINT LA$;A$;M1;D$;M2;M$;M3;S$;NS$
2390 LGN=ATN(YN/XN)
2400 IF XN>=0 THEN 2430
2410 IF YN>=0 THEN LGN=PI+LGN
2420 IF YN<0  THEN LGN=LGN-PI
2430 M0=LGN
2440 GOSUB 2600 'call raddms
2450 EW$="E"
2460 IF I2=-1 THEN EW$="W"
2470 PRINT
LO$;NW$;A$;M1;D$;M2;M$;M3;S$;EW$
2480 RNN=AN/SQR(1 -E2N*SIN(LTN)^2)
2490 HN=P/COS(LTN)-RNN
2500 PRINT HT$;NW$;A$;HN
2510 NSN=HN-H
2520 PRINT NG$;NW$;A$;NSN
2530 STOP:GOTO 1630
2540 END

2550 'subroutines
2560 'sub dmsrad
2570
M0=(ABS(M1)+(ABS(M2)+ABS(M3)/60)/60)*R
2580 M0=M0*I2
2590   RETURN

2600 'sub raddms
2610 I2=1
2620 IF M0<0 THEN I2=-1
2630 W1=ABS(M0/R)
2640 M1=INT(W1)
2650 W2=(W1-M1)*60
2660 M2=INT(W2)
2670 M3=(W2-M2)*60
2680   RETURN

2690 DATA "Ellipsoid" ,"Major axis ","1 / flatten-
ing","Delta X ","Delta Y "
2700 DATA "Delta Z ","Delta K ","Latitude-
","Longitude- ","deg, ","min, "
2710 DATA "sec. ","Omega   ","Epsilon ","Psi
","h- ","N- "
2720 DATA " old "," new ","= "
2730 END
```

□

11.7.9 RDED003x.BAS

RDED0031.BAS - Date: 01-27-1997

CONVERSION CO-ORDINATES RD TO TM AND VICE-VERSA UTM ZONE 31 - CM 3 E

Station :	*Lemsterland-4 - 1ST Order*			
OLD-RD or NEW-RD ?? [O/N]	**O**			
Conversion RD -> ED-31 [Y/N] :	**Y**			
RD1918-X :	**21599.42**			
RD1918-Y :	**76098.50**			
RD1918-X =	21599.42			
RD1918-Y =	76098.5			
i ...j =	18	24		
z1...z4 =	34	15	-8	-5
z1...z4 =	128	130	107	105
E and corr =	682468.042379598			.0005
N and corr =	5857969.490340485			.1146
ED50-E =	682468.042879598			
ED50-N =	5857969.604940485			
RD1918-X :	**<next case>**			

□

CONVERSION CO-ORDINATES RD TO TM AND VICE-VERSA UTM ZONE 31 - CM 3 E

OLD-RD or NEW-RD ?? [O/N] :	**O**			
Conversion RD -> ED-31 [Y/N] :	**N**			
ED50-E :	**682468.043**			
ED50-N :	**5857969.605**			
ED50-E =	682468.043			
ED50-N =	5857969.605			
i ...j =	18	24		
z1...z4 =	34	15	-8	-5
z1...z4 =	128	130	107	105
X and corr =	21599.42347818251			.0005
Y and corr =	76098.61457497817			.1146
RD1918-X =	21599.42297818251			
RD1918-Y =	76098.49997497817			
ED50-E :	**<next case>**			

□

RDED0032.BAS - Date: 01-27-1997

CONVERSION CO-ORDINATES RD TO TM AND VICE-VERSA
UTM ZONE 32 - CM 9 E

OLD-RD or NEW-RD ?? [O/N] :_________ **O**
Conversion RD -> ED-32 [Y/N] :_________ **Y**

RD1918-X : _________________________ **21599.42**
RD1918-Y : _________________________ **76098.50**
RD1918-X =_________________________ 21599.42
RD1918-Y =_________________________ 76098.50

i ...j =	18	24		
z1...z4 =	34	15	-8	-5
z1...z4 =	128	130	107	105

E and corr = _________________________ 278345.1201431439 .0005
N and corr = _________________________ 5859606.184388154 .1146

ED-50-E = _________________________ 278345.1206431439
ED-50-N = _________________________ 5859606.298988154

RD1918-X : _________________________ **<next case>**

□

CONVERSION CO-ORDINATES RD TO TM AND VICE-VERSA
UTM ZONE 32 - CM 9 E

OLD-RD or NEW-RD ?? [O/N] :_________ **O**
Conversion RD -> ED-32 [Y/N] :_________ **N**

ED50-E :_________________________ **278345.121**
ED50-N : _________________________ **5859606.299**
ED-50-E = _________________________ 278345.121
ED-50-N = _________________________ 5859606.299

i ...j =	18	24		
z1...z4 =	34	15	-8	-5
z1...z4 =	128	130	107	105

X and corr = _________________________ 21599.41499616882 .0005
Y and corr = _________________________ 76098.61426734213 .1146

RD1918-X =_________________________ 21599.41449616882
RD1918-Y =_________________________ 76098.49966734213

ED50-E :_________________________ **<next case>**

□

```
1000 ' "RDED0031.BAS
1010 CLS
1020 PRINT "RDED0031.BAS,A - Date: ";DATE$
1030 ' RDED0031.BAS,A - Date: 01-27-1997
1040 PRINT "--------------------------------------------"
1050 PRINT "CONVERSION CO-ORDINATES RD TO TM
                    AND VICE-VERSA "
1060 PRINT "        UTM ZONE 31 - CM 3 E"
1070               ©M. Hooijberg 1984-1997
1080 PRINT "--------------------------------------------"
1090 DEFDBL A-Z
1100 DEFINT I-L
1110 OPTION BASE 1
1120 DIM PE(31,33),PN(31,33)
1130 A$=" ="
1140 E$="ED50-E  "
1150 N$="ED50-N  "
1160 X$="RD1918-X  "
1170 Y$="RD1918-Y  "

1180 ' rd constants zone 31
1190 A0=5781778.67#
1200 A1=595071.95#
1210 A2=.999255381#
1220 A3=-.033094419#
1230 A4=-.061539#
1240 A5=-.195849#
1250 A6=-.212404#
1260 A7=.031163#
1270 A8=-.01259#
1280 A9=.07502#

1290 ' utm constants zone 31
1300 B0=465832.61#
1310 B1=86725.49#
1320 B2=.999648685#
1330 B3=.033107445#
1340 B4=.041732#
1350 B5=.201003#
1360 B6=.21375#
1370 B7=-.002847#
1380 B8=.0247#
1390 B9=-.07001#

1400 ' x, y constants zone 31
1410 Y1=465832.61#
1420 Y2=577404.33#
1430 Y3=580314.65#
1440 Y4=317522.71#
1450 X1=86725.49#
1460 X2=119438.51#
1470 X3=269564.37#
1480 X4=194845.62#

1490 ' e, n constants zone 31
1500 N1=5781778.67#
1510 N2=5894349.93#
1520 N3=5902277.8#
1530 N4=5637082.43#
1540 E1=595071.95#
1550 E2=624068.21#
1560 E3=774028.17#
1570 E4=707963.95#

1580 ' load data lookup-tables
1590 GOSUB 2650
1600 INPUT "OLD-RD or NEW-RD ?? [O/N] : ",V$
1610 RD$="OLD"
1620 IF V$="O" OR V$="o" THEN 1660
1630 IF V$="N" OR V$="n" THEN 1650
1640 GOTO 1600
1650 RD$="NEW"
1660 INPUT "Conversion RD -> ED-31 [Y/N] : ",V$
1670 IF V$="Y" OR V$="y" THEN 1700
1680 IF V$="N" OR V$="n" THEN 1950
1690 GOTO 1660
```

Important:

SUBROUTINE at line 1700 cont'd for both UTM zones 31/32

```
1700 INPUT "RD1918-X : ",X
1710 INPUT "RD1918-Y : ",Y
1720 PRINT X$;A$;X
1730 PRINT Y$;A$;Y
1740 IF RD$="NEW" THEN 1770
1750 X=X+155000#
1760 Y=Y+463000#
1770 P=Y-Y1
1780 Q=X-X1
1790 R=(P*(Y-Y2)-Q*(X-X2))*.00000001#
1800 S=(P*(X-X2)+Q*(Y-Y2))*.00000001#
1810 T=(R*(Y-Y3)-S*(X-X3))*.000001#
1820 U=(R*(X-X3)+S*(Y-Y3))*.000001#
1830 V=(T*(Y-Y4)-U*(X-X4))*.000001#
1840 W=(T*(X-X4)+U*(Y-Y4))*.000001#
1850 N=A0+P*A2-Q*A3+R*A4-S*A5+T*A6-
U*A7+V*A8-W*A9
1860 E=A1+Q*A2+P*A3+S*A4+R*A5+U*
A6+T*A7+W*A8+V*A9

1870 GOSUB 2210 ' bi-linear interpolation
```

```
1880 PRINT "E and corr = ";E;CE
1890 PRINT "N and corr = ";N;CN
1900 E=E+CE
1910 N=N+CN
1920 PRINT E$;A$;E
1930 PRINT N$;A$;N
1940 GOTO 1700 'next case ->
1950 INPUT "ED50-E : ",E
1960 INPUT "ED50-N : ",N
1970 PRINT E$;A$;E
1980 PRINT N$;A$;N
1990 Q=E-E1
2000 P=N-N1
2010 R=(P*(N-N2)-Q*(E-E2))*.00000001#
2020 S=(P*(E-E2)+Q*(N-N2))*.00000001#
2030 T=(R*(N-N3)-S*(E-E3))*.000001#
2040 U=(R*(E-E3)+S*(N-N3))*.000001#
2050 V=(T*(N-N4)-U*(E-E4))*.000001#
2060 W=(T*(E-E4)+U*(N-N4))*.000001#
2070 Y=B0+P*B2-Q*B3+R*B4-S*B5+T*B6-U*B7+V*B8-W*B9
2080 X=B1+Q*B2+P*B3+S*B4+R*B5+U*B6+T*B7+W*B8+V*B9
2090 GOSUB 2210
2100 IF RD$="NEW" THEN 2130
2110 X=X-155000#
2120 Y=Y-463000#
2130 PRINT "X and corr = ";X;CE
2140 PRINT "Y and corr = ";Y;CN
2150 X=X-CE
2160 Y=Y-CN
2170 PRINT X$;A$;X
2180 PRINT Y$;A$;Y
2190 GOTO 1950 'next case ->

2200 'bi-linear interpolation - eastings
2210 E5=X
2220 N5=Y
2230 DI=10000#
2240 E6=INT(E5/DI)*DI
2250 N6=INT(N5/DI)*DI
2260 XC=E5-E6
2270 YC=N5-N6
2280 I=INT(E6/DI)+1
2290 J=INT((N6-30*DI)/DI)+1
2300 IF I<1 OR I>30 OR J<1 OR J>32 THEN 2560
2310 Z1=PE(I,J)
2320 Z2=PE(I,J+1)
2330 Z3=PE(I+1,J+1)
2340 Z4=PE(I+1,J)
2350 PRINT "i ... j  = ";I;J
2360 PRINT "z1...z4 = ";Z1;Z2;Z3;Z4
2370 IF Z1=999 OR Z2=999 OR Z3=999 OR Z4=999 THEN 2600
2380 A=Z1
2390 B=(Z4-Z1)/DI
2400 C=(Z2-Z1)/DI
2410 D=(Z1-Z2+Z3-Z4)/DI^2
2420 CE=INT((A+C*YC+(B+D*YC)*XC)*10#)/DI
2430 ' bi-linear interpolation - northings
2440 Z1=PN(I,J)
2450 Z2=PN(I,J+1)
2460 Z3=PN(I+1,J+1)
2470 Z4=PN(I+1,J)
2480 PRINT "z1...z4 = ";Z1;Z2;Z3;Z4
2490 IF Z1=999 OR Z2=999 OR Z3=999 OR Z4=999 THEN 2600
2500 A=Z1
2510 B=(Z4-Z1)/DI
2520 C=(Z2-Z1)/DI
2530 D=(Z1-Z2+Z3-Z4)/DI^2
2540 CN=INT((A+C*YC+(B+D*YC)*XC)*10#)/DI
2550  RETURN
2560 PRINT "-> outside transformation limits
2570 CE=0#
2580 CN=0#
2590  RETURN
2600 PRINT "-> no correction applied
2610 CE=0#
2620 CN=0#
2630  RETURN
2640 END

2650 RESTORE 2790
2660 'read E-matrix
2670 FOR L=1 TO 33
2680 FOR K=1 TO 31
2690 READ PE(K,L)
2700 NEXT K
2710 NEXT L

2720 RESTORE 3120
2730 'read N-matrix
2740 FOR L=1 TO 33
2750 FOR K=1 TO 31
2760 READ PN(K,L)
2770 NEXT K
2780 NEXT L
```

This section is required for programs RDED0031.BAS (3° E) and RDED0032.BAS (9° E)

31×33 Matrix of Eastings

```
2790 DATA 999,999,999,999,999,999,999,999,999,999,999,999,999,999,999,999,999,999,-121,-71,010,100,190,999,999,999,999,999,999,999,999
2800 DATA 999,999,999,999,999,999,999,999,999,999,999,999,999,999,999,999,999,-130,-86,-40,030,100,180,999,999,999,999,999,999,999,999
2810 DATA 999,999,999,999,999,999,999,999,999,999,999,999,999,999,999,999,-140,-96,-50,-13,030,100,170,999,999,999,999,999,999,999,999
2820 DATA 999,999,999,999,999,999,999,999,999,999,999,999,999,999,999,999,-110,-70,-26,000,020,100,166,999,999,999,999,999,999,999,999
2830 DATA 999,999,999,999,110,999,999,999,999,999,999,999,999,999,999,999,-80,-40,-17,003,020,080,150,999,999,999,999,999,999,999,999
2840 DATA 999,999,070,074,070,030,-06,-40,999,999,999,999,999,999,-63,-60,-57,-40,-19,002,023,060,100,999,999,999,999,999,999,999,999
2850 DATA 999,020,024,025,020,-02,-27,-60,-100,-110,-90,-67,-55,-51,-50,-50,-50,-50,-40,-17,004,010,020,999,999,999,999,999,999,999,999
2860 DATA -90,-54,-16,-10,-08,-18,-40,-70,-107,-110,-96,-78,-67,-61,-51,-50,-50,-50,-51,-40,-28,-40,999,999,999,999,999,999,999,999,999
2870 DATA 999,-80,-50,-51,-36,-34,-60,-80,-100,-101,-90,-80,-70,-61,-52,-50,-50,-55,-60,-60,-60,-80,-83,999,999,999,999,999,999,999,999
2880 DATA 999,999,-85,-70,-64,-60,-71,-85,-97,-86,-77,-65,-58,-53,-50,-50,-55,-65,-76,-86,-103,-112,-103,999,999,999,999,999,999,999,999
2890 DATA 999,999,-98,-77,-60,-50,-60,-70,-72,-69,-60,-50,-44,-40,-44,-51,-64,-80,-104,-120,-135,-141,-110,999,999,999,999,999,999,999,999
2900 DATA 999,999,-80,-60,-40,-36,-39,-54,-60,-62,-60,-45,-30,-36,-47,-60,-80,-110,-130,-146,-162,-152,-110,-60,999,999,999,999,999,999,999
2910 DATA 999,999,999,-42,-28,-20,-24,-40,-52,-56,-60,-49,-33,-35,-51,-69,-100,-130,-150,-174,-170,-140,-90,-80,-113,-168,999,999,999,999,999
2920 DATA 999,999,999,999,-07,007,004,-10,-23,-34,-40,-40,-28,-30,-48,-80,-116,-147,-170,-170,-140,-100,-64,-100,-123,-170,-230,-290,999,999,999
2930 DATA 999,999,999,999,007,022,037,032,020,000,-05,001,-07,-20,-42,-80,-124,-151,-160,-117,-93,-60,-90,-120,-140,-180,-265,-350,999,999,999
2940 DATA 999,999,999,999,001,020,041,050,044,031,022,013,-02,-20,-40,-75,-103,-100,-120,-82,-63,-85,-107,-133,-160,-200,-313,-400,-490,999,999
2950 DATA 999,999,999,999,999,999,010,020,013,007,004,000,-09,-21,-30,-50,-50,-54,-100,-86,-86,-100,-127,-153,-186,-225,-320,-433,-530,999,999
2960 DATA 999,999,999,999,999,999,999,000,-03,-08,-12,-16,-20,-11,-07,-12,000,-10,-50,-99,-100,-115,-145,-180,-206,-255,-320,-443,999,999,999
2970 DATA 999,999,999,999,999,999,999,-02,-10,-23,-30,-30,-20,-04,014,020,040,050,010,-60,-100,-131,-160,-190,-227,-280,-330,-450,999,999,999
2980 DATA 999,999,999,999,999,999,999,999,-20,-34,-45,-36,-20,-02,030,060,076,068,020,-20,-60,-116,-164,-192,-230,-270,-320,-450,999,999,999
2990 DATA 999,999,999,999,999,999,999,999,999,-31,-36,-24,-13,003,035,076,090,030,-20,-60,-100,-130,-160,-190,-220,-247,-315,-435,999,999,999
3000 DATA 999,999,999,999,999,999,999,999,999,999,-08,000,007,018,034,077,082,030,-20,-70,-108,-130,-150,-180,-210,-220,-291,-420,999,999,999
3010 DATA 999,999,999,999,999,999,999,999,999,999,011,020,033,040,046,064,064,052,010,-30,-70,-110,-140,-171,-180,-177,-270,-380,-489,999,999
3020 DATA 999,999,999,999,999,999,999,999,999,999,020,031,040,045,050,048,045,034,-05,-30,-70,-97,-132,-137,-123,-120,-231,-333,-429,-507,999
```

```
3030 DATA 999,999,999,999,999,999,999,999,999,999,022,031,039,040,035,026,025,015,-08,-47,-70,-101,-
110,-90,-75,-90,-181,-279,-370,-458,999
3040 DATA 999,999,999,999,999,999,999,999,999,999,021,027,030,020,010,004,001,001,-10,-50,-40,-60,-
80,-60,-40,-40,-139,-220,-311,-400,999
3050 DATA 999,999,999,999,999,999,999,999,999,999,015,015,010,000,-10,-20,-24,-27,-34,-50,000,010,-
33,-30,005,005,-80,-170,-240,-290,999
3060 DATA 999,999,999,999,999,999,999,999,999,999,010,008,002,-17,-30,-40,-50,-60,-50,-
50,020,050,035,020,044,033,000,-89,-150,-159,999
3070 DATA 999,999,999,999,999,999,999,999,999,999,999,002,-15,-40,-50,-64,-80,-90,-90,-
57,000,033,065,070,082,066,030,000,-09,-20,999
3080 DATA 999,999,999,999,999,999,999,999,999,999,999,-20,-54,-70,-90,-104,-115,-122,-123,-90,-
30,017,050,100,130,110,094,136,130,999,999
3090 DATA 999,999,999,999,999,999,999,999,999,999,999,999,-110,-123,-140,-155,-165,-160,-115,-
60,005,064,130,180,200,250,280,266,999,999
3100 DATA 999,999,999,999,999,999,999,999,999,999,999,999,999,-152,-173,-190,-210,-200,-140,-80,-
10,060,150,220,296,374,410,999,999,999
3110 DATA
999,999,999,999,999,999,999,999,999,999,999,999,999,999,999,999,999,999,999,999,999,999,999,999,999,9
99,999,999,999,999,999
```

31×33 Matrix of Northings

```
3120 DATA 999,999,999,999,999,999,999,999,999,999,999,999,999,999,999,999,999,-60,-101,-141,-
180,-210,999,999,999,999,999,999,999,999
3130 DATA 999,999,999,999,999,999,999,999,999,999,999,999,999,999,999,999,060,015,-30,-74,-90,-
90,999,999,999,999,999,999,999,999
3140 DATA 999,999,999,999,999,999,999,999,999,999,999,999,999,999,999,999,110,077,050,020,030,030,
020,999,999,999,999,999,999,999,999
3150 DATA 999,999,999,999,999,999,999,999,999,999,999,999,999,999,999,125,094,066,047,080,114,
130,999,999,999,999,999,999,999,999
3160 DATA 999,999,999,999,155,999,999,999,999,999,999,999,999,999,999,141,113,089,084,124,160,
199,999,999,999,999,999,999,999,999
3170 DATA 999,999,000,086,134,164,196,185,999,999,999,999,999,999,120,140,160,136,118,129,170,213,
229,999,999,999,999,999,999,999,999
3180 DATA 999,-80,000,065,109,140,160,152,137,114,100,085,110,160,174,181,173,160,150,180,214,236,
250,999,999,999,999,999,999,999,999
3190 DATA -130,-59,010,049,100,120,131,130,119,103,091,080,111,140,160,174,174,160,166,196,222,250,
999,999,999,999,999,999,999,999,999
3200 DATA 999,-40,030,054,079,103,111,110,100,091,075,064,090,110,127,140,150,138,160,190,212,240,
289,999,999,999,999,999,999,999,999
3210 DATA 999,999,059,080,084,089,092,092,080,040,019,036,060,083,094,099,112,122,145,173,197,225,
280,999,999,999,999,999,999,999,999
3220 DATA 999,999,082,100,104,107,086,050,018,-20,-03,012,036,057,073,083,093,107,130,156,
180,216,280,999,999,999,999,999,999,999,999
3230 DATA 999,999,100,126,130,122,110,080,045,020,010,020,020,042,058,068,076,090,114,140,170,210,
276,350,999,999,999,999,999,999,999
3240 DATA 999,999,999,150,150,126,117,100,080,060,041,032,021,030,040,050,055,070,099,120,162,200,
270,323,370,420,999,999,999,999,999
3250 DATA 999,999,999,999,140,123,110,090,067,050,029,016,013,007,008,009,023,045,073,110,150,193,
250,295,340,390,435,479,999,999,999
3260 DATA 999,999,999,999,133,119,107,090,058,030,000,-25,-20,-24,-28,-32,-04,020,060,106,140,187
,230,270,310,354,390,420,999,999,999
3270 DATA 999,999,999,999,129,118,106,089,050,010,-23,-53,-59,-56,-60,-80,-60,-
20,030,090,135,173,210,249,279,294,320,350,380,999,999
```

```
3280 DATA 999,999,999,999,999,999,104,080,034,-05,-34,-60,-73,-81,-100,-130,-110,-
53,020,090,121,143,175,220,240,234,249,280,310,999,999
3290 DATA 999,999,999,999,999,999,999,070,040,004,-24,-50,-80,-89,-102,-127,-119,-
40,043,103,118,132,147,170,192,180,184,210,999,999,999
3300 DATA 999,999,999,999,999,999,999,090,066,040,010,-19,-49,-79,-90,-82,-89,-
70,009,090,107,117,124,142,140,130,119,130,999,999,999
3310 DATA 999,999,999,999,999,999,999,999,089,067,045,016,-10,-40,-40,-32,-34,-49,-
19,054,085,095,098,113,100,083,074,060,999,999,999
3320 DATA 999,999,999,999,999,999,999,999,999,089,070,044,020,000,004,010,020,020,009,050,073,080,
077,073,050,029,005,-70,999,999,999
3330 DATA 999,999,999,999,999,999,999,999,999,999,092,069,050,040,047,059,079,090,079,069,060,063,
063,055,033,008,-70,-189,999,999,999
3340 DATA 999,999,999,999,999,999,999,999,999,999,095,100,090,090,090,097,116,129,107,083,063,049,
046,040,021,002,-100,-230,-350,999,999
3350 DATA 999,999,999,999,999,999,999,999,999,999,100,096,092,091,096,107,123,128,105,083,062,043,
029,021,010,-14,-119,-214,-350,-483,999
3360 DATA 999,999,999,999,999,999,999,999,999,999,087,077,070,070,085,103,119,130,107,085,067,047,
026,010,010,012,-50,-186,-341,-499,999
3370 DATA 999,999,999,999,999,999,999,999,999,999,068,058,049,058,074,092,107,120,118,099,088,075,
040,042,040,042,040,-107,-290,-455,999
3380 DATA 999,999,999,999,999,999,999,999,999,999,050,036,029,043,059,077,094,107,117,117,130,119,
089,090,084,059,055,-50,-250,-420,999
3390 DATA 999,999,999,999,999,999,999,999,999,999,024,012,012,028,046,066,090,109,130,150,170,169,
150,135,120,080,028,-28,-200,-380,999
3400 DATA 999,999,999,999,999,999,999,999,999,999,999,-14,000,020,050,075,102,130,154,
176,190,201,190,183,150,100,050,-20,-170,-337,999
3410 DATA 999,999,999,999,999,999,999,999,999,999,999,-50,-29,000,020,050,080,125,169,
209,224,233,240,224,190,130,080,-12,-170,999,999
3420 DATA 999,999,999,999,999,999,999,999,999,999,999,999,999,-
29,010,059,105,154,200,240,270,280,289,260,229,170,090,-10,-170,999,999
3430 DATA 999,999,999,999,999,999,999,999,999,999,999,999,999,999,060,100,140,190,233,270,306,329,
336,330,300,210,100,-09,999,999,999
3440 DATA 999,999,999,999,999,999,999,999,999,999,999,999,999,999,999,999,999,999,999,999,999,999,
999,999,999,999,999,999,999,999,999
3450  RETURN
```

□

```
1000 ' "RDED0032.BAS
1010 CLS
1020 PRINT "RDED0032.BAS,A - Date: ";DATE$
1030 ' RDED0032.BAS,A - Date: 01-27-1997
1040 PRINT "-------------------------------------------"
1050 PRINT "CONVERSION CO-ORDINATES RD TO TM
                AND VICE-VERSA"
1060 PRINT "     UTM ZONE 32 - CM 9 E"
1070             M. Hooijberg 1984-1997
1080 PRINT "-------------------------------------------"
1090 DEFDBL A-Z
1100 DEFINT I-L
1110 OPTION BASE 1
1120 DIM PE(31,33),PN(31,33)
1130 A$=" ="
1140 E$="ED-50-E  "
1150 N$="ED-50-N  "
1160 X$="RD1918-X  "
1170 Y$="RD1918-Y  "

1180 ' rd constants zone 32
1190 A0=5790890.61#
1200 A1=184863.86#
1210 A2=.999470447#
1220 A3=.050233726#
1230 A4=-.07804#
1240 A5=.294143#
1250 A6=-.212926#
1260 A7=-.021184#
1270 A8=-.01963#
1280 A9=.07205#

1290 ' utm constants zone 32
1300 B0=465832.61#
1310 B1=86725.49#
1320 B2=.998008761#
1330 B3=-.050160262#
1340 B4=.032931#
1350 B5=-.302186#
1360 B6=.210869#
1370 B7=-.021515#
1380 B8=.00104#
1390 B9=-.07759#

1400 ' x, y constants zone 32
1410 Y1=465832.61#
1420 Y2=577404.33#
1430 Y3=580314.65#
1440 Y4=317522.71#
1450 X1=86725.49#
1460 X2=119438.51#
1470 X3=269564.37#
1480 X4=194845.62#

1490 ' e, n constants zone 32
1500 N1=5790890.61#
1510 N2=5900759.95#
1520 N3=5896096.39#
1530 N4=5637344.76#
1540 E1=184863.86#
1550 E2=223164.22#
1560 E3=373263.19#
1570 E4=285581.79#

1580 ' load data lookup-tables
1590 GOSUB 2650
1600 INPUT "OLD-RD or NEW-RD ?? [O/N] :",V$
1610 RD$="OLD"
1620 IF V$="O" OR V$="o" THEN 1660
1630 IF V$="N" OR V$="n" THEN 1650
1640 GOTO 1600
1650 RD$="NEW"
1660 INPUT "Conversion RD -> ED-32 [Y/N] : ",V$
1670 IF V$="Y" OR V$="y" THEN 1700
1680 IF V$="N" OR V$="n" THEN 1950
1690 GOTO 1660
```

Important:

SUBROUTINE at line 1700 cont'd for both UTM zones 31/32 of program RDED0031.BAS

↓

11.7.10 Footnotes on the Programs

In General

The computer subroutines were written in the most suitably economical form. Therefore, some subroutine listings may deviate from the basic equations given in this book. A short description of conversions and transformations [5], together with the full formulae [6.1.1; 7.1.1; 8.1.1; 9.1.1], for computing and the fundamental data used [3.4.3] are given.

Considerable space is devoted to the use of the routines, with fully worked out sample computations of all sections. Furthermore, each program listing is preceded by worked examples with *an intermediate output* for algorithm *testing*, because situations may arise in which a program fails to operate as expected due to incorrect signs, constants or arithmetic operators. Graphics are added describing the "round-trip errors" of the computations indicating when and how these reach certain limiting values.

(Vincenty, 1994) said:

" ...numerical and graphical presentations of running times and accuracies of programs should speak for themselves ... ".

A demonstration of the reliability of the results by calculating the round-trip errors is imperative. In this respect Figure 18 and Figure 48 may be helpful for use of correct signs. Suspicion may be placed on results that have *not* been checked. When a program known to be operating correctly is used for calculating the solution, checking should be concentrated on *all input* to the computer program to eliminate errors. Rounding off data variables is left to the user of the programs.

Be sure about their significance in the program before changing or removing any statement. Please bear in mind, the book is written chiefly for the reader who has studied, but is by no means familiar with the subject of map projections and their use for global and extended surveys.

Using BASIC

Double-precision numeric constants are stored with 17 digits of precision, and printed with as many as 16 digits. A double-precision constant is any numeric constant with a trailing number sign (#) e.g. 3490.0# or 3490.0d0. The trailing number signs # are removed here to clarify the listings and therefore are not shown in the listings, excepting [11.7.9]. The reader is expected to enter these number signs # into the programs, otherwise they do not work properly.

The structure of a program listing - in fact a subroutine - is not fully utilised in order to permit wider use of the (sub) routines :

- *input* and *output* routines are kept as simple as possible
- REMind statements (') are entered
- several *superfluous print statements* are left in the routines
- some other statements are left in the listings for clarity.

360° Sexagesimal System, and 400^g Centesimal System

For any 400^g centesimal system, such as used in FRANCE, exchange following statements:

- ```
  470 ' input 360° sexagesimal system
  480 INPUT "Lat. Std Par (D,M,S):",M1,M2,M3
  490 PRINT LP$;A$;M1;D$;M2;M$;M3;S$;NS$
  500 GOSUB 3730 ' call dmsrad
  510 LAT=M0
  ↓
  ↓
  2310 ' output 360° sexagesimal system
  2320 M0=LAT
  2330 GOSUB 3820 ' call raddms
  2340 PRINT LP$;A$;M1;D$;M2;M$;M3;S$;NS$
  ↓
  ↓
  3730 ' sub dmsrad
  3740 M0=
     (ABS(M1)+(ABS(M2)+ABS(M3) / 60) / 60)*RD

  3750 M0=M0*I2
  3760   RETURN
  ↓
  ↓
  3820 ' sub raddms
  3830 I2=1
  3840 IF M0<0 THEN I2=-1
  3850 W1=ABS(M0/RD)
  3860 M1=INT(W1)
  3870 W2=(W1-M1)*60
  3880 M2=INT(W2)
  3890 M3=(W2-M2)*60
  3900   RETURN
  ```

- ```
  470 ' input 400g centesimal system
  480 INPUT "Lat. Std Para (G.g):",M1
  490 PRINT LP$;A$;M1;G$;NS$
  500 GOSUB 3730 ' call grdrad
  510 LAT=M0
  ↓
  ↓
  2310 ' output 400g centesimal system
  2320 M0=LAT
  2330 GOSUB 3820 ' call radgrd
  2340 PRINT LP$;A$; M1;G$
  ↓
  ↓
  3730 ' sub grdrad
  3740 M0=
     (ABS(M1)*.9)*RD

  3750 M0=M0*I2
  3760   RETURN
  ↓
  ↓
  3820 ' sub radgrd
  3830 I2=1
  3840 IF M0<0 THEN I2=-1
  3850 M1=ABS(M0/.9/RD)
  3860   RETURN
  ```

Remarks about the Routines

Verification of historical data and output of other programs is *not* a function of these (sub)routines. Co-ordinates of historical data points must be taken at face value, with the realisation that such co-ordinates could be significantly in error (Floyd, 1985).

BDG00000.BAS program

Computing of intermediate points in *the Direct Problem* of BDG00000.BAS requires to REM (or ') line 2130. All intermediate points are computed in lines 2140-2270. Results (accuracy) depend on the setting of the integrations (number).

GBD00000.BAS program

Computing intermediate points in *the Inverse Problem* of GBD00000.BAS. This computation requires the insertion of lines between 2310 and 2450, such as given in program BDG00000.BAS: lines 2140-2270. Accuracy of the intermediate points computed depends on the setting of the integrations (number).

In order to solve *the Inverse Problem* it is advisable that the position of point P_1 is situated south of point P_2. Furthermore, the Line P_1-P_2 should not run exactly east-west or exactly north-south. In case this may occur during data processing, the computer will issue an error message without further notice!

See [2.2] , Error Messages, or the *BASIC-Handbook*, which explains the various error conditions. Some calculations in the Western or Southern Hemisphere can be mirrored and calculated in the N/E Hemisphere. See Figure 16.

GK000000.BAS program

GK000000.BAS program can also be put to use with the Meridional Arc formulae as shown in [3.4.3], formulae (3.03 - 3.08) of the Spheroidal Mapping Equations section. It may speed up the computing time by a factor 1.9.

Recasting the original NGS 5 formula may speed up the computing time by a factor 0.9, thus it is *slightly slower* than the present algorithm (Hooijberg, 1996; Vincenty, 1984a).

LCC00000.BAS program

Comments on using the Lambert algorithm as given in [11.7.6]:

Using the algorithm as given in the calculation of *isometric latitude* for the Oblique Mercator program OM000000.BAS, the latitude can be obtained *without iteration* as shown in the LCC00000.BAS program.

Belgium Lambert1972 System.

Remember to use the lines "Belgium only".
Thus, for every " '@ " REMind statement showing a line:

<linenumber>'@ <statement> '[Belgium ...], e.g. lines 1860-1870:

- `1860 SS=(LOG(WU*COS(LTL)/(WL*COS(LTU))))/(QU-QL)`
- `1870 '@SS=.7716421928'` *[belgium]*

(1860) REM the preceding <line number> if appropriate
(1870) remove " '@ ". Consequently the line should read:

<linenumber> <statement> '[Belgium ...]

Consequently, the lines 1860-1870 should read:

- `1860` *REM* `SS=(LOG(WU*COS(LTL)/(WL*COS(LTU))))/(QU-QL)`
- `1870 SS=.7716421928'` *[Belgium]*

Do not use the Belgium transformation statements in the (sub)routine for any other country.

OM000000.BAS program

The constants for the formulae in OM000000.BAS to calculate the isometric latitude are adapted by Berry and Burkholder. The equations are recasted by the author in 1994.

TRM00000.BAS program

Comments on using the S-transformation formulae in TRM00000.BAS are given in [9.1]. Transformed curvilinear co-ordinates can be obtained without iteration. However, situations may arise in which iterative correction is required due to a *large ellipsoidal height*, e.g. satellite tracking at 20 000 km height + 6 378 km (earth's radius)=26 400 km height. Schuhr's algorithm can provide corrections to such data (Schuhr, 1996).

RDED0003x.BAS program

Bi-linear interpolation. TDN-data matrix given is 30×32 (integers). For calculation, one column and one row of zero data (given as: 999) are added to each matrix. Hence, in the program the data-matrix used for calculation is 31×33 (integers).

RDED00031.BAS and RDED00032.BAS subroutines are continued at line 1700 for both UTM zone 31 and UTM zone 32, using the matrix 31×33.

12. Bibliography and Indices

12.1 Index of Subjects

A

B

C

S

T

U

V

W

X

Z

12.2 Index of Authors

V

W

X

Z

12.3 Bibliography

Achilli, V., 1994. Comparison between GPS and IGM Co-ordinates in the Italian Area. Bollettino di Geodesia e Scienze Affini, Anno LIII - N.1., 1994. Publicazione dell' Istituto Geografico Militare, Florence: pp 1-23

ACIC, 1957. Geodetic Distance and Azimuth Computations for Lines over 500 Miles. ACIC TR Nr. 80. Geosciences Branch Aeronautical Chart and Information Centre, Missouri

Adams, O.S. / Deetz, C.H., 1921 (1990). Elements of Map Projection. 5th Edition. Special Publication No. 68. US Government Printing Office, Washington, DC

Adams, O.S., 1949. Latitude Developments Connected with Geodesy and Cartography. Special Publication No. 67. US Government Printing Office, Washington, DC

Admiralty, 1965. Manual of Hydrographic Surveying, Volume One. Hydrographer of The Navy. HMSO, London.

Agajelu, S.I., 1987. On Conformal Representation of Geodetic Positions in Nigeria. Survey Review 29, No. 223, Jan 1987: pp 3-12

AGS, 1995. Australian Geodesy Subcommittee - National Committee for Solid Earth Sciences. Australian Academy of Science. National Report 1991-1995. Presented at IAG, General Assembly XXI of I.U.G.G., Bouldery Colorado, Jul 1995

Airy, G.B., 1830. Encyclopaedia Metropolitana. Trigonometry on The Figure of Earth, Tides and Waves. Observatory, Scientific Department of Cambridge University

Al-Bayari, O / Capra, A. / Radicioni, F. / Vittuari, L., 1996. GPS Network for Deformation Control in the Mount Melbourne Area (Antarctica): Preliminary Results of Third Measurements Campaign. Reports on Surveying and Geodesy, DISTART, Università degli Studi di Bologna Dipartimento di Ingegneria delle DISTART: pp 64-71

Alberda, J.E. / Krijger, B.G.K. / Meerdink, E.F., 1960. The Adjustment of UELN as executed at Delft. Bulletin Géodésique, No. 55, Mar 1960: pp 41-53

Alberda, J.E., 1963. Report on the Adjustment of the United European Levelling Net and Related Computations. New Series, Volume 1, Number 2. Netherlands Geodetic Commission, Delft

Alberda, J.E., 1978 (1991). Inleiding Landmeetkunde. 4e druk. Delftse Uitgevers Maatschappij, Delft

Allan, A.L. / Hollwey, J.R. / Maynes, J.H.B., 1968 (1975). Practical Field Surveying and Computations. Heinemann, London

Anzidei, M. / Baldi, P. / Casula, G. / Riguzzi, F. / Surace, L., 1995. La Rete Tyrgeonet. Supplement to Bollettino di Geodesia e Scienze Affini, Anno LIV - N.2. Publicazione dell' Istituto Geografico Militare, Florence

Arnold, K.,1970. Methoden der Satelitengeodäsie. Akademie-Verlag, Berlin

Arrighi, A., 1994. L'effetto dell'introduzione di un Fattore di Riduzione di Scala nella Rappresentazione di Gauss. Bollettino di Geodesia e Scienze Affini, Anno LIII - N.4, 1994. Publicazione dell' Istituto Geografico Militare, Florence: pp 401-417

Ashkenazi, V. et al, 1986a. The 1980 Readjustment of the Triangulation of the United Kingdom and the Republic of Ireland OS(SN)80. Ordnance Survey Professional Paper No. 31. Ordnance Survey, Southampton

Ashkenazi, V., 1986b. Co-ordinate Systems: How to get Your Position Very Precise and Completely Wrong. Journal of Navigation, Vol. 39, No. 2, 88: pp 269-278

Ashkenazi, V. / Storey, J., 1991. The Co-ordinate Datum Problem and its Solution. In: *ION GPS-91, 1991: pp 387-392*

Ashkenazi, V. / Hill, C.J. / Nagel, J., 1992. Wide Area Differential GPS: A Performance Study. In: *ION GPS-92, 1992: pp 589-598*

Ashkenazi, V. / Moore, T., 1993. GPS Co-ordinates as a GIS Attribute. In: *ION GPS-93, 1993: pp 85-88*

Ayres, J.E., 1995. Private Communication. NIMA, US Department of Defense, Washington, DC

Ayres, J.E., 1996. Private Communication - UTM Notes. NIMA, US Department of Defense, Washington, DC

Baarda, W., 1981. S-Transformations and Criterion Matrices. Publications on Geodesy, New Series Volume 5, Number 1. Second revised edition. Netherlands Geodetic Commission, Delft

Baarda, W., 1995. Linking Up Spatial Models in Geodesy - Extended S-Transformations. Publications on Geodesy, New Series Number 41. Netherlands Geodetic Commission, Delft

Baeschlin, C.F., 1948. Lehrbuch der Geodäsie. Orell Füssli Verlag, Zürich

Beers, B.J., 1995. FRANK - the Design of a New Land Surveying System using Panoramic Images - Thesis. Delft University of Technology. Delft University Press, Delft

Benhallam, A / Rosso, R., 1996. Performance Comparisons between GPS and GLONASS Transmission Systems. In: *NTM, 1996: pp 423-430*

Berry, R.M. / Bormanis, V., 1970. Plane Co-ordinate Survey System for the Great Lakes Based Upon the Hotine Skew Orthomorphic Projection. US Lake Survey Miscellaneous Paper 70-4, Department of the Army, US Corps of Engineers, Lake Survey District, Detroit, MI

Berry, R.M., 1971.The Michigan Co-ordinate System. Seminar presented to Michigan Society of Registered Land Surveyors at the University of Michigan, Ann Arbor, MI

Beser, J., 1992. GPS and GLONASS Visibility Characteristics and Performance Data of the 3S Navigation R-100 Integrated GPS / GLONASS Receiver. In: *ION GPS-92, 1992: pp 187-205*

Bhattacharji, J.C., 1962. The Indian Foot-Metre Ratio and its Adoption in the Indian Geodetic System. Empire Survey Review, No. 119, Jan 1962: pp 13-18

Bigourdan, G., 1912. Grandeur et Figure de la Terre - J.B.J. Delambre. Gauthier-Villars, Imprimeur-Libraire, Paris

Bjerhammer, A., 1986. Relativistic Geodesy. NOS 118 NGS 36. National Geodetic Survey, Silver Spring, MD

Bjork, A. / Dahlquist, G., 1974. Numerical Methods. Prentice Hall, Englewood Cliffs, New Jersey

Bogdanov, V.I. / Taybatorov, K.,1995. Some Results of the Russian Research in the Frame of the International Baltic Sea Level Project. In: *Vermeer, 1995: pp 31-41*

Böhme, R., 1993. Inventory of World Topographic Mapping, Volume 3. The International Cartographic Association, Elsevier Applied Science Publishers, London-New York

Bomford, G., 1977. Geodesy. Third Edition. Oxford University Press, Oxford

Bordley, R.F. / Calvert, C.E., 1985. The Horsnet Project. The Hydrographic Journal, No. 36, Apr 1985: pp 5-10

Boucher, C., 1979 (1981). Les Présentations Planes Conformes de Lambert / Projections en Usage pour la France Métropolitaine (NT/G No. 13). Institut Géographique National, St. Mandé.

Bowring, B.R., 1971. The Normal Section - Forward and Inverse Formulae at Any Distance. Survey Review No. 161, Jul 1971: pp 131-136

Bowring, B.R., 1976. Transformation from Spatial to Geographical Co-ordinates. Survey Review 23, No. 181, Jul 1976: pp 323-327

Bowring, B.R. / Vincenty, T., 1979. Use of Auxiliary Ellipsoid in Height Controlled Spatial Adjustments. Technical Memorandum NOS NGS 23. US Government Printing Office, Washington, DC

Bowring, B.R., 1981. The Direct and Inverse Problems for Short Geodesic Lines on the Ellipsoid. Surveying and Mapping, Vol. 41, No. 2: pp 135-141

Brazier, H.H. / Hotine, M., 1950. Projection Tables for British Commonwealth Territories in Borneo. Directorate of Colonial Surveys, Teddington, Middlesex

Brouwer, F.J.J. / Buren, J. van / Gelder, B.H.W. van, 1989. GPS - Navigatie en Geodetische Puntsbepaling met het Globale Positioning System. Delftse Universitaire Pers, Delft

Buchar, E., 1962. Determination of Some Parameters of the Gravity Field of the Earth from Rotation of the Nodal Line of artificial satellites. Bulletin Géodésique, No. 65, Sep 1962: pp 269-271

Burkholder, E.F., 1984. Geometrical Parameters of the Geodetic Reference System 1980. Surveying and Mapping, Vol. 44, No. 4: pp 339-340

Burkholder, E.F., 1985 (1990). State Plane Co-ordinates on the NAD83. 1985 ASCE Spring Convention, Denver, CO: pp 34

Burkholder, E.F., 1993. Design of a Local Co-ordinate System for Surveying, Engineering, and LIS / GIS. Surveying and Land Information System, Vol. 53, No. 1: pp 29-40

Burkholder, E.F., 1995a. Consideration of a 3-Dimensional Model for 3-D Data. GIS-T 95 Sparks, Nevada, Apr 1995

Burkholder, E.F., 1995b. GIS Application of GPS Technology via Local Co-ordinate Systems: Past, Present, and Future. ACSM / ASPRS Annual Convention and Exposition, Charlotte, NC, Feb / Mar 1995

Burša, M., 1966. Fundamentals of the Theory of Geometric Satellite Geodesy. Geofysikalni Sbornik 1966. Akademia Praha: pp 25-74

Burša, M., 1992. Parameters of Common Relevance of Astronomy, Geodesy, and GeoDynamics. In: *(GH [The Geodesist's Handbook], 1992)*

Burton, E.L., 1996. In: *Ayres, 1996*

Calvert, C.E., 1995. Private Communication. Ordnance Survey, Southampton

Chovitz, B.H., 1989. Datum Definition. In: *Schwartz, 1989: pp 81-85*

Claire, C.N., 1968. State Plane Co-ordinates by Automatic Data Processing. Special Publication No. 62-4. US Government Printing Office, Washington, DC

Clark, D., 1976 (1923). Plane and Geodetic Surveying for Engineers, Volume 2: Higher Surveying. 6th edition revised and rewritten. Constable, London

Clarke, A.R., 1880. Geodesy. Clarendon Press, Oxford

Clarke, F.L., 1973. Zone to Zone Projection on the Transverse Mercator Projection. The Australian Surveyor, Vol. 25, No. 4. Dec 1973: pp 293-302

Codd, J.F., 1995 (1996). Private Communication. Ordnance Survey of Northern Ireland, Belfast

Cory, M.J., 1995 (1996). Private Communication. Ordnance Survey of Ireland, Dublin

Craig, Th., 1882. A Treatise on Projections. Unites States Coast and Geodetic Survey, US Government Printing Office, Washington, DC

DA (Department of the Army), 1958. Universal Transverse Mercator Grid. Technical Manual No. TM 5-241-8. Headquarters of the Army, Washington, DC

Daly, P. / Raby, P. / Riley, S., 1992. GLONASS Status and Initial C/A and P-Code Ranging Tests. In: *ION GPS-92, 1992: pp 145-151*

Daly, P. / Raby, P., 1993. Using the GLONASS System for Geodetic Survey. In: *ION GPS-93, 1993: pp 1129-1138*

Daly, P. / Riley, S., 1994. GLONASS P-Code Data Message. In: *NTM, 1994: pp 195-202*

De Min, E.J., 1996. De geoïde van Nederland. Ned. Commissie voor Geodesie, Publicatie 34. Delft

Dewhurst, W.T., 1990. NADCON. The Application of Minimum Curvature-Derived Surfaces in the Transformation of Positional Data from the North American Datum of 1927 to the North American Datum of 1983. NOS NGS 50. US Department of Commerce, National Geodetic Survey, Silver Spring, MD

DGPS '91, 1991. First International Symposium Real Time Differential Applications of the Global Positioning System, Sep 1991, TÜV Rheinland GmbH, Cologne

DMA, 1973 (1977), see (NIMA, 1973-1977)

DMA, 1991 (1995), see (NIMA, 1991-1995)

Doodson, A.T., 1960. Mean Sea Level and Geodesy. Bulletin Géodésique, No. 55, Mar 1960: pp 69-88

Dracup, J.F., 1994. Local Plane Co-ordinate Systems: An Overall View. Surveying and Mapping, Vol. 54, No. 3: pp 168-180

Dupuy, M., 1954. La Détermination des Dimensions de la Terre pour les Travaux Géodésiques en U.R.S.S. Bulletin Géodésique, No. 31, Mar 1954: pp 55-66

Ehrnsperger, W., 1989. Das Europäische Datum 1987 (ED87) und sein österreichischer Anteil. ÖZfVuPh., Heft 2: pp 47-90, 77. Jg, Heft 4: pp 192-193

Ekman, M. / Mäkinen, J.,1995a. Mean Sea Surface Topography in a Unified Height System for the Baltic Sea Area. In: *Vermeer, M., 1995: pp 53-57*

Ekman, M., 1995b. What is the Geoid? In: *Vermeer, 1995: pp 49-51*

Epstein, E.F. / Duchesneau, T.D., 1984. The Use and Value of a Geodetic Reference System. National Geodetic Information Center, NOAA, Rockville, MD

Field, N.J., 1980. Conversions between Geographical and Transverse Mercator Co-ordinates. Survey Review 25, No. 195, Jan 1980: pp 228-230

Fischer, F., 1845. Lehrbuch der Höheren Geodäsie. Carl Wilhelm Leske, Darmstadt

Fischer, I.K., 1959. The Hough Ellipsoid or The Figure of the Earth from Geoidal Heights. Bulletin Géodésique, No. 54, Dec 1959: pp 45-52

Fischer, I.K., 1972. The Geoid 4M-710-E-010-010. Defense Mapping School, Ft. Belvoir

Fister, F.I., 1980. Some Remarks on Landsat MSS Pictures. 14th Congress of the International Society of Photogrammetric, Commission IV. Int. Arch. Photogramm.: pp 214-222

Floyd, R.P., 1985. Co-ordinate Conversion for Hydrographic Surveying. NOS 114. US Department of Commerce, National Geodetic Survey, Silver Spring, MD

Geodetic Glossary, 1986. US Department of Commerce, NOAA-NGS, Silver Spring, MD

GH (The Geodesist's Handbook 1992), 1992. International Association of Geodesy. Bulletin Géodésique, 1992

GH (The Geodesist's Handbook 1996), 1996. International Association of Geodesy. Journal of Geodesy, 1996. Springer-Verlag

Glasmacher, H. / Krack, K., 1984. Umkehrung von vollständigen Potenzreihen mit zwei Veränderlichen, Schriftenreihe Wiss. Studiengang Vermessungswesen, Hochschule der Bundeswehr, München, Heft 10: pp 49-69, München

Göhler, H., 1991. Zu Anforderungen an das Vermessungswesen in den neuen Bundesländern. Allgemeine Vermessungs Nachrichten, 5 / 1991: pp 149-157

Gouzhva, Y.G. / Gevorkyan, A.G. / Bogdanov, P.P., 1991. Frequency and Time Support of GLONASS and GPS Realisation of Differential Fixing. In: *DGPS '91, 1991: pp 355*

Gouzhva, Y.G. / Gevorkyan, P. / Bogdanov, P.P., 1995. Getting in Sync: GLONASS Clock Synchronisation. GPS World, Apr 1995: pp 48-56

GPE (Geodesy and Physics of the Earth), 1993. Proceedings of IAGS-1993, Springer-Verlag, Berlin-Heidelberg

Graaff-Hunter, J. de, 1960. The Shape of the Earth's Surface Expressed in Terms of Gravity at Ground Level. Bulletin Géodésique, No. 56, Jun 1960: pp 191-200

Grafarend, E.W. / Krumm, F. / Schaffrin, B., 1985. Criterion Matrices of Heterogeneously Observed Three-dimensional Networks. Manuscripta Geodaetica, 1985, No. 10: pp 3-22

Grafarend, E.W., 1995a. The optimal Universal Transverse Mercator Projection. Manuscripta Geodaetica, 1995, No. 20: pp 421-468

Grafarend, E.W. / Krumm, F. / Okeke, F., 1995b. Curvilinear Geodetic Datum Transformations. Zeitschrift für Vermessungswesen, No. 7, 1995: pp 334-350

Grafarend, E.W. / Syffus, R. / You, R.J., 1995c. Projective Heights in Geometry and Gravity Space. Allgemeine Vermessungs Nachrichten, 102. Jg, Heft 10, Oct 1995: pp 382-403

Grafarend, E.W. / Syffus, R., 1995d. The Oblique Azimuthal Projection of Geodesic Type for the Bi-axial Ellipsoid: Riemann Polar and Normal Co-ordinates. Journal of Geodesy, 1995, No. 70: pp 13-37

Grafarend, E.W. / Engels, J., 1995e. The Oblique Mercator Projection of the Ellipsoid of Revolution $E^2_{a,\,b}$. Journal of Geodesy, 1995, No. 70: pp 38-50

Grafarend, E.W. / Krarup, R. / Syffus, R., 1996a. An Algorithm for the Inverse of an Multivariate Homogeneous Polynomial of Degree n. Journal of Geodesy, 1996, No. 70: pp 276-286

Grafarend, E.W. / Kampmann, R., 1996b. $C_{10}^{(3)}$: The Ten Parameter Conformal Group as a Datum Transformation in Three-dimensional Euclidean Space. Zeitschrift für Vermessungswesen, 121 Jg, Heft 2, Feb 1996: pp 68-77

Grafarend, E.W. / Syffus, R., 1996c. The Optimal Mercator Projection and the Optimal Polycylindric Projection of Conformal Type - Case Study Indonesia. International Conference About Geodetic Aspects of the Law of the Sea (GALOS). Denpassar / Indonesia: pp 1-12

Grafarend, E.W. / Shan. J., 1997a. Closed-form Solution of P4P or the Three-Dimensional Resection Problem in Terms of Möbius Barycentric Co-ordinates. Journal of Geodesy, 1997, No. 71: pp 217-231

Grafarend, E.W. / Shan. J., 1997b. Closed-form Solution to the Twin P4P or the Combined Three-Dimensional Resection-Intersection Problem in Terms of Möbius Barycentric Co-ordinates. Journal of Geodesy, 1997, No. 71: pp 232-239

Grafarend, E.W. / Ardalan, A.A., 1997c. W_0. - an Estimate in the Finnish Height Datum N60, Epoch 1993.4, from Twenty-five GPS Points of the Baltic Sea Level Project -. Journal of Geodesy, In Print, 1997

Greenfield, J.S., 1992. Development of Projection Tables for a Transverse Mercator Projection of NAD83. Surveying and Land Information Systems, Vol 52, No. 4: pp 219-226

Gretschel, H., 1873. Lehrbuch der Karten-Projektion. Bernhard Friedrich Voigt, Weimar

Groten, E., 1974. Zur Genauigkeit im Südhessisischen Teil des Deutschen Haupthöhennetzes. Zeitschrift für Vermessungswesen, 10 / 1974: pp 431-443

Hager, J.W. et al., 1990. Datums, Ellipsoids, Grids, and Grid Reference Systems. DMA (now NIMA) Technical Manual, DMA TM 8358-1. National Imagery and Mapping Agency, Washington, DC

Harper, D., 1976 (1977). Eye in the Sky - Introduction to Remote Sensing. Multiscience Publications Limited, Montréal

Harsson, B.G., 1995. Private Communication. Norwegian Mapping Authority, Hønefoss, Norway

Hartman, R.G. / Brenner, M.A. / Kant, N.M. / Fowler, B., 1991a. GPS / GLONASS Flight Test, Lab Test and Coverage Analysis Tests. In: *ION GPS-91, 1991: pp 333-344*

Hartman, R.G. / Brenner, M.A. / Kant, N.M. / Fowler, B., 1991b. Results from GPS / GLONASS Flight and Static Tests. In: *DGPS '91, 1991:* pp 95-106

Hartman, R.G., 1992. Joint US / USSR Satellite Navigation. GPS World, Feb 1992: pp 26-36

Heiskanen, W.A., Moritz, H., 1967. Physical Geodesy. W.H. Freeman, San Francisco.

Helmert, F.R., 1880 (1962). Die Mathematischen und Physikalischen Theorieen der Höheren Geodäsie, Part 1, 1884, Part 2. Minerva GmbH, Leipzig (Frankfurt / Main)

Herrewegen, M. van der, 1989. Referentiesystemen en Transformatieformules in Gebruik in België. Nationaal Geografisch Instituut, Brussels

Heuvelink, Hk.J., 1918. De Stereografische Kaartprojectie in hare Toepassing bij de Rijksdriehoeksmeting. Technische Boekhandel J. Waltman Jr, Delft

Hirvonen, R.A., 1962. The Reformation of Geodesy. Bulletin Géodésique, No. 65, Sep 1962: pp 197-214

Hooijberg, M., 1979. Toepassing van Elektronische Zakcomputers in de Geodesie. Geodesia, Feb 1979: pp 52-65

Hooijberg, M., 1984. Report on: Application of Bi-linear Interpolation - Dutch Continental Shelf. To: Dienst der Hydrografie (Royal Neth. Navy), Nov 1984

Hooijberg, M., 1996. Map Projections - Revisiting Latitude Developments for Ellipsoids. Workshop Hydrographic Society - Benelux Branch, Feb 1996, Amsterdam

Hopfner, F., 1949. Grundlagen der Höheren Geodäsie. Springer-Verlag, Vienna

Hotine, M., 1946-1947. The Orthometric Projection of the Ellipsoid. Empire Survey Review, Vol. 8, 9. Nos. 62, 63, 64, 65 and 66

Hotine, M., 1969. Mathematical Geodesy. ESSA Monograph 2. US Department of Commerce. Washington, DC

Hristov, V.K., 1968. Developpement de l' Aplatissement et de la Pesanteur Terrestres jusqu'aux Termes du Troisieme Ordre Inclusivement. Bulletin Géodésique, No. 90, Dec 1968: pp 435-457

Husti, G.J., 1996. Getallenvoorbeeld voor de Transformatie tussen het ETRS89 en het RD / NAP Stelsel. GPS Nieuwsbrief No. 2, 1996: pp 41-42

IAG (International Association of Geodesy), 1967. Geodetic Reference System 1967. Bulletin Géodésique, No. 85, Sep 1967

IAG, 1971. Geodetic Reference System 1967. Bulletin Géodésique, Publ. Spéc. No. 3

IERS (International Earth Rotation Service), 1993a. IERS Annual Report. Central Bureau of IERS - Observatoire de Paris, Paris

IERS (International Earth Rotation Service), 1993b. IERS Technical Note 17: Earth Orientation, Reference Frames and Atmospheric Excitation Functions Submitted for the 1993 IERS Annual Report. VLBI, LLR, GPS, SLR and AAM Analysis Centres, Sep 1994, Central Bureau of IERS, Observatoire de Paris, Paris

IGN (Institut Géographique National), 1963. Tables des Constantes Numériques des systèmes de projections Lambert en Usage á l'Institut Géographique National. Eyrolles, Paris

IGN (Institut Géographique National), 1986. Note Complementaire sur les Projections Lambert Utilisees pour la Carte de France Type 1922 (NT/G No. 52). Institut Géographique National, St. Mandé.

IGN (Institut Géographique National), 1994. Private Communication. Institut Géographique National, St. Mandé

Ihde, J., 1981. Zu den Beziehungen zwischen traditionellen astronomisch-geodetischen Netzen und satelliten-geodäti-schen Bezugssysteme. Vermessungstechnik. 29. Jg, 1981, Heft 5: pp 163-167

Ihde, J., 1991. Geodätische Bezugsysteme. Vermessungstechnik. 39. Jg, 1991, Heft 1: pp 13-15, Heft 2: pp 57-63

Ihde, J., 1993. Some Remarks on Geodetic Reference Systems in Eastern Europe in Preparation of a Uniform European Geoid. Bulletin Géodésique, Vol. 67, No. 2, 1993: pp 81-85

Ihde, J. / Schoch, H. / Steinich, L., 1995. Beziehungen zwischen den geodätischen Bezugssystemen Datum Rauenberg, ED50 und System 42. DGK Reihe B, Heft No. 298. Verlag des Instituts für Angewandte Geodäsie, Frankfurt am Main

ION GPS-91, 1991. Proceedings of the Fourth International Technical Meeting of the Satellite Division of the Institute of Navigation, Albuquerque Convention Centre, Albuquerque, New Mexico. Institute of Navigation, Alexandria, VA, USA

ION GPS-92, 1992. Proceedings of the Fifth International Technical Meeting of the Satellite Division of the Institute of Navigation, Albuquerque Convention Centre, Albuquerque, New Mexico. Institute of Navigation, Alexandria, VA, USA

ION GPS-93, 1993. Proceedings of the Sixth International Technical Meeting of the Satellite Division of the Institute of Navigation, Salt Palace Convention Center, Salt Lake City, Utah. Institute of Navigation, Alexandria, VA, USA

ION GPS-94, 1994. Proceedings of the Seventh International Technical Meeting of the Satellite Division of the Institute of Navigation, Salt Palace Convention Center, Salt Lake City, Utah. Institute of Navigation, Alexandria, VA, USA

ISG (Integrated Survey Grid Tables), 1972. Australian National Spheroid. Technical Subcommittee, Survey Integration Committee, Government Printer, Sydney, New South Wales

Ivanov, N. / Salischev, V., 1992. The GLONASS System - An Overview. The Journal of Navigation, Vol. No. 45, 1992: pp 175-182

Izotov, A.A., 1959. Reference-Ellipsoid and the Standard Geodetic Datum adopted in the USSR. Bulletin Géodésique, No. 53, Sep 1959: pp 1-6

J/E/K (Jordan / Eggert / Kneissl / Ledersteger), 1956 (1969). Handbuch der Vermessungskunde, Band V. Tenth Edition. J.B. Metzlersche Verlagsbuchhandlung, Stuttgart.

J/E/K (Jordan / Eggert / Kneissl), 1959. Handbuch der Vermessungskunde, Band IV / 1 (1958); Band IV / 2 (1959). Tenth Edition. J.B. Metzlersche Verlagsbuchhandlung, Stuttgart

Jank, W. / Kivioja, L.A., 1980. Solution of the Direct and Inverse Problems on Reference Ellipsoids by Point-by-Point Integration Using Programmable Pocket Calculators. Surveying and Mapping, Vol. XL: pp 325-337

Jivall, L., 1995. GPS for Geodetic Control Surveying in Sweden. In: *Vermeer, 1995: pp 93-102*

Johnston, G.T., 1993 Results and Performance of Multi-Site Reference Station Differential GPS. In: *ION-GPS 93, Vol. I: pp 677-689*

Kakkuri, J., 1995. The Baltic Sea Level Project. In: *Vermeer, 1995: pp 43-47*

Kakkuri, J., 1996. Private Communication. Finnish Geodetic Institute, Masala

Kaula, W.M., 1962. Cospar - IAG Symposium on the Use of Artificial Satellites for Geodesy. Bulletin Géodésique, No. 65, Sep 1962: pp 193-196

Kazantsev, V.N. / Karnauhov, V., 1991. GLONASS Satellite Radionavigation System. In: *DGPS '91: pp 67-77*

Kazantsev, V.N. / Cheremisin, V.F. / Kozlov, A.G. / Reshetnev, M.F., 1992. Current Status, Development Program and Performance of the GLONASS System. In: *ION GPS-92, 1992: pp 139-144*

Kivioja, L.A., 1971. Computation of Geodetic Direct and Indirect Problems by Computers Accumulating Increments from Geodetic Line Elements, Bulletin Géodésique, No. 99, Mar 1971: pp 55-63

König, K. / Weise, K.H., 1951. Mathematische Grundlagen der Höheren Geodäsie und Kartographie. Band I. Springer-Verlag, Berlin

Kouba, J. / Popelar, J., 1994. Modern Geodetic Reference Frames for Precise Satellite Positioning and Navigation. In: *Proceedings of the International Symposium on Kinematic Systems in Geodesy. Geomatics and Navigation, Banff, Canada, Sep 1994: pp 79-85*

Krack, K., 1981. Die Umwandlung von gaußschen konformen Koordinaten in geographische Koordinaten des Bezugsellipsoides auf der Grundlage des transversalen Mercatorentwurfes. Allgemeine Vermessungsnachrichten 88. Jahrgang, Heft 5, Mai 1981: pp 173-178

Krack, K., 1982. Zur direkten Berechnung der geographischen Breite aus der Bogenlänge auf Rotationsellipsoiden. Allgemeine Vermessungs Nachrichten, 3 / 1982: pp 122-125

Krack, K., 1983. Private Communication. FAF University, München

Krack, K., 1995. Private Communication. FAF University, München

Krüger, L., 1919. Formeln zur konformen Abbildung des Erdellipsoids in der Ebene. Im Selbstverlage, Berlin

Kukkamäki, T.J. / Honkasala, T., 1954. Measurement of the Standard Baseline of Buenos Aires with Väisälä Comparator. Bulletin Géodésique, No. 34, Dec 1954: pp 355-362

Kukkamäki, T.J., 1978. Väisälä Interference Comparator. Finnish Geodetic Institute, Helsinki

Laplace, P.-S., 1799. Mécanique Celèste, Bd. II, Book III. Paris

Laurila, S.H., 1976. Electronic Surveying and Navigation. John Wiley & Sons.

Le Pape, M., 1994. Private Communication. Institut Géographique National, St. Mandé

Lechner, W. / Vieweg, S., 1993. Realisation of GNSS: Results of Combined GPS / GLONASS Data Processing. In: *ION GPS-93, 1993: pp 409-418*

Lechner, W. / Vieweg, S., 1994. Competitors or Initial Elements of a Future Civil GNSS. In: *NTM, 1994: pp 503-509*

Ledersteger, K., 1954. Die einheitliche Begründung der metrischen Höhendefinition (A Synthesis of the Different Metric Definitions of Altitude). Bulletin Géodésique, No. 32, Dec 1954: pp 108-145

Lee, L.P., 1962. The Transverse Mercator Projection on the Entire Ellipsoid. Empire Survey Review, No. 123, Jan 1962: pp 208-217

Leick, A. / Gelder, H.W. van, 1975. On Similarity Transformations and Geodetic Network Distortions Based on Doppler Satellite Observations. Ohio State Univ., Report Dept. Geodetic Science, Columbus 1975, No. 235

Leick, A., 1990 (1995). GPS Satellite Surveying. John Wiley & Sons, New York

Levallois, J.-J., 1970. Géodésie Generale, Tome 2 - Géodésie Classique Bidimensionelle. Eyrolles, Paris

Linden, J.A., 1985. Over RD, ED en WGS; het Verband tussen de Coördinatensystemen op de Nederlandse Topografische Kaarten. NGT Geodesia, Okt 85: pp 342-347

Listing, J.B., 1873. Nachrichten von der Königlichen Gesellschaft der Wissenschaften und der Universität zu Göttingen, 1873: pp 33-98

Logsdon, T., 1992. The NavStar Global Positioning System. Van Nostrand Reinhold, New York

Luse, J.D. / Malla, R., 1985. Geodesy from Astrolabe to GPS, A Navigator's View. Navigation: Journal of the Institute of Navigation, Vol. 32, No. 2, Summer 1985

Luymes, J.L.H., 1924. Kaartprojecties Beschouwd uit een Hydrographisch Oogpunt. Algemeene Landdrukkerij, 's-Gravenhage

Macdonald, A.S. / Christie, R.R., 1991. From Miles to Millimetres. Survey Review, Vol. 31, No. 241, Jul 1991

Maling, D.H.,1992. Co-ordinate Systems and Map Projections. Second Edition. Pergamon Press, Oxford

Malys, J. / Slater, S., 1994. Maintenance and Enhancement of the World Geodetic System 1984. In: *ION GPS-94, 1994: pp 17-24*

Meade, B.K., 1987. Program for Computing the Universal Transverse Mercator (UTM) Co-ordinates for Latitudes North or South and Longitudes East or West. Surveying and Mapping, Vol. 47, No. 1: pp 37-40

Misra, P.N., 1992. GLONASS Data Analysis: Interim Results. Navigation: Journal of the Institute of Navigation, Vol. 39, No. 1. Spring 1992: pp 93-109

Misra, P.N. / Abbot, R.I., 1994. SGS85 - WGS84 Transformation. Manuscripta Geodaetica, 1994, No. 19: pp 300-308

Mitchell, H.C. / Simmons, L.C., 1945 (1979). Special Publication No. 235. US Government Printing Office, Washington, DC

Molenaar, M., 1981. A further Inquiry into the Theory of S-Transformations and Criterion Matrices. New Series Volume 7, Number 1. Netherlands Geodetic Commission, Delft

Moritz, H., 1968. The Geodetic Reference System 1967. Allgemeine Vermessungs Nachrichten, 1 / 1968: pp 2-7

Moritz, H., 1979. Fundamental Geodetic Constants, Report of Special Study Group No. 5.39 of IAG, presented at XVII General Assembly of I.U.G.G., Canberra

Moritz, H., 1980. Geodetic Reference System 1980. International Association of Geodesy 1980. Bulletin Géodésique, The Geodesist's Handbook 1980, Vol 54, No. 3: pp 395-405

Moritz, H., 1981. Das Geodätische Bezugssystem 1980. Vermessungstechnik. 29. Jg, 1981, Heft 9: pp 292-294

Moritz, H., 1992. Geodetic Reference System 1980. In: *GH (The Geodesist's Handbook), 1992*

Moskvin, G.I. / Sorochinsky, V.A., 1990. GLONASS Satellite Navigation System, Navigational Aspects. GPS World, Jan / Feb 1990: pp 50-54

Murphy, D.W., 1981. Direct Problem Geodetic Computation Using a Programmable Pocket Calculator. Survey Review, Vol. 26, No. 199, Jan 1981: pp 11-15

NIMA, 1973 (1977). Basic Geodesy. Report 4MF1-F-010-010. Defense Mapping School, Ft. Belvoir

NIMA, 1991 (1995). Department of Defense - World Geodetic System 1984. Technical Report DMA TR 8350.2-B. Second Edition-1991 (1995). Defense Mapping Agency, Washington, DC

NMC (The National Mapping Council of Australia), 1986. The Australian Geodetic Datum. Technical Manual. Special Publication No. 10. Australian Government Publishing Service, Canberrra

NOAA-C&GS, 1952. Formulas and Tables for the Computation of Geodetic Positions. Special Publication No. 8, 7th Edition reprint. US Government Printing Office, Washington, DC

NOAA-C&GS, 1954. Plane Co-ordinate Projection Tables for Hawaiian Islands. Special Publication No. 302. US Government Printing Office, Washington, DC

NOAA-C&GS, 1961. Plane Co-ordinate Intersection Tables for Alaska Zone 1. Special Publication No. 65-1 Part 49. US Government Printing Office, Washington, DC

NTM (Proceedings of the 1994 National Technical Meeting), 1994. Navigating the Earth and Beyond. San Diego, Jan 1994. Institute of Navigation, Alexandria, VA, USA

NTM (Proceedings of the 1996 National Technical Meeting), 1996. Technology and Operations: Partnership for Success in Navigation. Santa Monica, Jan 1996. Institute of Navigation, Alexandria, VA, USA

Ollikainen, M., 1995. The Finnish Geodetic Co-ordinate Systems and the Realisation of the EUREF89 in Finland. In: *Vermeer, 1995: pp 151-164*

Olson, A.C., 1977. Graphic Analysis of Resources by Numerical Evaluation Techniques (GARNET). Computer and Geosciences, Vol. 3: pp 539-545

OS, 1950 (1975). Constants, Formulae and Methods used in Transverse Mercator Projection. Ordnance Survey, Her Majesty's Stationary Office, London.

OS, 1995a. The Ellipsoid and the Transverse Mercator Projection. Geodetic Information Paper No. 1, 9 / 1995. Ordnance Survey, Southampton.

OS, 1995b. National Grid / ETRF89 Transformation Parameters. Geodetic Information Paper No. 2, 8 / 1995. Ordnance Survey, Southampton

OSI, 1995. Private Communication. Ordnance Survey of Ireland, Dublin

OSNI, 1994. The Trigonometrical Survey Network, Height Above Mean Sea Level and the Irish Grid. Leaflet No. 34. Ordnance Survey of Northern Ireland, Belfast, Oct 1994

Osterhold, M., 1993. Landesvermessung und Landinformationssysteme in den Vereinigten Staaten von Amerika. Allgemeine Vermessungs Nachrichten, 8-9 / 1993: pp 287-295

Otero, J. / Sevilla, M., 1990. On the Optimal Choice of the Standard Parallels for a Conformal Conical Projection. Bollettino di Geodesia e Scienze Affini, Anno XLIX - N.1. Publicazione dell' Istituto Geografico Militare - Florence: pp 1-14

Paggi, G. / Stoppini, A. / Surace, L., 1994a. Tecniche per l'Inserimento di Rilievi GPS nella Cartografia Esistente. ASIT Bolletino delle Associazione Italiana topografi, 25-26

Paggi, G. / Stoppini, A. / Surace, L., 1994b. Trasformazioni di Co-ordinate nei Rilievi GPS. Bollettino di Geodesia e Scienze Affini, Anno LIII - N.3. Publicazione dell' Istituto Geografico Militare, Florence: pp 285-312

Parkinson, B.W. / Spilker Jr., J.J., 1996. Global Positioning System: Theory and Applications. American Institute of Aeronautics and Astronautics, Inc. Washington, DC

Pearson II, F., 1990. Map Projections. Theory and Applications. CRC Press Inc., Boca Raton, FL

Pfeifer. L., 1984. The use of Bowring's Algorithms for Hydrography and Navigation. The Hydrographic Journal, No. 31, Jan 1984: pp 21-23

Poder, K. / Hornik, H., 1989. The European Datum of 1987 (ED87). Publication No. 18. Report IAG, Lisbon and Munich 1988

Poutanen, M. / Vermeer, M. / Mäkinen, J., 1996. The Permanent Tide In GPS Positioning. Journal of Geodesy, 1996, Vol. 70: pp 499-504

Poutanen, M., 1995. Some Practical Aspects of High-Precision GPS. In: *Vermeer, 1995: pp 113-123*

Price, W.F., 1986. The New Definition of the Metre. Survey Review, Vol. 28, No. 219, Jan 1986: pp 276-279

Rapp, R.H., 1981. Transformation of Geodetic Data Between Reference Datums. Geometric Geodesy, Vol. III, Ohio State University, Columbus: pp 53-57, 66

Rens, J. / Merry, C.L., 1990. Datum Transformation Parameters in Southern Africa. Survey Review, Vol. 30, No. 236, Apr 1990: pp 281-293

Rizos, C. / Fu, W.X. / Subsuantaeng, S., 1990. Antarctic GPS Surveying with the WM101 Receiver: Relative Positioning using Pseudo-Range Data. Aust. J. Geod. Photogram. Surv. No. 52, June 1990: pp 57-82

Rizos, C. / Morgan, P. / Ching-Mei C., 1992. GPS Orbit Computations in Australia within the International GPS GeoDynamics service: Should we? Could we? In: *GPE, 1993, pp 28*

Robbins, A.R., 1950. Length and Azimuth of Long Lines on the Earth. Empire Survey Review, No. 84, Oct 1950: pp 268-274

Robbins, A.R., 1962. Long Lines on the Ellipsoid. Empire Survey Review, No. 125, Jul 1962: pp 301-309

Rotter, F., 1984. Vom Erdmeridian zum Lichtzeitmeter. ÖZfVuPh., 72. Jg 1984, Heft 1: pp 1-10

Rüeger, J.M., 1994 (1996). Private Communication. University of New South Wales, Sydney

Rune, G.A., 1954. Some Formulae Concerning the Transverse Mercator Projection (Gauss Conformal Projection). Bulletin Géodésique, No. 34, Dec 1954: pp 309-317

Sacks, R., 1950. The Projection of the Ellipsoid. Empire Survey Review, No. 78, Oct 1950: pp 369-375

Salzmann, M.A., 1996. GPS, De HTW en Transformatie Parameters. GPS Nieuwsbrief No. 2, 1996: pp 33-37

Schmidt, H.H., 1966. Reformatory and Revolutionary Aspects in Geodesy. Bulletin Géodésique, No. 80, Jun 1966: pp 141-150

Schödlbauer, L., 1982. Transformation Gaußer konformer Koordinaten von einem Meridianstreifen in das benachbarte unter Bezugsnahme auf strenge Formeln der querachsigen sphärischen Mercator projektion. Allgemeine Vermessungs Nachrichten, 1 / 1982

Schreiber, O., 1866. Theorie der Projectionsmethode der Hannoverschen Landesvermessung. Hahn'sche Hofbuchhandlung, Göttingen

Schuhr. P., 1996. Transformationen zwischen kartesischen und geographischen Koordinaten. Allgemeine Vermessungs Nachrichten, 3 / 1996: pp 111-116

Schut, T.G., 1991. Transformatie Parameters voor RD naar WGS84, GPS Nieuwsbrief, October 1991, Ned. Commissie voor Geodesie, Delft

Schutz, B. / Bevis, M. / Taylor, F. / Kuang, D. / Abusali, P. / Watkins, M. / Recy, J. / Perin, B. / Peroux, O., 1993. The South West Pacific GPS - Project: Geodetic Results from Burst 1 of the 1990 Field Campaign. Bulletin Géodésique, Vol. 67, No. 4: pp 224-240

Schwartz, C.R., 1989. North American Datum of 1983. NOAA Professional Paper NOS 2. US Department of Commerce, National Geodetic Survey, Silver Spring, MD

Seeber, G., 1993. Satellite Geodesy. de Gruyter, Berlin

Seppelin, T.O., 1974. The Department of Defense World Geodetic System 1972. Technical Paper: Headquarters, DMA, Washington, DC, and also: Can. Surv., 1974, No. 28: pp 496-506

Seymour, W.A., 1980. A History of the Ordnance Survey, Wm. Dawson & Sons Ltd, Folkestone

Sjöberg, L.E. / Ming, P., 1993. Baltic Sea Level project with GPS. Bulletin Géodésique, Vol. 67, No. 1, 1993: pp 51-59

Sluiter, P.G., 1995. Geodetic Dual-Frequency GPS Receivers under Anti-Spoofing. Publications on Geodesy, New Series, Number 42. Netherlands Geodetic Commission, Delft

Sodano, E.M., 1958. Determination of Laplace Azimuth between Non-intervisible Distant Stations by Parachuted Flares and Light Crossings. Bulletin Géodésique, No. 49, 1958: pp 16-32

Stem, J.E., 1989a (1994). State Plane Co-ordinate System of 1983. NOAA Manual NOS NGS 5. US Department of Commerce, National Geodetic Survey, Silver Spring, MD

Stem, J.E., 1989b. User Participation and Impact. In: *Schwartz, 1989: pp 237-248*

Stoppini, A., 1996. Private Communication. Universita degli Studi Facoltà di Ingegneria, Perugia, Italy

Strang van Hees, G.L., 1996. Private Communication. Ned. Commissie voor Geodesie, Delft

Strang van Hees, G.L., 1997. Globale en Lokale Geodetische Systemen. 3e druk. Ned. Commissie voor Geodesie, Publicatie 30, Delft

Strasser, G.L., 1957. Ellipsoidische Parameter der Erdfigur (1800-1959). DGK, Reihe A, No. 19. Verlag der Bayerischen Akademie der Wissenschaften, Munich

Strasser, G.L., 1966. Heinrich Wild's Contribution to the Development of Modern Survey Instruments. Survey Review, Vol. XVIII, No. 140, Apr 1966: pp 263-268

Strauss, R., 1991. Lagebezugssysteme in Deutschland im Wandel. Allgemeine Vermessungs Nachrichten, 4 / 1991: pp 130-137

Surace, L., 1995 (1996). Private Communication. Istituto Geografico Militare, Florence

Tarczy-Hornoch, A. / Hristov, V.K., 1959. Tables for the Krassovsky-Ellipsoid. Akademiai Kiado, Budapest.

Tardi, P., 1934. Traité de Géodésie (fascicule 1). Gauthier-Villars, Éditeur, Paris

Thomas, P.D., 1952 (1978). Conformal Projections in Geodesy and Cartography. Special Publication No. 251. US Government Printing Office, Washington, DC

Torge, W., 1989. Gravimetry. Walter de Gruyter & Co, Berlin-New York

Torge, W., 1991. Geodesy. Second Edition. Walter de Gruyter & Co, Berlin-New York

Urmajew, N.A., 1955 (1958). Sphäroidische Geodäsie. VEB Verlag Technik, Berlin (Redaktion-Verlagsabteilung des militar-topographischen Dienstes), Moscow

Väisälä, Y., 1923. Die Anwendung der Lichtinterferenz zu Längenmessungen auf grösseren Distanzen. Publication of the Finnish Geodetic Institute, No. 2, Helsinki

Väisälä, Y., 1930. Die Anwendung der Lichtinterferenz bei Basismessungen. Publication of the Finnish Geodetic Institute, No. 14, Helsinki

Vermeer, M., 1995. Co-ordinate Systems, GPS, and the Geoid. 95:4 - Report of the Finnish Geodetic Institute, Helsinki: Proceedings of the Nordic Academy for Advanced Study (NorFA), Espoo, Finland, Jun 27-29, 1994

Vincenty, T., 1971. The Meridional Distance Problem for Desk Computers. Survey Review No. 161, Jul 1971: pp 136-140

Vincenty, T., 1976a. Determination of North American Datum of 1983 - Co-ordinates of Map Corners (Second Prediction). Technical Memorandum NOS NGS 16. US Government Printing Office, Washington, DC

Vincenty, T., 1976b. Direct and Inverse solutions of Geodesics on the Ellipsoid with applications of nested equations. Survey Review No. 176, Apr 1976: pp 88-93

Vincenty, T., 1984a. Transverse Mercator Projection Formulas for GRS80 Ellipsoid. NOAA-NGS (unpublished file report, Aug 1984)

Vincenty, T., 1984b. Oblique Mercator Projection Formulas for GRS80 Ellipsoid. NOAA-NGS (unpublished file report, Sep 1984)

Vincenty, T., 1985a. Lambert Conformal Conical Projection Formulas for GRS80 Ellipsoid, NOAA-NGS (unpublished file report, Mar 1985)

Vincenty, T., 1985b. Precise Determination of the Scale Factor from Lambert Conformal Conical Projection Co-ordinates. Surveying and Mapping, Dec 1985, Vol. 45, No. 4: pp 315-318

Vincenty, T., 1986a. Lambert Conformal Conical projection: Arc-to Chord Correction. Surveying and Mapping, Jun 1986, Vol. 46, No. 2: pp 163-167

Vincenty, T., 1986b. Use of Polynomial Coefficients in Conversions of Co-ordinates on the Lambert Conformal Conical projection. Surveying and Mapping, Mar 1986, Vol. 46, No. 1: pp 15-18

Vincenty, T., 1994. Private Communication. NOAA, Silver Spring, MD

Waalewijn, A., 1986. Der Amsterdamer Pegel (NAP). ÖZfVuPh., 74. Jg 1986, Heft 4: pp 264-270

Waalewijn, A., 1987. The Amsterdam Ordnance Datum (NAP). Survey Review, Vol. 29, No. 226, Oct 1987: pp 197-204

Wadley, T.L., 1957. The Tellurometer System of Distance Measurement. Empire Survey Review, Vol. XIV, No. 105, 106

Wilford, J.N., 1981 (1982). The Mapmakers. First Vintage Books Edition. Reprint. Random House, Inc., New York-Toronto

Williams, O.W. / Iliff, R.L. / Tavenner, M.S., 1966. Lasers and Satellites: A Geodetic Application. Bulletin Géodésique, No. 80, Jun 1966: pp 151-156

Wittke, H., 1949 (1958). Geodätische Briefe. 3. Auflage. Hanseatische Verlagsanstalt GmbH, Hamburg

Wolf, H., 1987. Datums-Bestimmungen im Bereich der Deutschen Landesvermessung. Zeitschrift für Vermessungswesen, 8 / 1987: pp 406-413

Xiang, J., 1988. Ellipsoidal Geodesy. Beijing. (Chinese text)

Zeger, J., 1991. 150 Jahre Bessel Ellipsoid, 1841-1991. ÖZfVuPh., Heft 4: pp 337-340

Zhu, H.T., 1986. The Establishment of Geodetic Co-ordinate Systems. Beijing. (Chinese text)

Zilkoski, D.B. / Richards, J.H. / Young, G.M., 1992. Results of the General Adjustment of the North American Vertical Datum of 1988. Surveying and Mapping, Vol. 52, No. 3: pp 133-149